AADCU PUBLICATION
WWW.AADCU.ORG
INFO@AADCU.ORG

ACKNOWLEDGEMENTS
This publication has been made possible with the help and cooperation of many individuals and institutions. Grateful acknowledgement is made to Vicente Guallart for his inspiring work and for his kind support in the preparation of book on Re-Naturalisation for the AADCU book series of the world's most emerged practices in the field of digital, multi-media and net based architecture design.

PROJECT DIRECTOR: BRUCE Q. LAN
COORDINATOR: XIAOQU LUO/Stanford University
CURATOR/EDITOR IN CHIEF: BRUCE Q. LAN
BOOK DESIGN: °C Office®/BRUCE Q. LAN(principle)
EDITED AND PUBLISHED BY: Beijing Office/United Asia Art&Design Cooperation/bj-info@aadcu.org
IN COLLABORATION WITH: Vicente Guallart/www.guallart.com; d-Lab&International Architecture Research; °C Office®

ISBN: 978-7-5609-4861-4

编辑与出版：亚洲艺术与设计协作联盟\美国\info@aadcu.org
协同合作：Vicente Guallart\西班牙;
国际建筑研究与设计中心\美国;零度事务\美国

主编及学术项目总策划：蓝青
国际总协调：罗小渠\斯坦福大学
书籍设计：°C Office®\蓝青(艺术总监)
翻译：李璠\清华大学建筑学院

※美国亚洲艺术与设计协作联盟的建筑出版项目旨在关注国际前沿建筑学研究和实践，由于诸多因素的制约和学术环境的差异，中文阅读建议敬请读者最大程度参阅相对应之英文，并致以歉意和谢意。※

RE-NATURALISATION

A ADCU 2008 ARCHITECTURE DESIGN MONOGRAPH SERIES:
西班牙当代建筑实践 VICENTE GUALLART, BARCELONA, SPAIN

重构自然
山峦 · 媒介 · 建筑

媒介、山峦与建筑是三种可栖居的自然域——数码、自然与人工——的交界面。这三种被人们所感知的体系需要一种新的组织方式来重组他们之间的交互关系：物质世界被数字世界的出现而改变；建成环境被曾经所谓“自然”的东西改变；数字世界自发地构成了其内部的规律并建造了自身的操作系统。在面对任何人类行为的时候，这三种产物同时产生了波动。

THE RE-NATURALISATION OF TERRITORY
区域性的自然重构

THE HISTORY OF THE FUTURE
The cities and lands had grown physically at the same time of their economical growth.
Soon the urban Barcelona will be totally occupied.
The metropolitan Barcelona is running out of available free land.
If Catalunya keeps the same growing rate than the last 150 years it will be completely occupied by the year 2375.
Before the artificial world of our civilization we it existed a virginal geography, virgen and natural, made of plains and mountains .
Land, a necessary but limited good has the key for it's re-foundation...

未来的历史
在经济增长的同时，城市和土地也在物质性地增长。
很快，巴塞罗那市区就将被全部占领，而不再拥有可利用的土地。
如果加泰罗尼亚继续维持过去这150年的增长速度，那么它将在2375年被完全占据。
在人类文明创造出的人工世界存在之前，存在着一个由平原和高山组成的未经碰触的地球环境，它完全自然，没有经过改造。
土地是一种必要且有限的资源，并且只有它自身才能启动自身的重塑功能。

THE BIRTH OF A NEW SPECIE
Montserrat, formed by the conglomeration that joins the rests of the sediments of the Mountains of its surroundings, has the genetic code of the lands that surround it.
Facing this urban invation, it starts a proces of auto-clonation.

一个新物种的诞生
将环绕在其周边群山的沉积物凝聚在一起的蒙特塞拉特岛，具有其周围土地的基因码。
面对这样的城市侵占，它开始了一系列的自我克隆行为。

THE RE-NATURALISATION OF TERRITORY

First, Mountains discover highways, that block any possibility of a natural continuity.

In San Cugat, they devour and reform corporative centers, creating new Mountain-buildings capable of being auto-sufficient.

The continuity of the landscape is now a public golf court over the A-7 highway.

More Land

In Tarragona the Mountains detect some chemical agents that block the possibility to live the coasts. They isolate, they move them towards the interior and unfold new topographies to organize a new city.

More coasts, more activities

In Barcelona, the expansion of the XIX century starts the growth of the city and becomes a key of attraction to people, avoiding the territorial expansion. The energy of the Mountains is so big that they decide to cretae a set of islands in front of the city.

区域性的自然重构

首先，高速公路将山体割裂，切断了自然延续的可能性。

在San Cugat，人们吞噬并重新组成了综合性的中心，创造了能够自给自足的新山体式建筑。

景观的连续性现在由7号高速公路上的一个公共高尔夫球场来完成。

更多的土地

Tarragona的山体发现了一些阻碍海岸线生存的化学元素。它们将其分离然后推向内部，并展开新的地形来组织新的城市。

更多的海岸，更多的活动

在巴塞罗那，XIX世纪的扩建引发了城市的增长，并且在避免领地扩张的情况下成为了新的名胜。山体的能量足够大，以至于在城市的前沿创造出了一系列新的岛屿。

EPILOGUE

In Barcelona a new «Camp Nou».

In the A-7 highway, a new continuity.

In Tarragona, biotechnology and cinema.

In El Perelló, techno-agricultural micro-villages

2020 The Mountains have established in land a new code natural-artificial.

A new attitude that allows the growth towards the interior of the cities, and a more natural way to live the established landscape.

The laws have changed. It exists a three-dimensional urbanism and the law of the forth factor is used: To get the double Benefit with half resources.

结论

巴塞罗那的新《Camp Nou》。

A7高速公路的一种新的延续性。

Tarragona的生物技术与电影院。

El Perelló的技术化农业微型村庄。

2020年，山峦就已经为土地建立了一种新的自然人工规则。

一种容许城市内向增长的新态度以及延续现有景观的更加自然的方式。

法则被改变了。一种三维的都市主义生成了，外部因素的法则被运用其中，以达成事半功倍的资源利用目标。

DENIA MOUNTAIN
DENIA山峦

Denia is a town on the Mediterranean coast whose principal economic activity is tourism. In the centre of the town is a small hill topped by a castle of Arabic origin that was converted to agricultural use in the late 19th century. In the early years of the 20th century stone began to be quarried from the north face of the hill in order to extend the port. During the Spanish Civil War a refuge tunnel was dug into the hillside.

The Denia Castle Cultural Park is a town project, the object of which is the reordering of the castle and its surroundings and the creation of the cultural, civic and social amenities that will make it a territorial reference point promoting the urban recycling and compact development of the town.

The principal projects to be carried out are:

1. The conservation and protection of the castle
2. The creation of a cultural centre in the Plaça del Consell square
3. The extension of the pedestrian tunnel
4. The construction of a multifunctional centre
5. The protection of the Hort de Morand, which contains the remains of the Roman town.

The principal new construction project is a multifunctional centre (with facilities, services and open space) in the quarry adjacent to the castle that will restore the original volumetry of the hill.

Given that, in the last analysis, the best way to preserve the heritage is to increase it, this project sets out to conserve the historical remains, at the same time creating a new architectural landmark that will act as a focus of attraction for culture, tourism and economic activity.

The project, originally a public initiative, will be developed with the participation of the private sector.

位于地中海沿岸的Denia镇的支柱产业是旅游业。小镇中心的山丘上，伫立着在19世纪后期改做农业用途的阿拉伯式城堡。20世纪初，山丘的背面开始进行采石活动以扩充港口。在西班牙内战期间，山体的一侧又挖掘了一个避难隧道。

Denia城堡文化公园作为一个市镇项目，其工作目的为重新组织城堡及其周边环境，以创造一个具有文化性、本土性以及社会性的休闲娱乐场所，并成为该地区推行城市循环的根据地，并压缩市镇的发展。

即将实施的主要项目是：

1. 城堡的维护与保护；
2. 在Placa del Consell广场创建文化中心；
3. 行人通道的扩充；
4. 建造一个多功能中心；
5. 对包含有罗马市镇遗迹的Hort de Morand进行保护。

新的建设项目中，最为主要的是通过在城堡附近采石场建造多功能中心（硬件设施、服务以及开放空间）来还原山丘原有的体量。以此为前提，同时建造一个新的建筑性地标以发挥其作为文化、旅游和经济活动的聚焦点的作用。

这一项目以公共需求为出发点，将同时吸纳私有因素的参与。

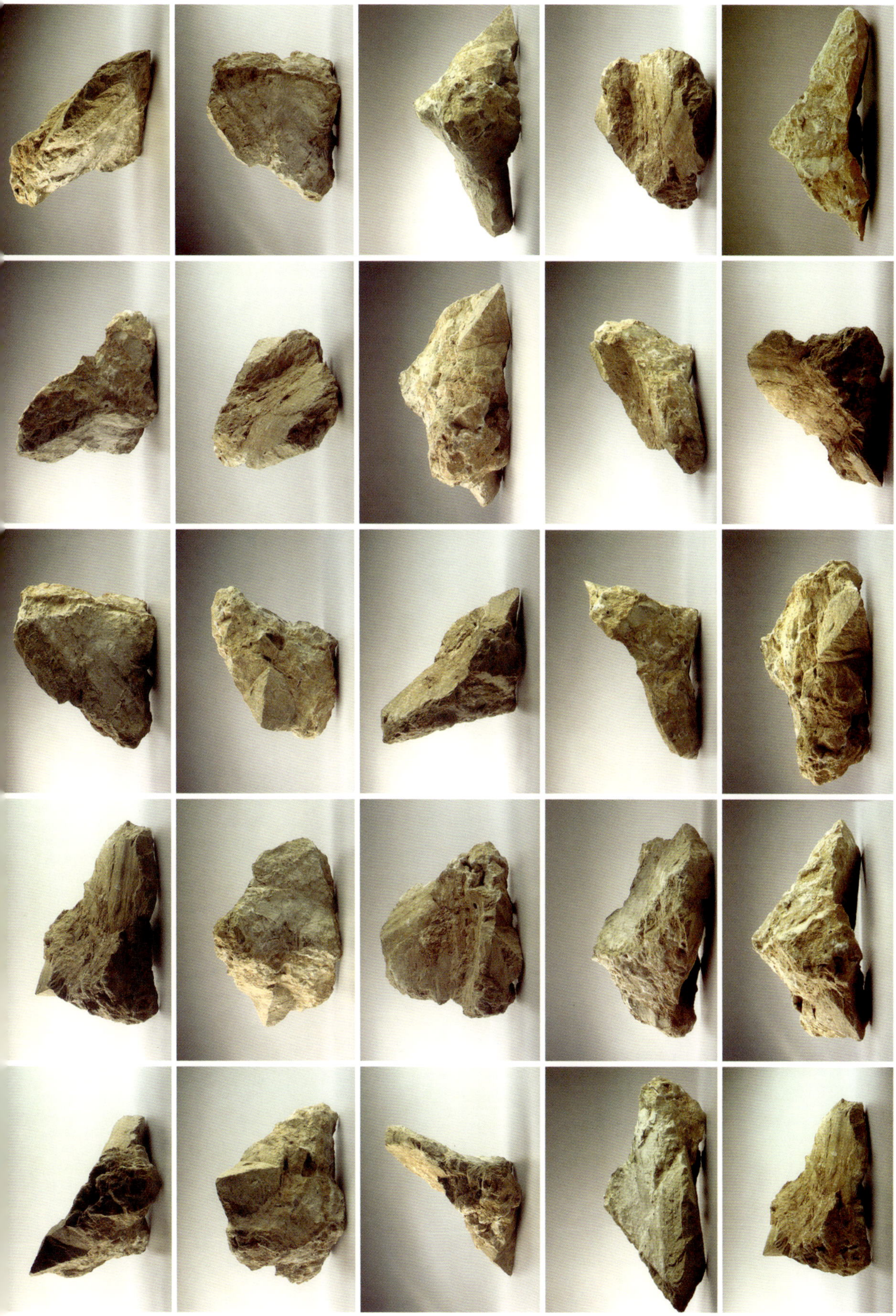

怎样制造一座山

Denia城堡山丘的重新生成

这个项目应当被归入哪一类？重建一座山的原型能够仅仅是一个简单的表面行为么？

怎样将建筑的第五立面转化成为一座公园？

Denia采石场的未来建设将控制在一个结构性的逻辑下。

山体的微观、中等以及大尺度上的地理结构为我们提供了推进矿石系统的规则，并以此指导其运行。

山体的莱因石以及其菱形的方解石晶体使我们能够从多个尺度层次来为项目构思一个透明的基础，从结构到外部边界均对单一的晶体系统进行回应。通过这一方法，表皮就像山上的土壤一样，直接反映了物质的内在逻辑和与环境的交互关系。

在这样一个案例中，菱形系统生成了一个六角形的几何体来充当“基因”（一个能完全由其表面构成的六角微型）的几何基座，这个基座将激发建造表皮的过程。这个基因将会启动山体的重新生成过程。建造一座山并将其投入使用的另一个关键问题是：是否假定横穿路线应当来自山体假定界限上挖掘出来（就像传统的通向隐居之处的路线）；是否山体本身应该建造其自己的路线（就像在露天采石场中所见的那样）。在这一案例中着力进行的是开发一个能够对允许表面上的多种运动的内在结构逻辑产生回应的系统，这样的系统可能包含了很多的斜坡，并且通过这样的方式创造出一个包含多种巡回路线的系统。这样创造一个附带路径的山体，山体本身就成为了路径。

室内

室内以定义社会性交流的三个功能焦点为中心进行组织：

1. **一座市场**，围绕其组织商业区域、电影院以及餐馆区域；
2. **一座礼堂**，一个供召开会议和市民集会的场所；
3. **一个水疗中心**，其中包括一个盐水湖。

整个项目围绕这些宽敞空旷的参照空间布置，并采用方解石的三种结晶特质系统来塑造其形式。这些晶体形式中的每一个个体都从矿石之间相互反应的基本系统当中生长出来：双晶、并排以及聚合体。

HOW TO MAKE A MOUNTAIN

The re-generation of Denia's Castle hill

To what category does this project correspond?

Can reconstructing the topography of a hill be a simple epidermic action?

How to transform the façade-roof of the building into a park?

The future construction of the quarry in Denia will be governed by a structural logic.

The geological structure of the hill, on the micro, the medium and the macro scales, offers us rules with which to put forward a mineralogical system that will guide its functioning.

The limestone of the hill and its rhombohedric crystals of calcite enabled us to conceive, at multiple scales, a crystalline genesis for the project. A coherent system, from the structure itself to its outer limit that responds to a single system of crystallization. In this way, the skin, like soil in the hills, directly reflects the internal logic of the mass and its interaction with the environment.

In this case, the rhombohedric system generates a hexagonal geometry that will be the geometric base of the 'gene' (a hexagonal micro-topography that can be combined all over its faces) that will initiate the process of constructing the skin. This gene will set in motion the re-generation of the hill.

The other key issue, that of constructing a hill and trying to make it usable, is the question of whether the hypothetical roads that traverse it should come from a system excavated out of the hypothetical limits of the hill (like the traditional paths to hermitages), or whether the hill itself ought to construct its own roads (as in the case of open quarries).

In this case the effort has been concentrated on developing a system that will respond to the logic of the internal structure that permits multiple movements across the skin, following various slopes, and in this way creating a system of multiple itineraries. Thus, instead of defining a hill with paths, the hill itself is the path.

THE INTERIOR

The interior is organized on the basis of the proposed programme around three focuses of functional attraction that define spaces for social interaction:

1. **AN AGORA**, around which is organized a commercial zone, cinemas and a restaurant zone;
2. **AN AUDITORIUM** and a zone for congresses and civic meetings;
3. **A BATHING SPA**, which includes a salt-water lake

These large empty reference spaces, around which the programme is organized, take on form on the basis of three systems of crystallization characteristic of calcite. Each of these forms of crystallization grows out of the fundamental systems of interaction between minerals: macles, juxtaposition and aggregation.

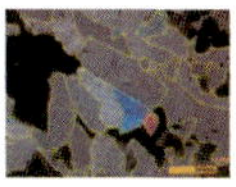

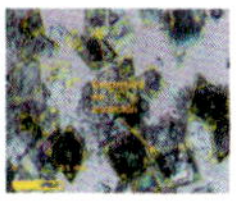

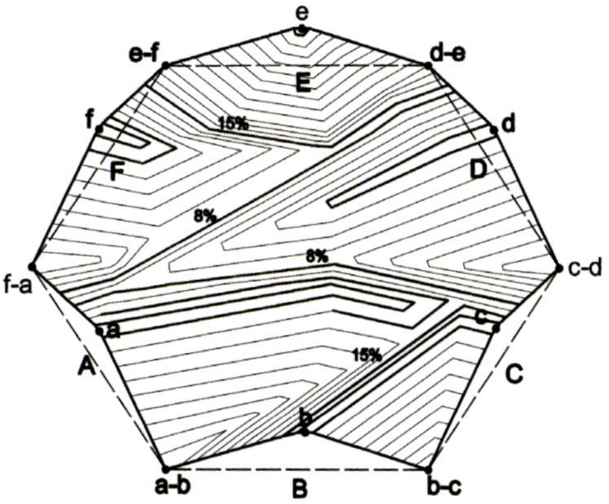
e
e-f
d-e
E
f
15%
d
F
D
8%
8%
f-a
c-d
a
c
15%
A
C
b
a-b
B
b-c

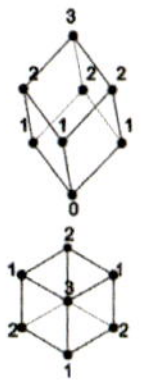

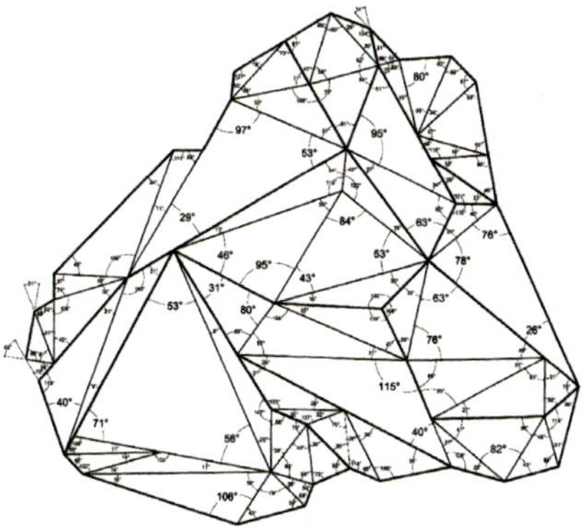

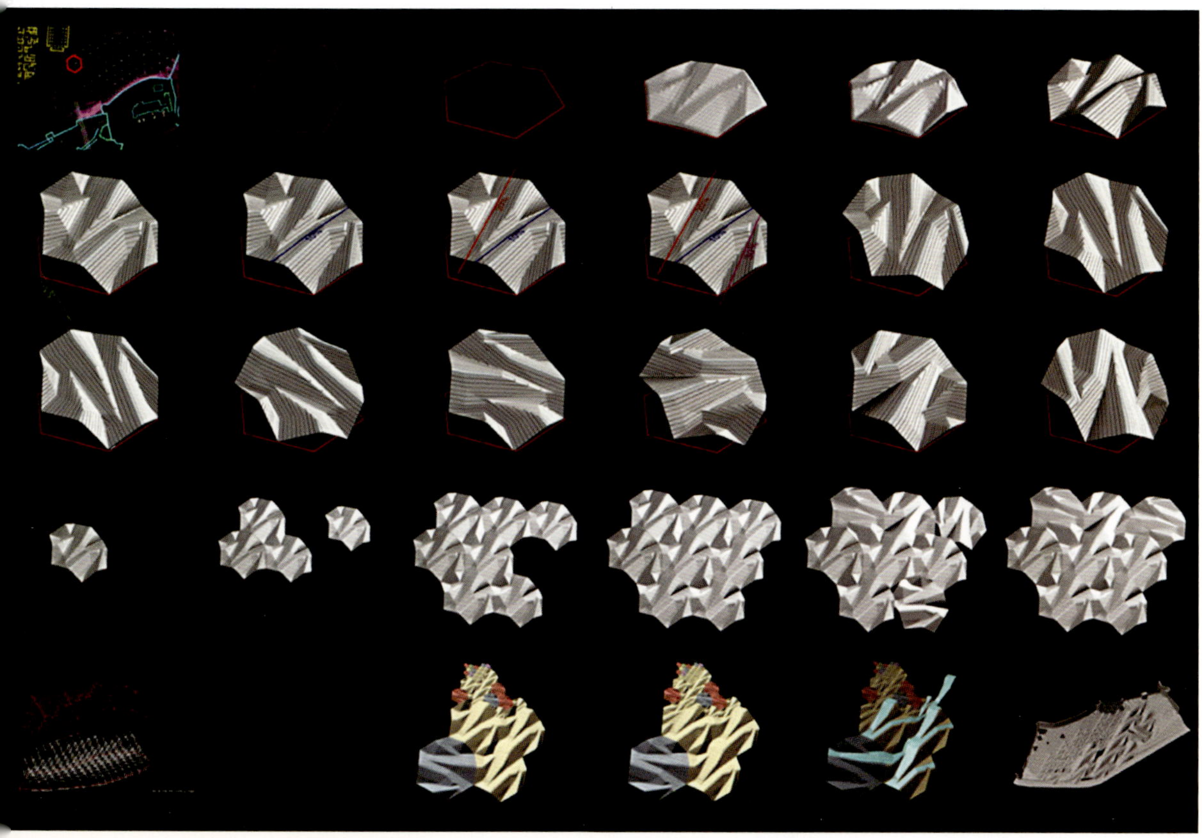

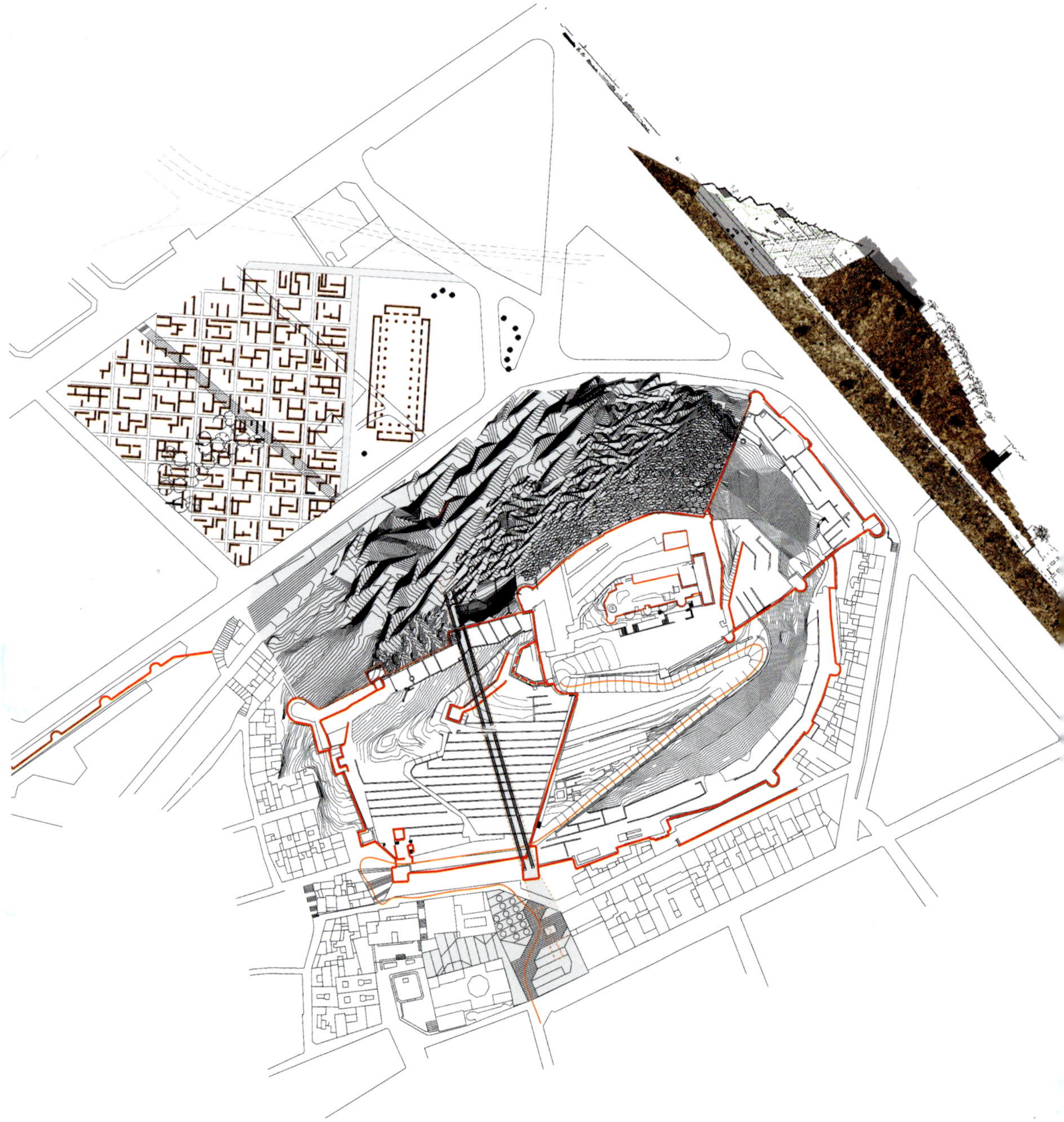

PIEZA GEN
PENDIENTES DE MOBILIDAD
15%
8%
8%
15%
COMBINACIONES EPIDERMICAS
0°
60°
120
180°
240°
300°
GEOMETRIA
CANTERA/ESTRUCTURA PIEL

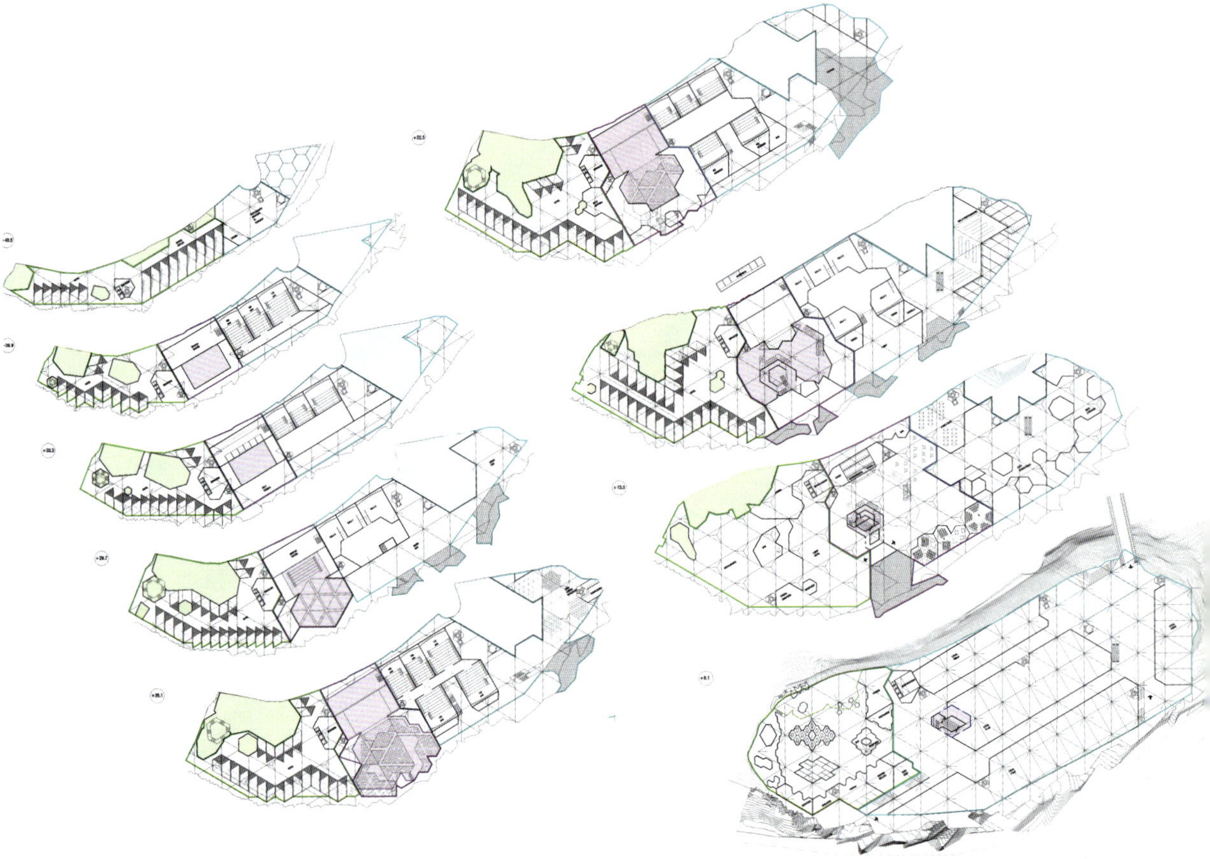

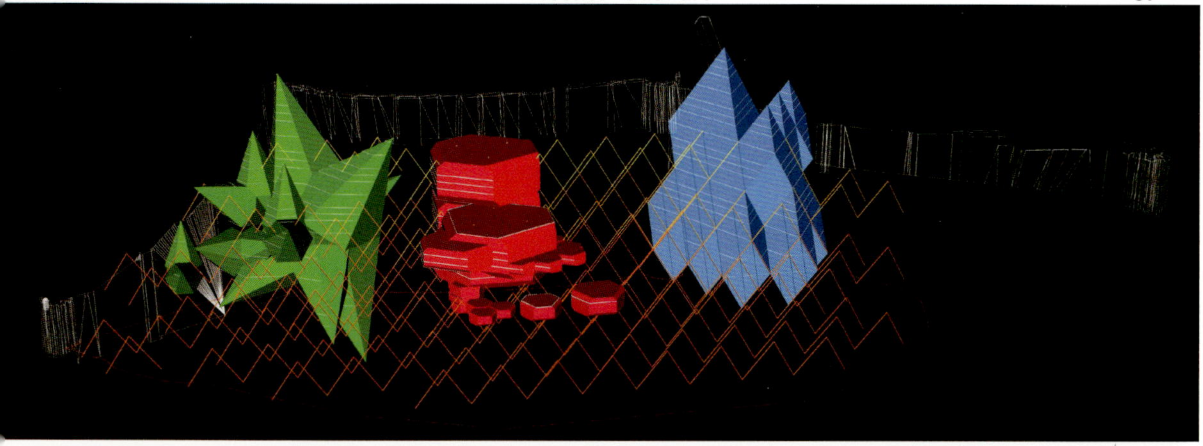

The PHANTOM of the OPERA
MAJESTIC THEATRE

EXPO 2012 WROCLAW MOUNTAIN

2012世博会，WROCLAW山

The Universal and International Expositions, like the Olympics, are events that tend to be staged either in emerging countries that are consolidating a position in the international panorama primarily in political o economic terms, or in central locations engaged in reasserting their international leadership. Wroclaw, in the south of Poland, near the German border, belongs to the first category, as a place that is undergoing major political and economic transformation and hopes to attract the world's attention with a great event. In this endeavour it is competing with Tangiers in Morocco and Yeosu in Korea to host the forthcoming Expo 201. In order to choose the maste plan for the laying out of its Expo and a scheme for an emblematic building, the city organized an international competition, in which our proposal was selected for the design of the building.
An Expo is an event that has no real identity in its own right. This being so, the ones that have gone down in history are those that have been very clearly symbolized by a landmark building that has come to constitute an icon of the city, such as Paxton's Crystal Palace in London, the Eiffel Tower in Paris, the Atomium in Brussels or Buckminster Fuller's Dome in Montreal.
Wroclaw stands at a crossroads, a territory occupied throughout the course of history by different cultures and empires, in which the rivers Oder and Olawa meet. In the light of this we saw crossing and cross-fertilization as fundamental attributes of the icon of this Expo, the key theme of which will be 'the culture of leisure in the world's economies.' Two of Poland's outstanding buildings of the 20th century are the Hala Ludowe or Centennial Hall by Max Berg in Wroclaw (completed in 1913, and in its day the largest concrete dome in the world) and the Tower c Culture and Science in Warsaw, a potent symbol of a foreign power in the centre of Warsaw that neverthe less represents a surprising typology of high-rise, containing cinemas, exhibition spaces and events venues, offices and other culture-related services. The Centennial Hall was built under German dominance, and the Warsaw Tower under de facto Soviet influence.
Poland is starting on a new phase with its full integration into the European Union, and is due to receive a massive amount of funds with which to con struct infrastructures and amenities.
In view of this, this Expo, its urban setting and its principal buildings are an important opportunity to define the identity of a new turn, political, economic and social, just as previously occurred in countries such as Spain.
Our building is located, volumetrically speaking, between the Centennial Hall and the Tower of Culture and Science in Warsaw. In its initial phase it was conceived as an accumulation of generic and as yet imprecise programmes that could include offices, ar

auditorium, a leisure centre and exhibition spaces that generated its identity on the basis of its mountain-like form. The building's activities would be allowed to expand onto exterior terraces, created directly by the topography of the building, and onto its roof, directly accessible by way of a cable car from the park in which the building stands.

The topography of the mountain is such that its south face forms an angle to accommodate a bank of photovoltaic panels on the façade which will generate part of the energy consumed by the building.

就像奥林匹克运动会一样，在国际性的世界博览会引人注目
主办国永远是那些寻求在政治或经济的国际舞台上取得重
位置的国家，或者是那些想要重申其国际性的领导地位的
心地区。波兰南部靠近德国边境的Wroclaw正属于上述的
一类，这是一个正在经历重大的正式和经济变迁的地区，并
望通过举办重大活动来吸引全世界的目光。在这一次冒险中，
Wroclaw与摩洛哥的Tangier和韩国的Yeosu竞争即将到
的2012年世界博览会的主办权。为了给世博会的布展选择总
以及一个标志性建筑的设计方案，Wroclaw组织了一次国
性的竞赛，而其中我们的建筑设计方案被最终采纳。

博览会是一种本身不具备实体的活动。在这样的前提下，那
历史上被记住的博览会通常是以其特有的建筑为标志，从而
为一个城市的标志，诸如Paxton设计的伦敦水晶宫，巴黎的
菲尔铁塔，布鲁塞尔的Atomium，或者是蒙特利尔的Buck
minster Fuller穹顶等。

Wroclaw矗立于这样一个十字路口，纵观整个历史，它被不
的文化和帝国所统治，而它所在的地区则是oder河与olawa
交汇的地方。在这样的背景之下，我们将跨越文化和跨越性
孕育过程都视为这一届博览会标志性的基础，同时其关键
主题则是"世界经济当众的休闲文化"。20世纪波兰最为杰
的两座建筑是位于Wroclaw的由Max Berg所设计的Hal
ludowe即百年纪念堂（建成于1913年，同时期世界上最大
混凝土穹顶）以及位于Warsaw的文化科学大厦——一座位
Warsaw中心外权象征，毫无悬念地表达着一种令人惊异
罪囚高度的原型，其中包含了电影院、展览陈列空间和集会
地，办公室以及其他相关文化服务设施。百年纪念堂建成于
国统治时期，而Warsaw大厦则受到了苏联实体的影响。

波兰正在以其对于欧盟的完全融入展开一个新的阶段，并且
于接受了大量的基金会资助而正在开展基础设施和公众休
娱乐设施的建造。

由此看来，这次展览会的都市背景条件以及其中的建筑都将
一个政治上，经济上以及社会上的新转折的良机，正如在诸
西班牙这样的国家不久前发生的那样。

我们的建筑从体量上讲定位于百年纪念堂和Warsaw文化
学大厦之间。在其初始阶段，它被设计成为一种对于诸如办
室、礼堂、休闲中心和展览空间这样的使用项目的广泛而不
确的积累，这种积累在其类似于山的外形下生成了其独特
识别性。这栋建筑的活动将会被允许拓展到室外台地上，
这类台地则是由建筑的外型直接创造出来的；活动也同样
以延续到建筑的屋顶这个可以通过公园内的电车直接到达
场所。

山的原型南面形成一个可以负荷立面上的光合作用板的承
面，而建筑所消耗的能源正来自于这些光合作用板。

EXPO 2012
WROCLAW

EXPO 2012
WROCLAW

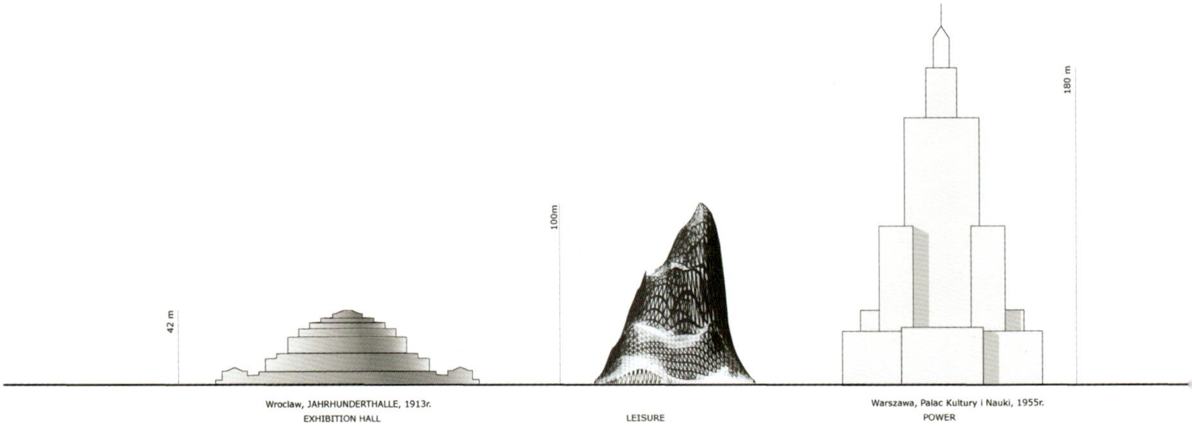
42 m
100m
180 m
Wroclaw, JAHRHUNDERTHALLE, 1913r.
EXHIBITION HALL
LEISURE
Warszawa, Palac Kultury i Nauki, 1955r.
POWER

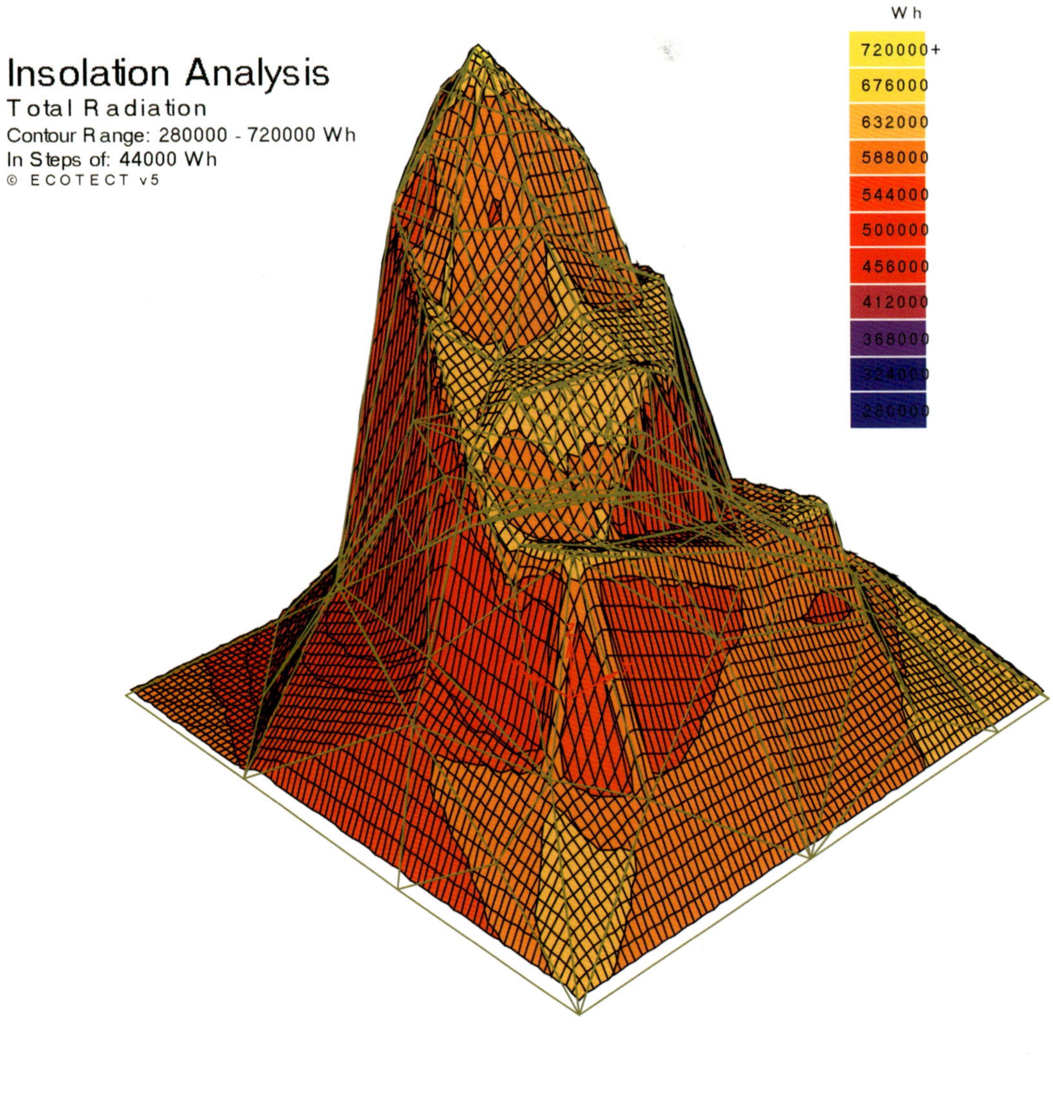
Insolation Analysis
Total Radiation
Contour Range: 280000 - 720000 Wh
In Steps of: 44000 Wh
© ECOTECT v5
Wh
720000+
676000
632000
588000
544000
500000
456000
412000
368000
324000
280000

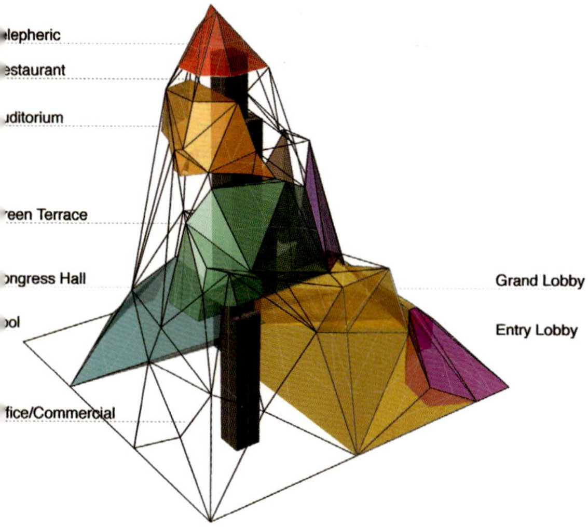
lepheric
estaurant
uditorium
een Terrace
ongress Hall
ol
fice/Commercial
Grand Lobby
Entry Lobby

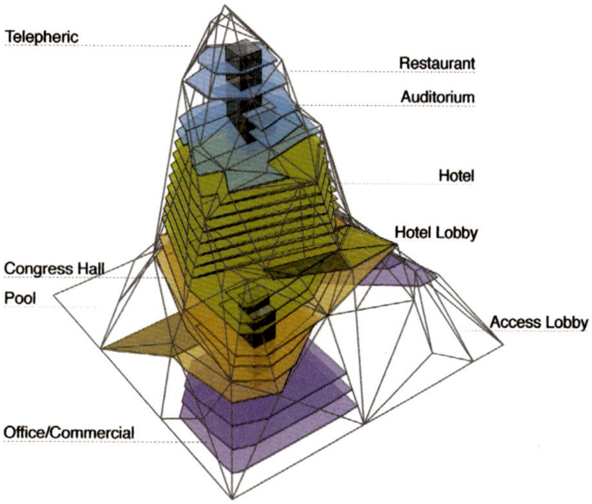
Telepheric
Restaurant
Auditorium
Hotel
Hotel Lobby
Congress Hall
Pool
Access Lobby
Office/Commercial

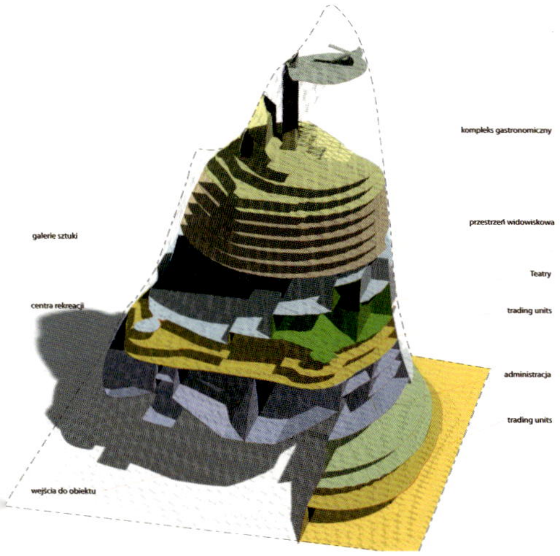
kompleks gastronomiczny
galerie sztuki
przestrzeń widowiskowa
Teatry
centra rekreacji
trading units
administracja
trading units
wejścia do obiektu

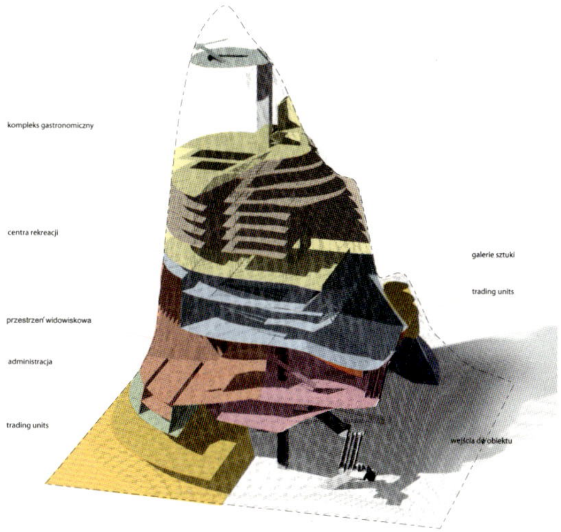
kompleks gastronomiczny
centra rekreacji
galerie sztuki
trading units
przestrzeń widowiskowa
administracja
trading units
wejścia do obiektu

Expo 2012
WROCLAW 2012
The Culture Of Leisure In World Economies
Kultura Czasu Wolnego w Gospodarkach Świata

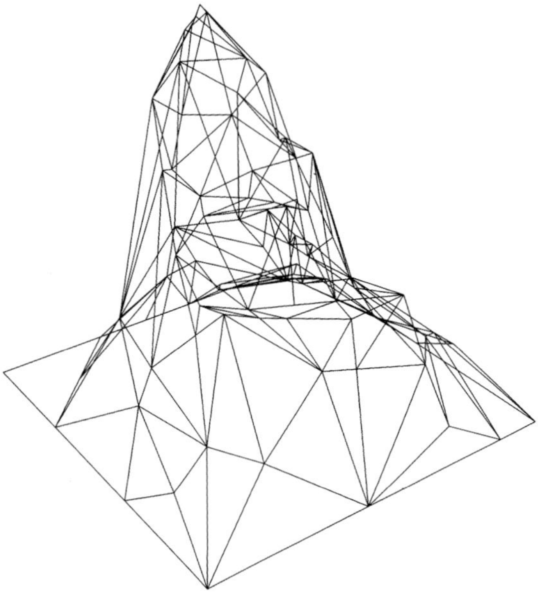

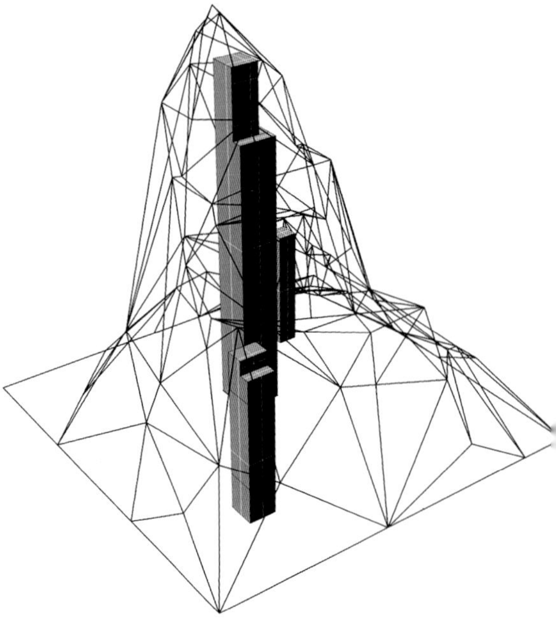

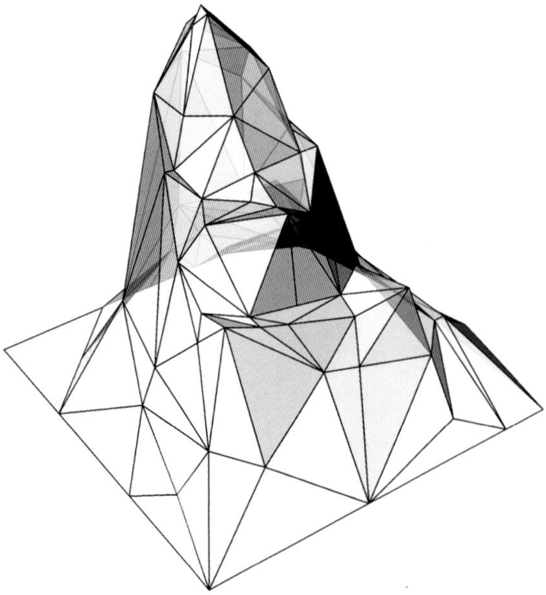

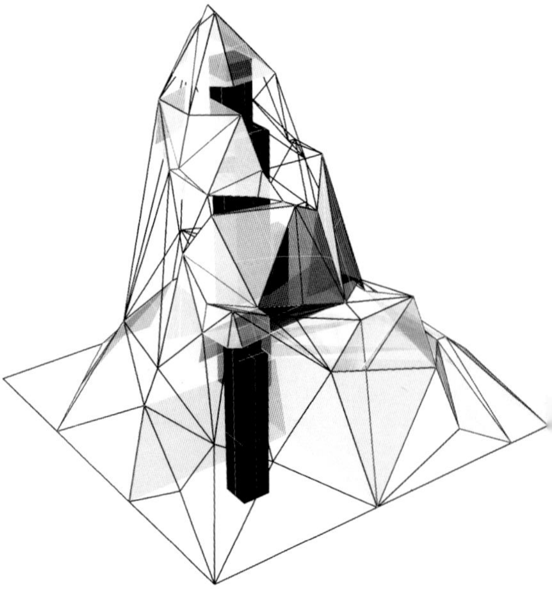

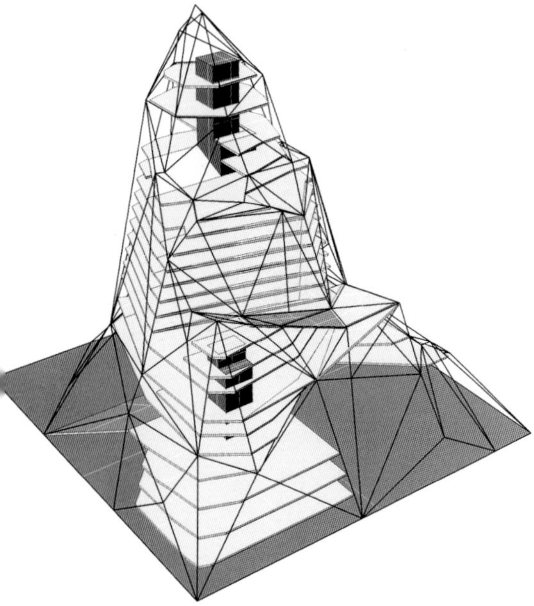

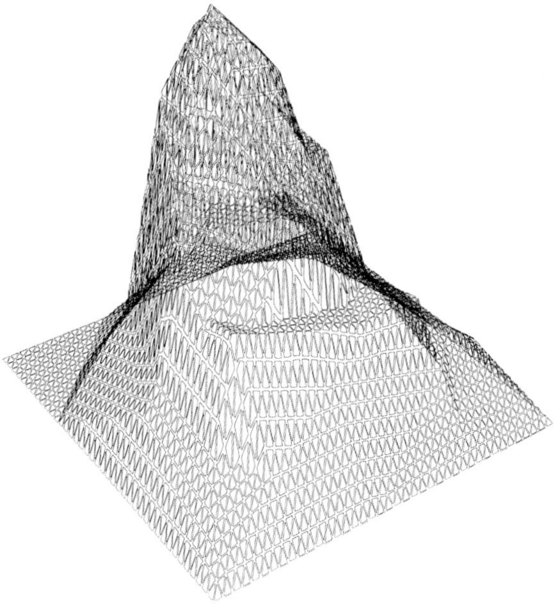

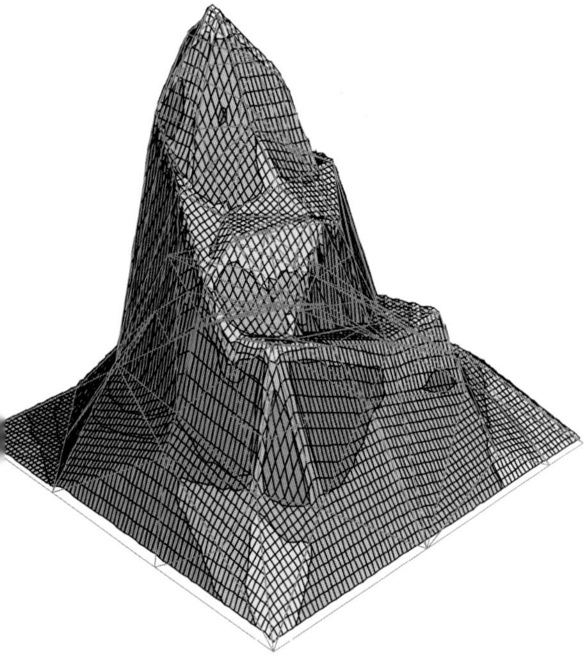

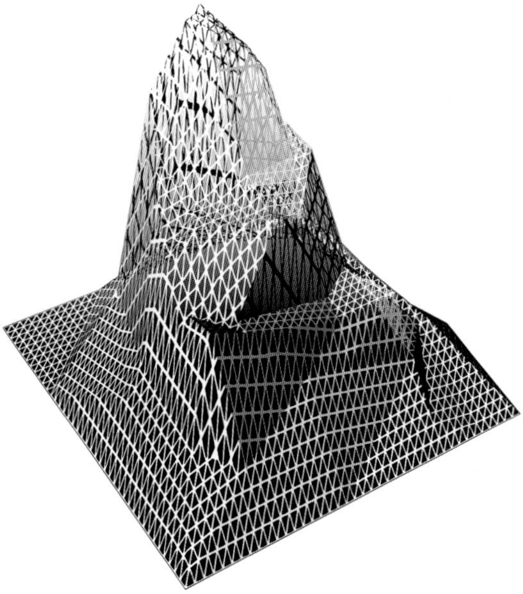

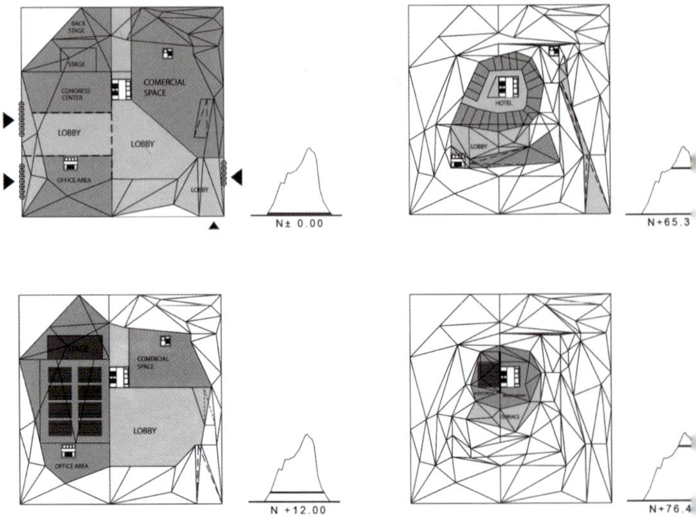
BACK STAGE
STAGE
CONGRESS CENTER
COMERCIAL SPACE
LOBBY
LOBBY
OFFICE AREA
LOBBY
N± 0.00
HOTEL
LOBBY
N+65.3
STAGE
COMERCIAL SPACE
LOBBY
OFFICE AREA
N +12.00
N+76.4

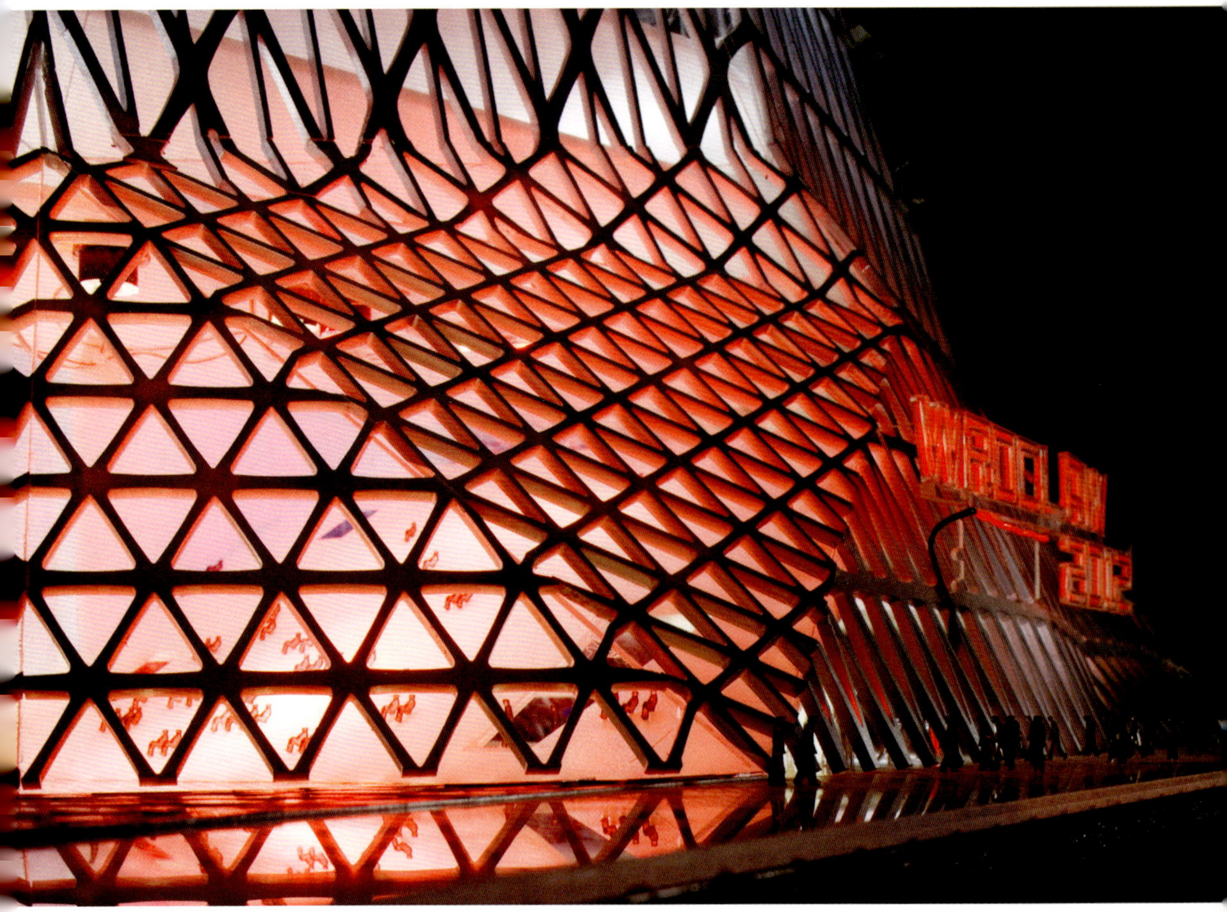

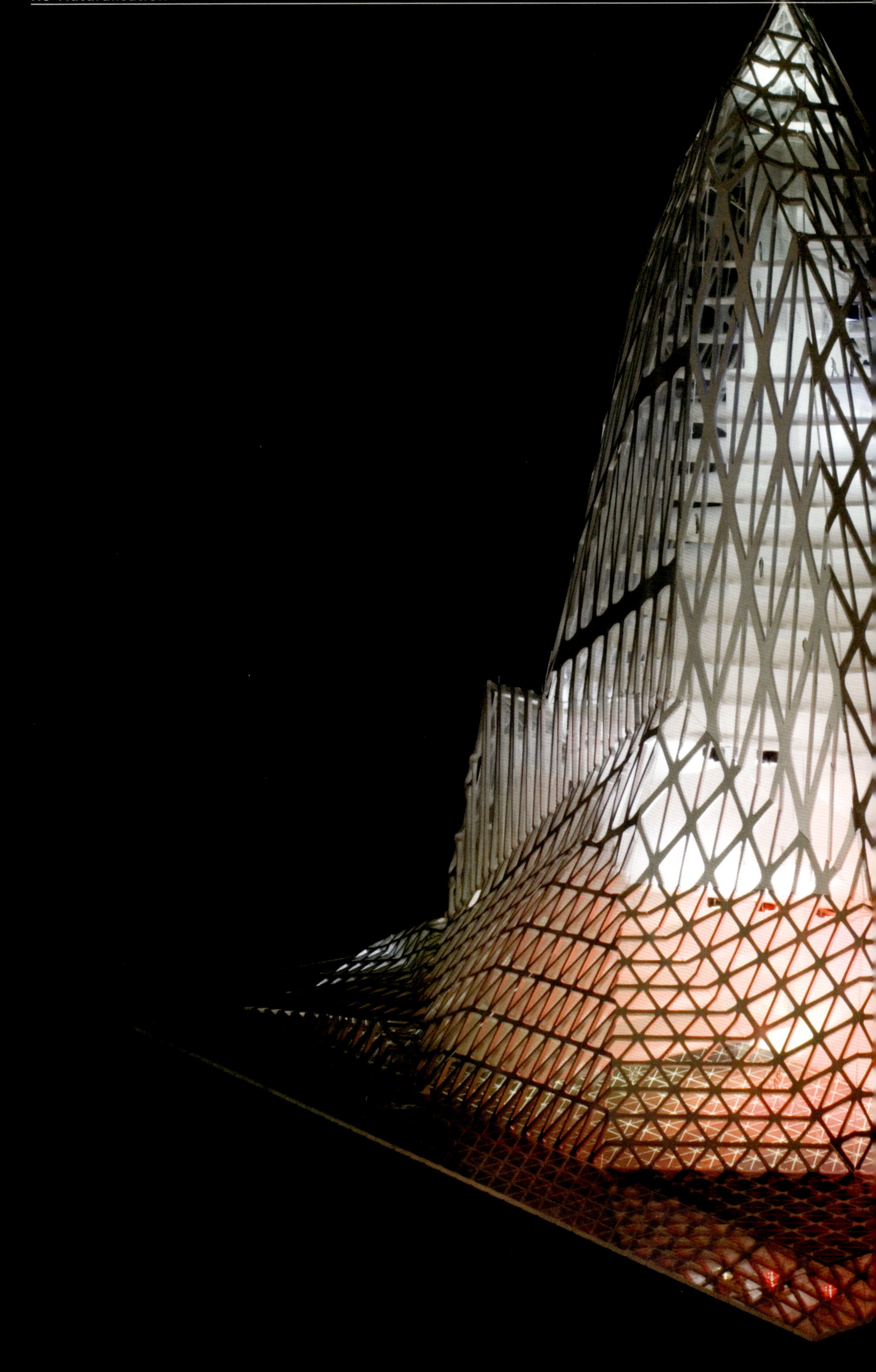

WROCLAW
2012

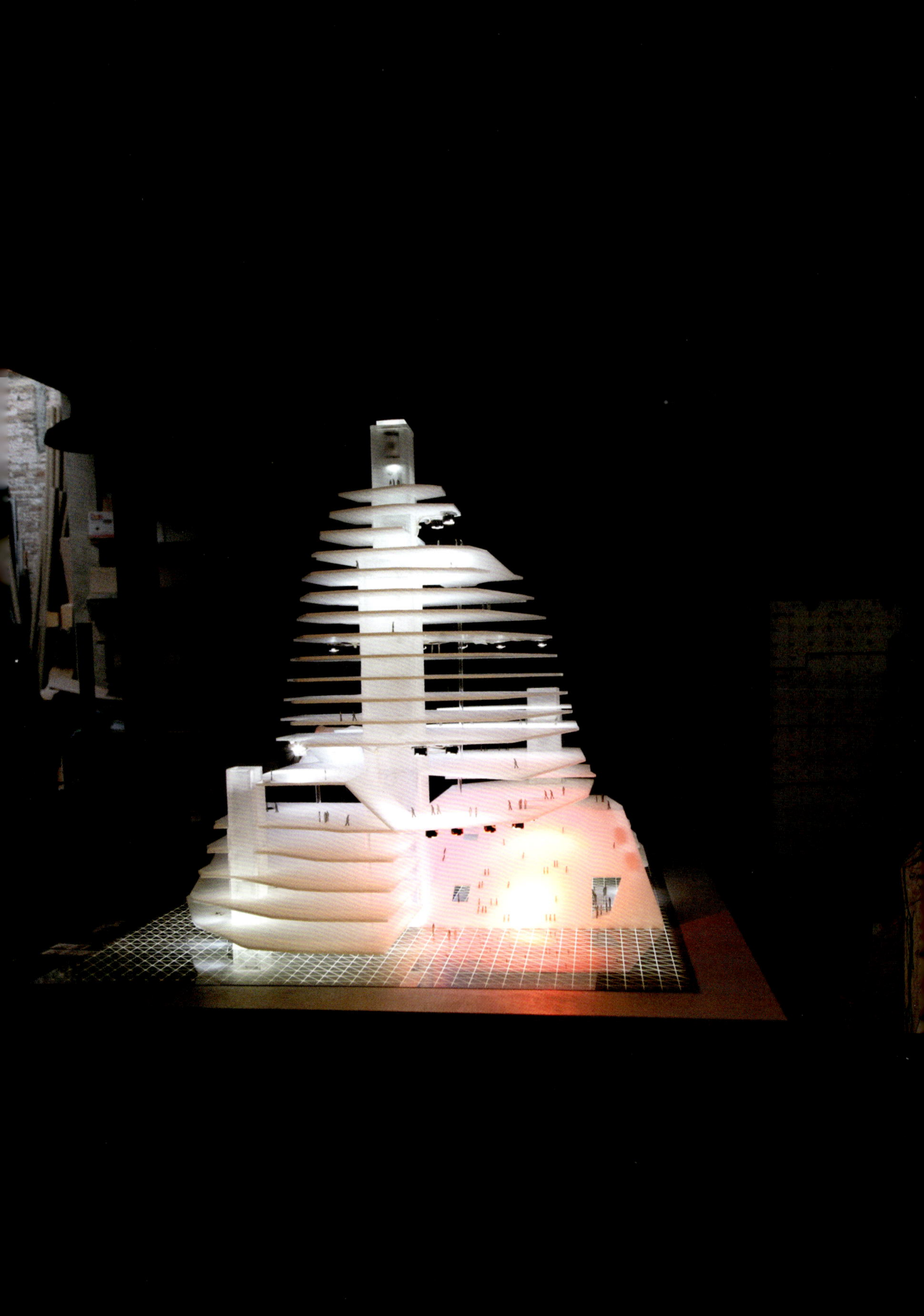

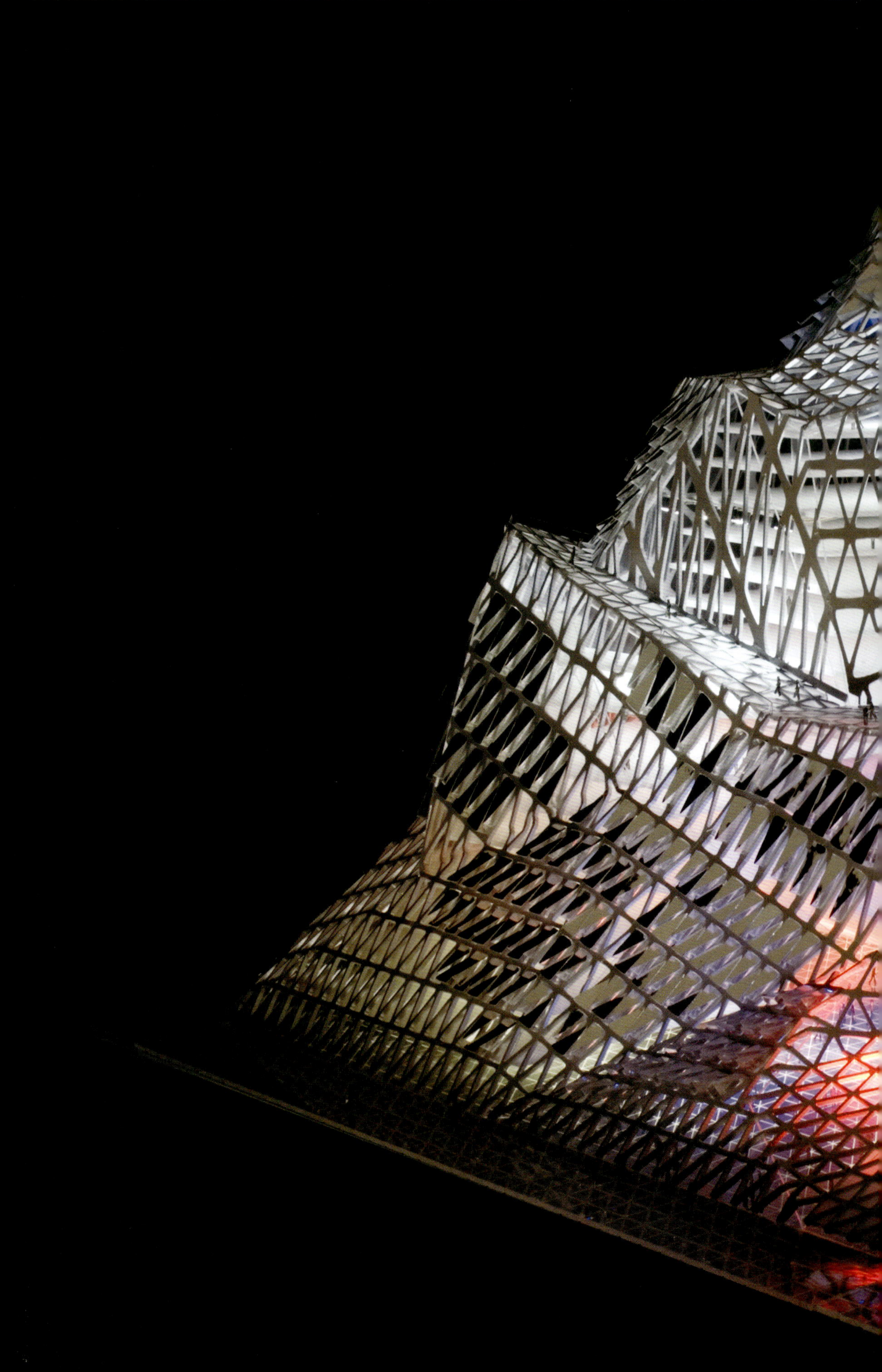

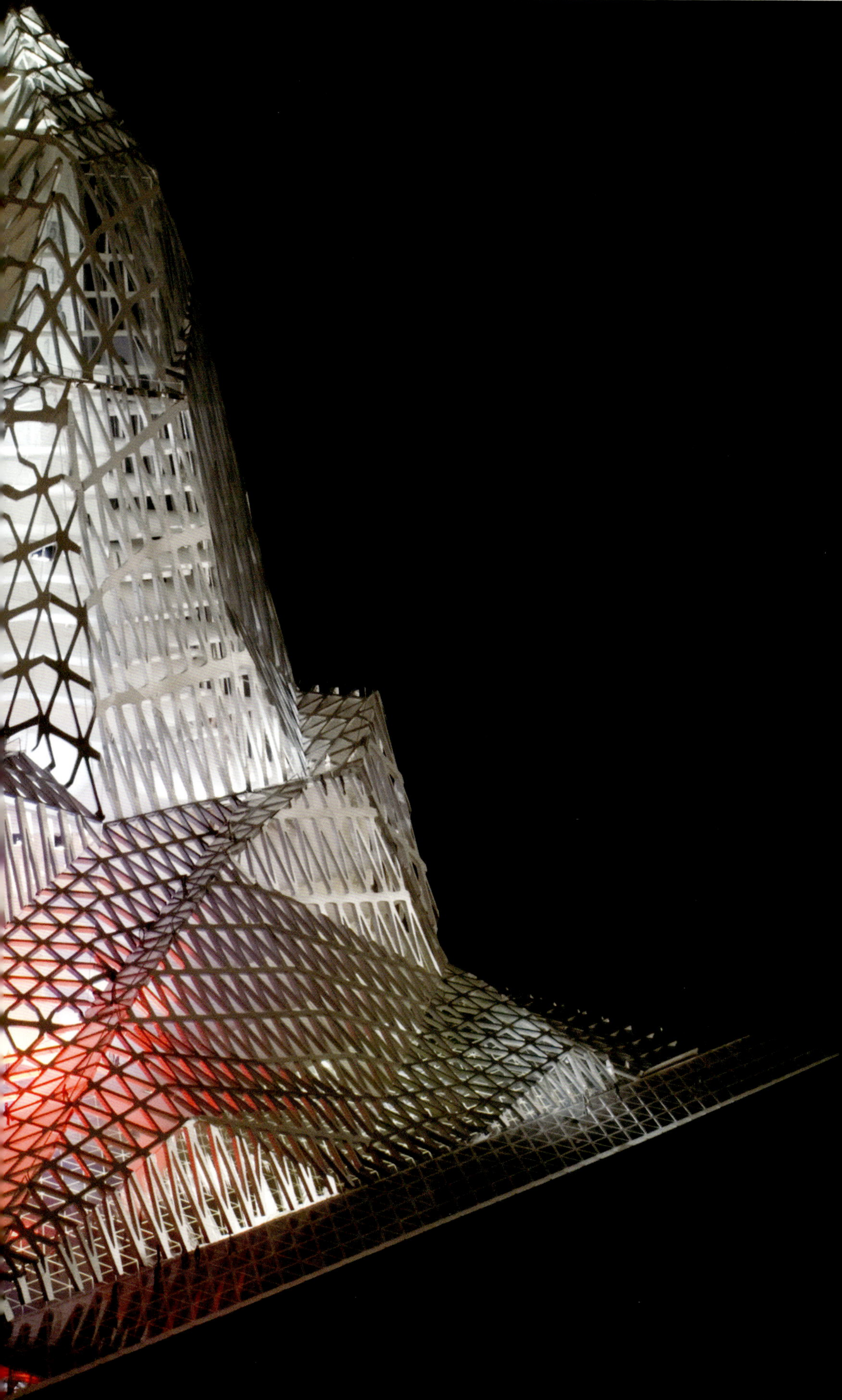

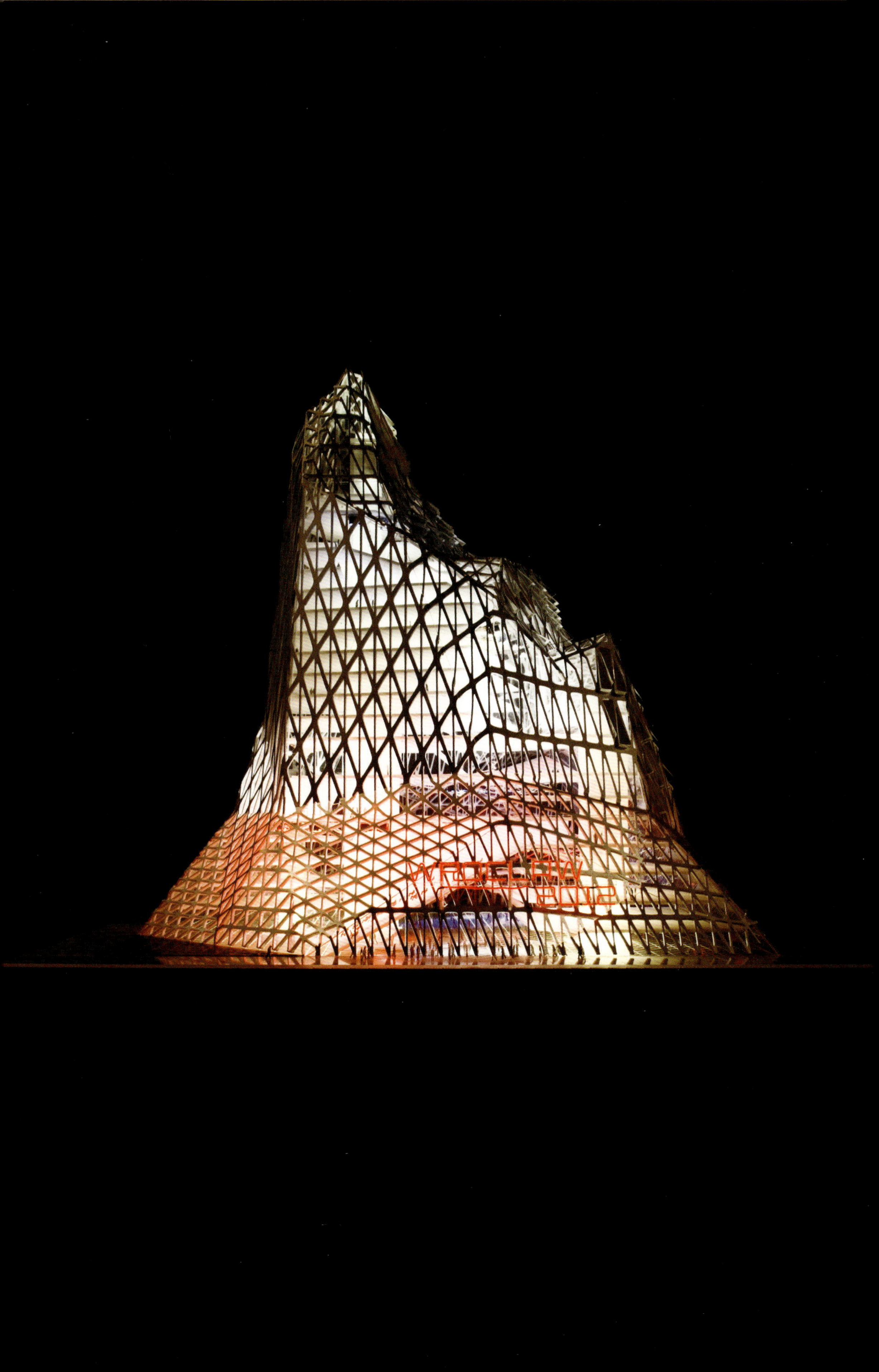
WROCLAW
2012

My projects explore people's capacity to construct an inhabitable environment by assuming this new three way interaction. Starting from the hypothesis that the end purpose of architecture is not (solely) to build but also to define environments where human life can be organized, I work as much through action as through omission ; with stones as well as with glass and "bits" ; in urban and rural territories ; and in the physical and virtual worlds.

我的方案假设一种在三种方式之间探寻新的互动，正是人类对于某个宜居环境的掌控力。假设建筑的终极目的并不（仅仅）是建造——而是来定义人类生活，以组织起来的环境为起点，我的工作对采取行动与放弃干预一视同仁；在使用石材与玻璃的同时，我也使用“比特”；我的工作地点在都市也在乡村，不仅是在物质世界当中也在虚拟世界当中。

VINAROS BAR
VINAROS酒吧

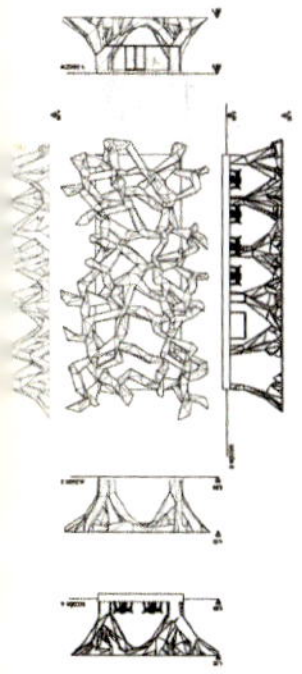

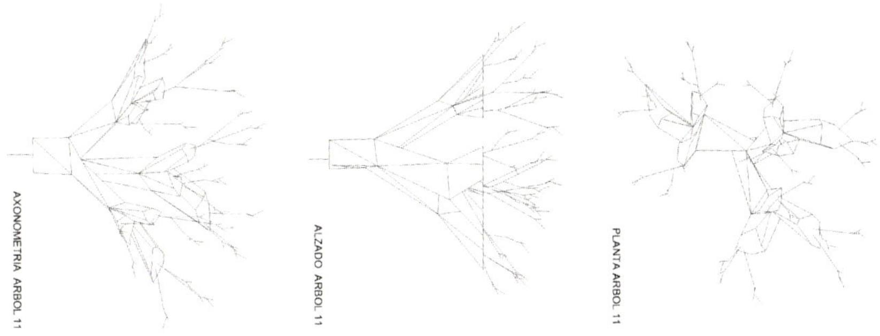
AXONOMETRIA ARBOL 11
ALZADO ARBOL 11
PLANTA ARBOL 11

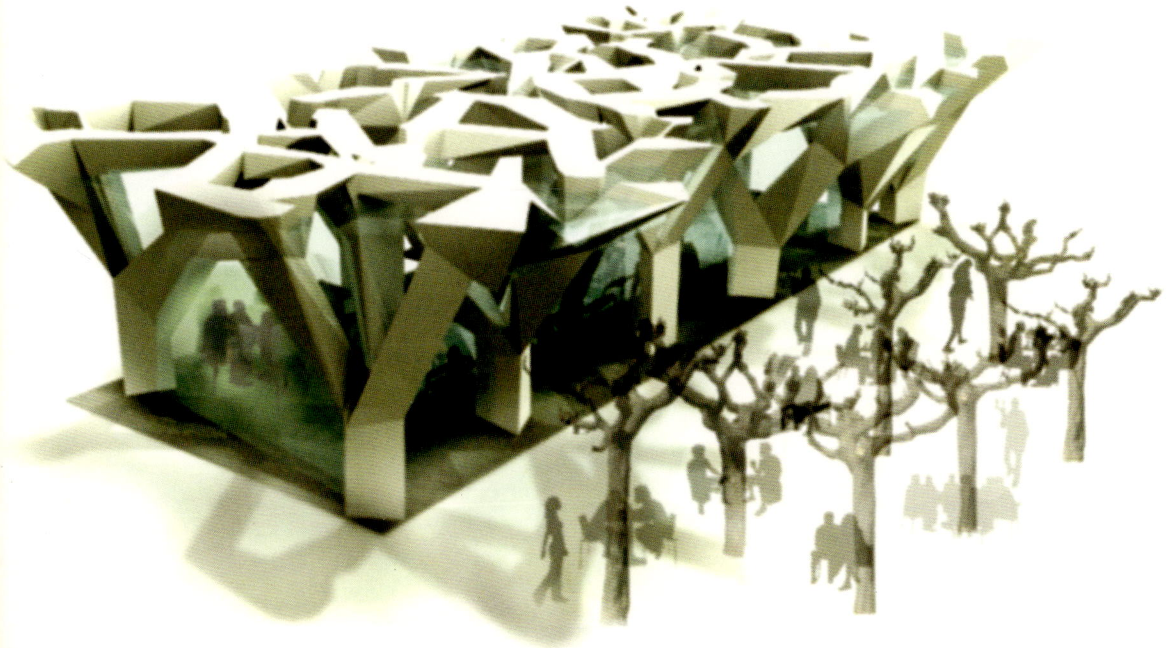

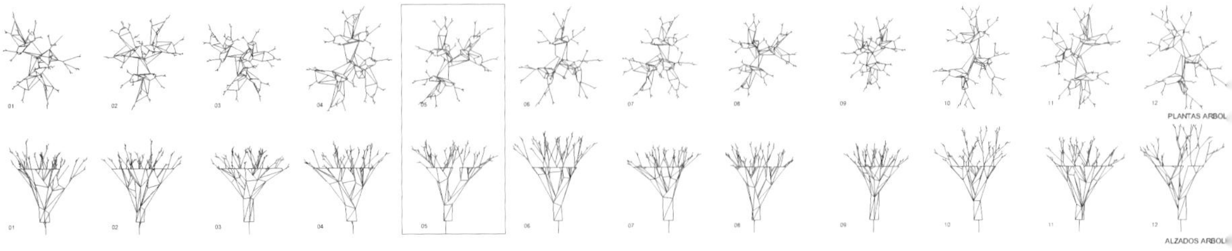

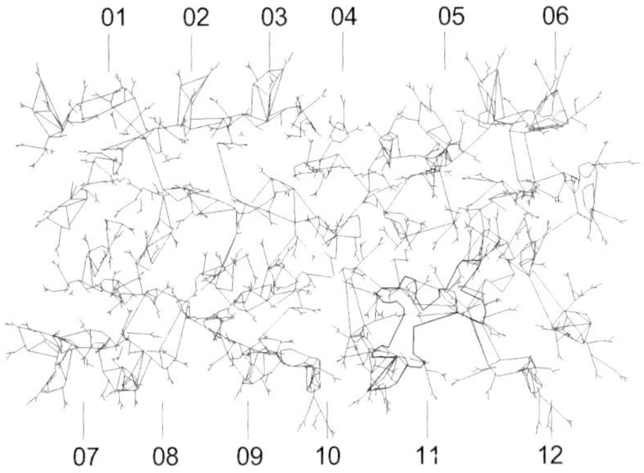

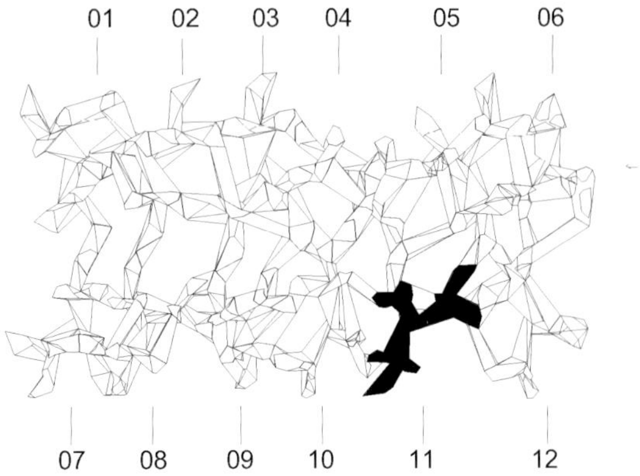

PLANTA GENERAL

PLANTA ARBOLES

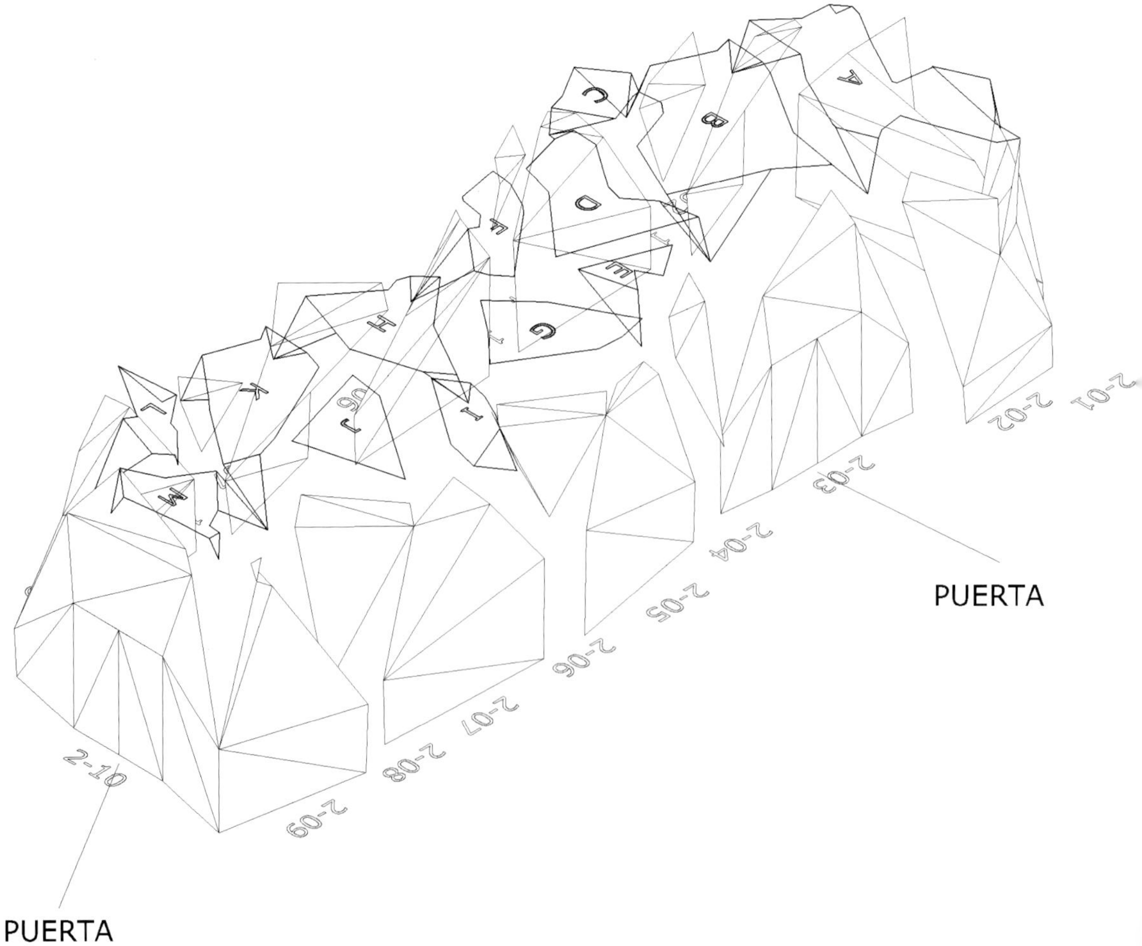

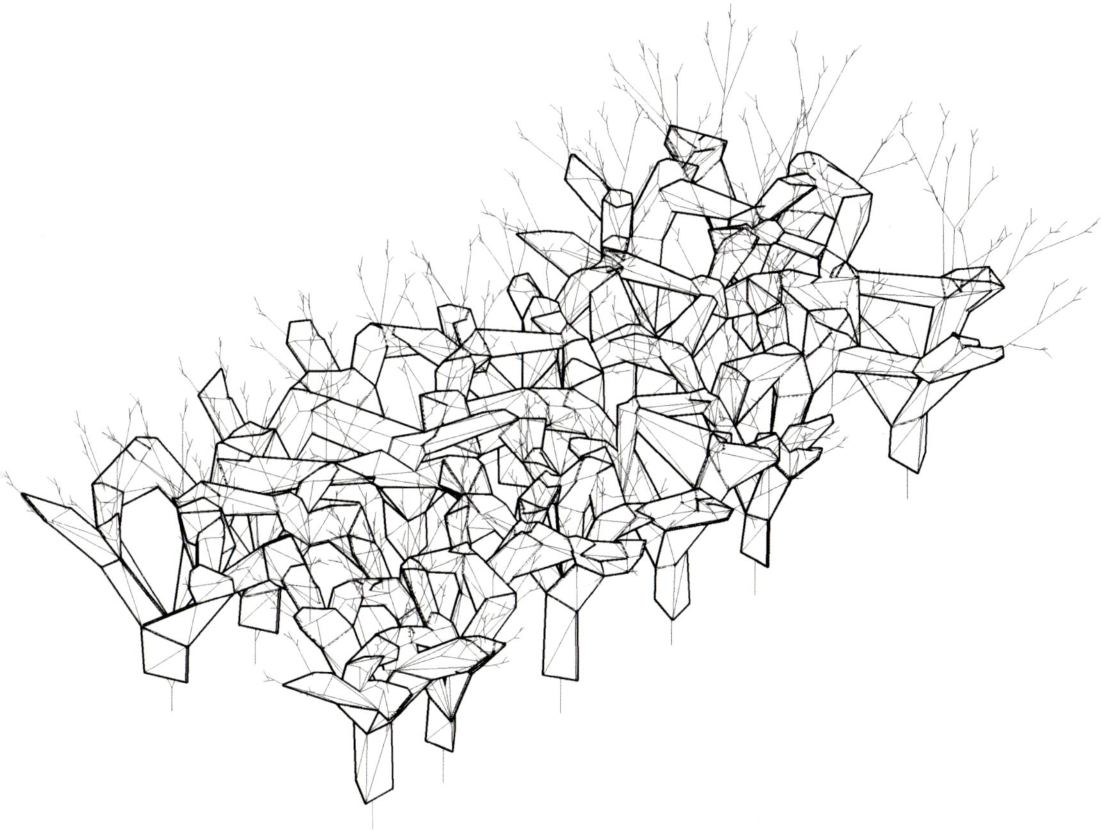

DESPIECE ARBOL 05

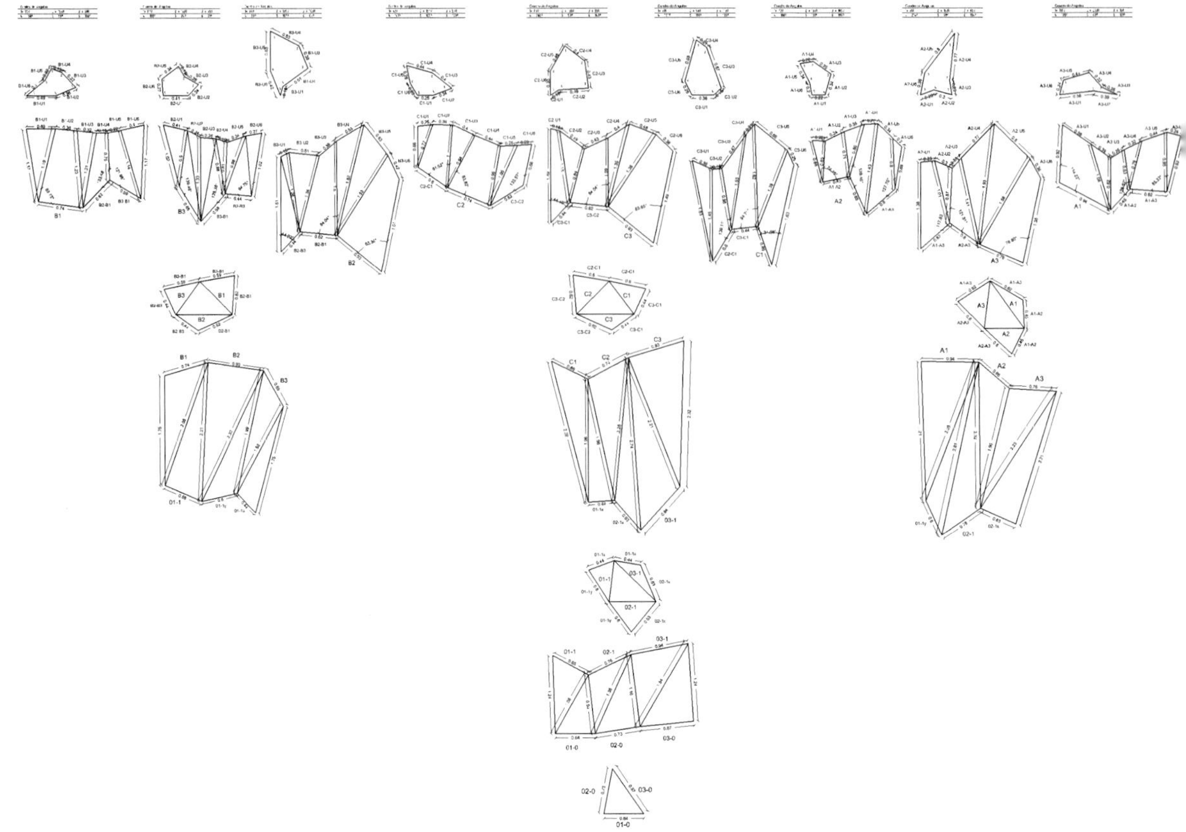

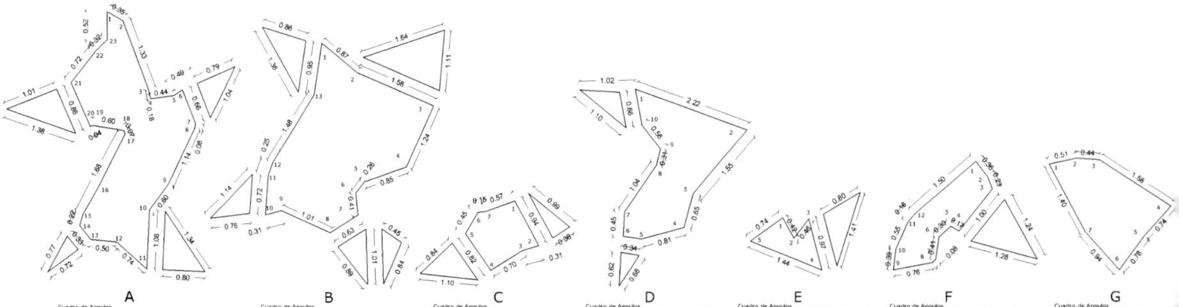

A

Cuadro de Angulos

1= 60º	2 = 141º	3 = 173º
4 = 264º	5 = 205º	6 = 78º
7 = 145º	8= 167º	9 = 168º
10 = 215º	11 = 52º	12 = 216º
13 = 135º	14 = 124º	15 = 167º
16= 183º	17 = 228º	18 = 241º
19= 216º	20 = 87º	21 = 117º
22= 172º	23 = 227º	

B

Cuadro de Angulos

1= 61º	2 = 195º	3 = 87º
4 = 134º	5 = 211º	6 = 226º
7 = 115º	8= 125º	9 = 118º
10 = 72º	11 = 164º	12 = 168º
13 = 203º		

C

Cuadro de Angulos

1= 99º	2 = 76º	3 = 196º
4 = 94º	5 = 113º	6 = 130º
7 = 191º		

D

Cuadro de Angulos

1= 60º	2 = 68º	3 = 205º
4 = 121º	5 = 156º	6 = 96º
7 = 149º	8= 201º	9 = 232.º
10 = 153º		

E

Cuadro de Angulos

1= 84º	2 = 292º	3 = 53º
4 = 54º	5 = 56º	

F

Cuadro de Angulos

1= 89º	2 = 163º	3 = 119º
4 = 99º	5 = 253º	6 = 219º
7 = 145º	8= 144º	9 = 73º
10 = 161º	11 = 163º	12 = 173º

G

Cuadro de Angulos

1= 74º	2 = 164º	3 = 1
4 = 79º	5 = 186º	6 = 8
7 = 165º		

DESPIECE DE CRISTALES 6+6 CUBIE

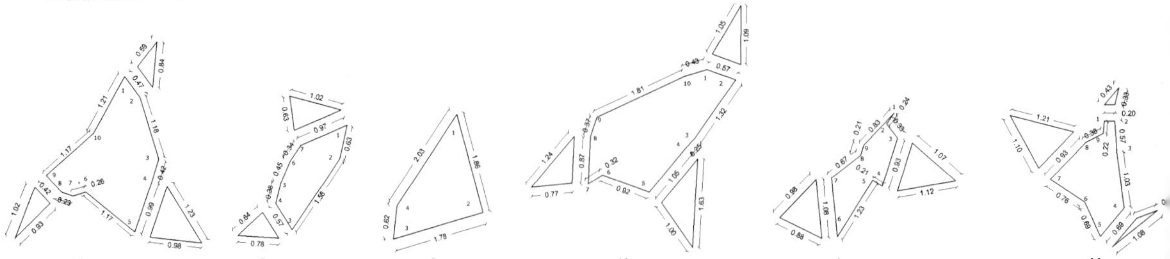

H

Cuadro de Angulos

1= 66º	2 = 161º	3 = 145º
4 = 179º	5 = 71º	6 = 234º
7 = 146º	8= 151º	9 = 87º
10 = 200º		

I

Cuadro de Angulos

1= 54º	2 = 159º	3 = 68º
4 = 143º	5 = 168º	6 = 169º
7 = 147º		

J

Cuadro de Angulos

1= 50º	2 = 91º	3 = 67º
4 = 152º		

K

Cuadro de Angulos

1= 144º	2 = 73º	3 = 172º
4 = 176º	5 = 110º	6 = 304º
7 = 128º	8= 164º	9 = 133º
10 = 172º		

L

Cuadro de Angulos

1= 9º	2 = 254º	3 = 123º
4 = 77º	5 = 269º	6 = 38º
7 = 122º	8= 206º	9 = 146º

M

Cuadro de Angulos

1= 95º	2 = 94º	3 = 181º
4 = 132º	5 = 66º	6 = 211º
7 = 79º	8= 161º	9 = 128º

DESPIECE DE CRISTALES 6+6 CUBI

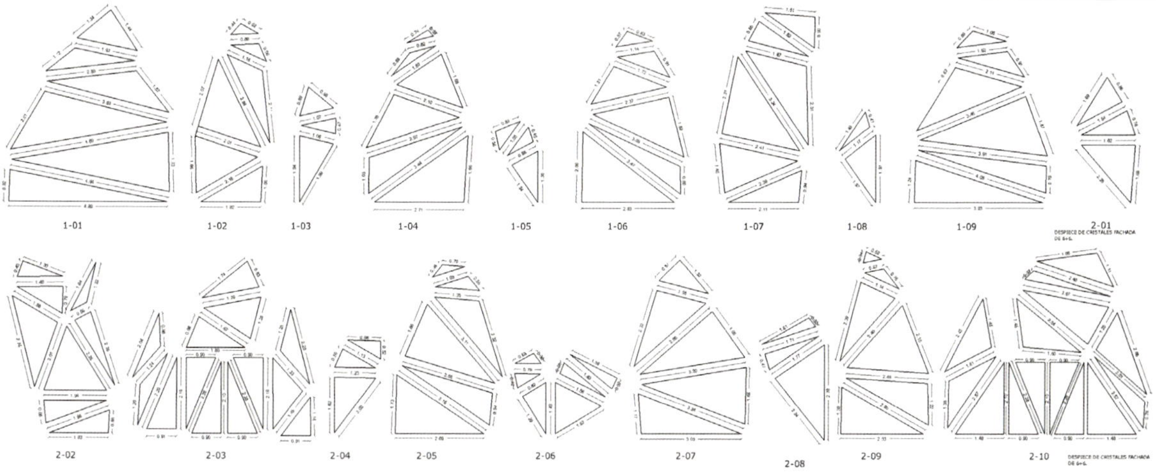
1-01
1-02
1-03
1-04
1-05
1-06
1-07
1-08
1-09
2-01
DESPIECE DE CRISTALES FACHADA
DE 6+6.
2-02
2-03
2-04
2-05
2-06
2-07
2-08
2-09
2-10
DESPIECE DE CRISTALES FACHADA
DE 6+6.

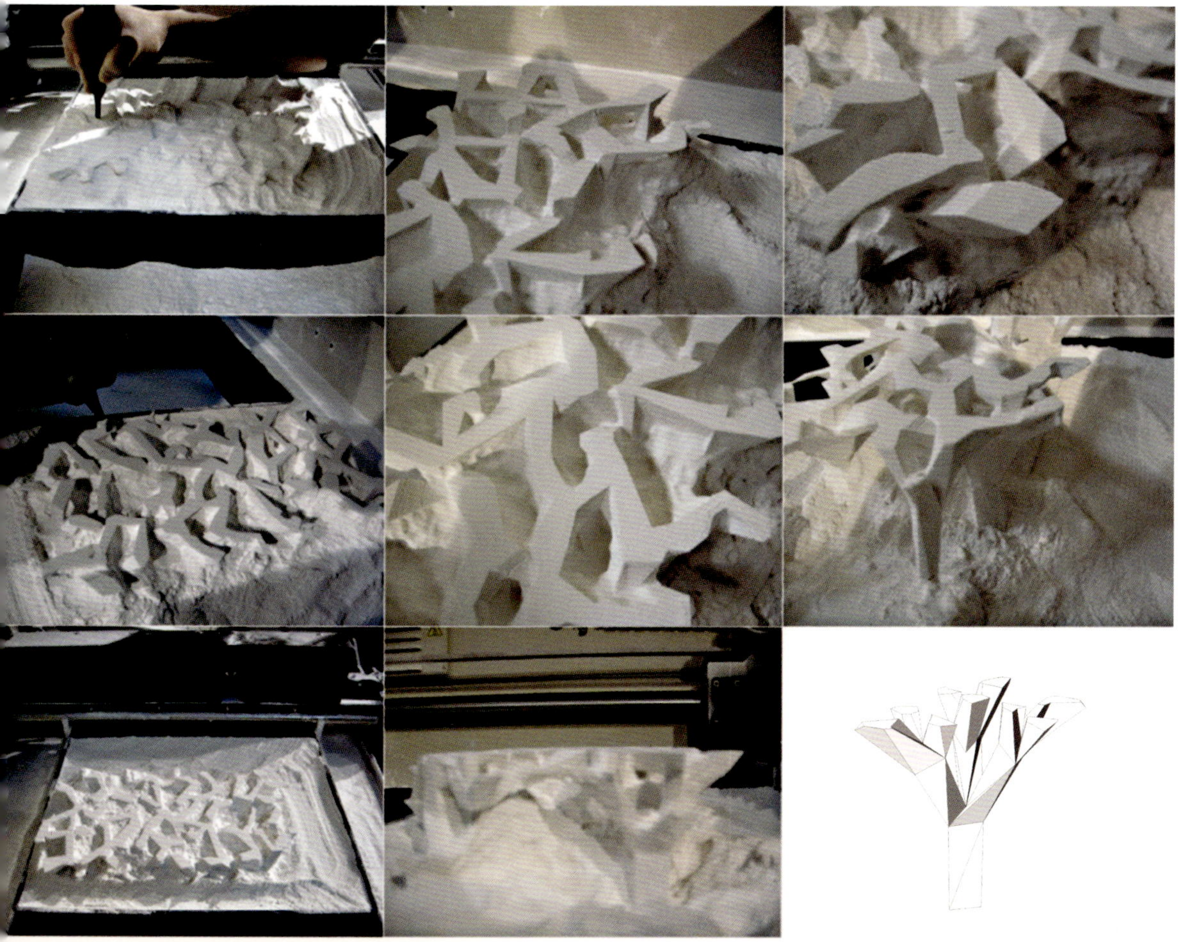

SHANGHAI EXPO 2010
2010上海世博会

The Spanish Pavilion for the Shanghai Expo project itself as a recognizable space which connects with the culture of Spain and generates synergies in the Chinese context. The pavilion is clearly iconic and develops the theme of Expo —'Better city, Better Life'— to propose that cities be developed on the basis of a formal and functional hybridization of the natural and the artificial. The scheme creates a plantation of artificial trees, surrounded by the fruits of Spanish agriculture, especially orange and lemon trees. The trees, of laser-cut sheet metal, wil constitute the physical structure of the building, and will also contain its metabolic systems. Photovoltaic panels on the tops of these trees will provide the building with energy, which will be distributed through the structural elements to the spotlights that will indirectly illuminate the interior spaces. The dimensions of the structure (which in places reaches an internal height of five metres) ar sufficient to house the air conditioning and water vapour systems. The elevators are also incorporate into these structures. This makes it possible to create large empty spaces to accommodate the activities of the Expo. The same principle has been applied to a series of smaller pavilions to be constructed all over Spain.

The project serves to explore the development of a parametric arboreal structure based on the L-system in which a constant section is applied to th three-dimensional structure. The proliferation of trees is sufficient to allow them to be grouped into hypostyle structure.

上海世博会西班牙馆的项目本身是一个与西班牙文化相联系同时在中国的文脉下生成了增效作用的可识别空间。该馆具清晰的标志性，同时也顺应本次世博会的主题“更好的城市更好的生活”提出城市发展应当以自然与人工在形式和功能的交融为基础。这一设计创造了一种人工树木植被，馆体被班牙特产的果树诸如橙树和柠檬树所包围。这些激光裁切金属片所构成的树木将会组成建筑的物理构架，同时也包含自身的新陈代谢系统。这些树木树冠处的光合作用面板将会建筑提供能源，并通过结构性元素分散到聚光灯处来间接为室内提供照明。结构（内部高度为五米）的尺度足够放置调和水蒸气系统。电梯也被嵌入到这些结构当中。这使得创大的空旷空间来容纳世博会活动成为可能。同样的原则也被用于一系列建造在西班牙馆主体周围的小型馆当中。

这一设计是人们为了对于一种全景式球状结构的探索，该结所基于的L系统中，一个恒定的切面被应用于三维结构当中树木的繁殖使得它们能够被分组纳入一个多柱式结构之中。

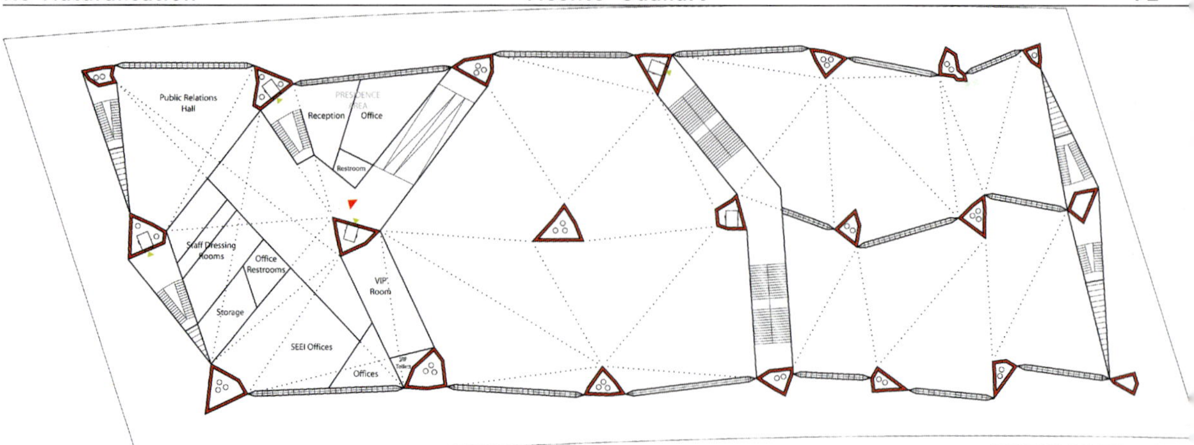

OFFICES PLAN LEVEL +6.00

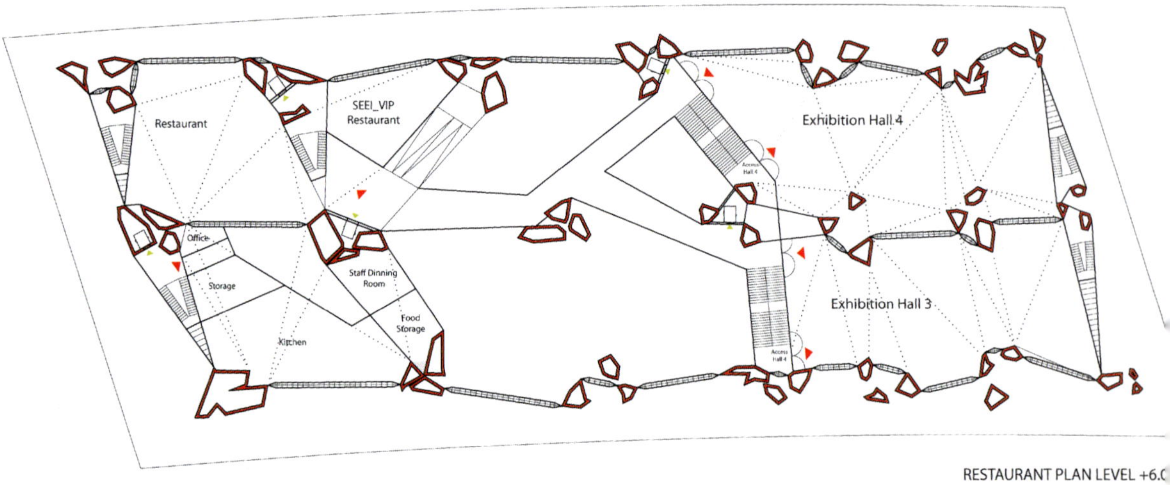

RESTAURANT PLAN LEVEL +6.(

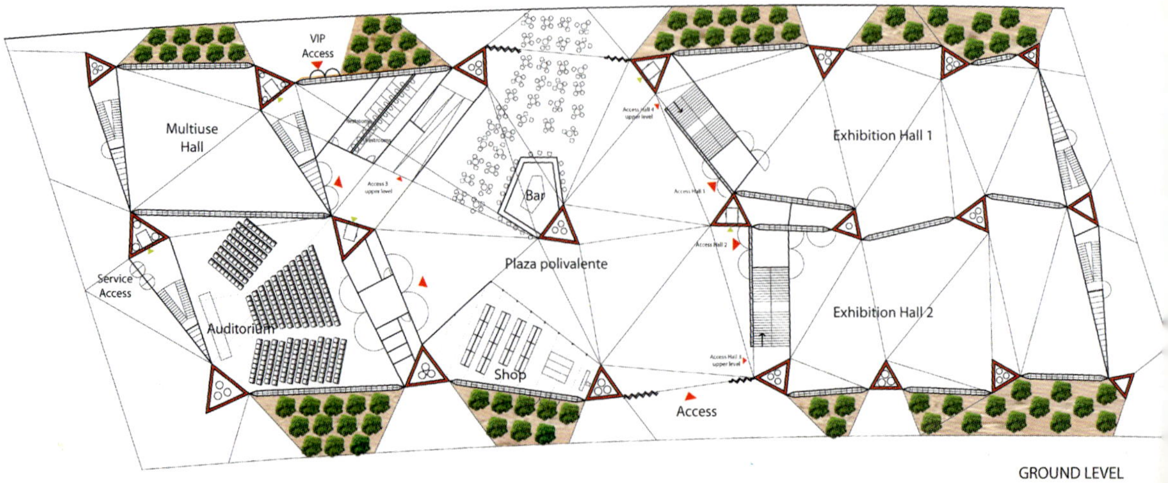

GROUND LEVEL

As a traditionally constructed interface, architecture must thus redefine its aims in the face of the emergence of the digital world, which encompasses and transforms everything, and through which people become immersed in a new reality. This situation might in addition give rise to an historical gap between a dazzling, luminous, thrilling and creative world on the one hand, and a deteriorated, dark and decadent physical world on the other.

作为一种传统的建造表面，面对数字世界的出现，建筑必须要重新确定其自身的目标，而这也催生并转化了一切，而人们正是通过这些活动逐渐沉浸到一个新的现实当中的。这种情形更可能加大一个光彩熠熠充满令人兴奋的创造力的世界与历史之间的鸿沟，而这个世界另一方面看上去更可能是一个恶化的，昏暗的并且不断衰退着的物质世界。

VINAROS COAST
VINAROZ海岸

RIBERA´S PROMENADE IN THE SOUTH COAST OF VINAROZ

The coast line on the Valencia's community has 474 km. An integrated project, like the one we are proposing will allow a continuous connection, for pedestrians and bicycles of the littoral space from Vinaroz until Pilar de la Horadada. This connection could adopt in it's different stages adequate morphologies decided by the local government.

As Vinaroz location is towards the north part, it has to adopt the compromise to start this territorial program with a very creative, innovative and audacious way, valuating the quality of their marine space for their inhabitants, visitors and tourists.

The growing activity around the coasts makes necessary that next to the traditional marine promenades of the urban nucleus and the littoral beaches, to organize them to allow their use and enjoyment of the rocky zones and the difficult access zones.

Also, next to this architectural and urban actions it would be needed to make certain actions of an ecological character in order to avoid coast erosion. Th could be solved by bulding an artificial breakwater (by building rock walls that protect the natural terrai from the direct erosion of the waves.

The blue footpath intents to assure the continuity in the pedestrian or cycling mobility along the whole coast. It's a path with contents. A route that advocates the richness of the local territory, this is why it carefully deals with its immediate surroundings, creating a continuous route with resting areas and observation points, which allow the user to take advantage of the most amazing urban and littoral landscapes.

In the south coast of Vinaroz we find stone beaches rocks and small cliffs that altogether can add up 2.892 meters.

The recognition of this characteristic was the previous step to decide the following actions with the following catalog of elements:

Pedestrian promenade

All along the coast one would be due to be able to circulate very slowly, walking, jogging or in the bike in a continuous way. The bicycle lane will be of basa tic rock to allow the transpiration of the rained wate

Automobiles and Bicycles parking

In the areas that we can dispose of space, it will be necessary to organize parking zones from the highway, in order to organize the circulation around the beaches. The deck will be made of gravel and sand avoiding the prolongation of asphalt towards the beach.

Access

Seen the interest of people towards the coasts for the human enjoyment and for bathing, it will be

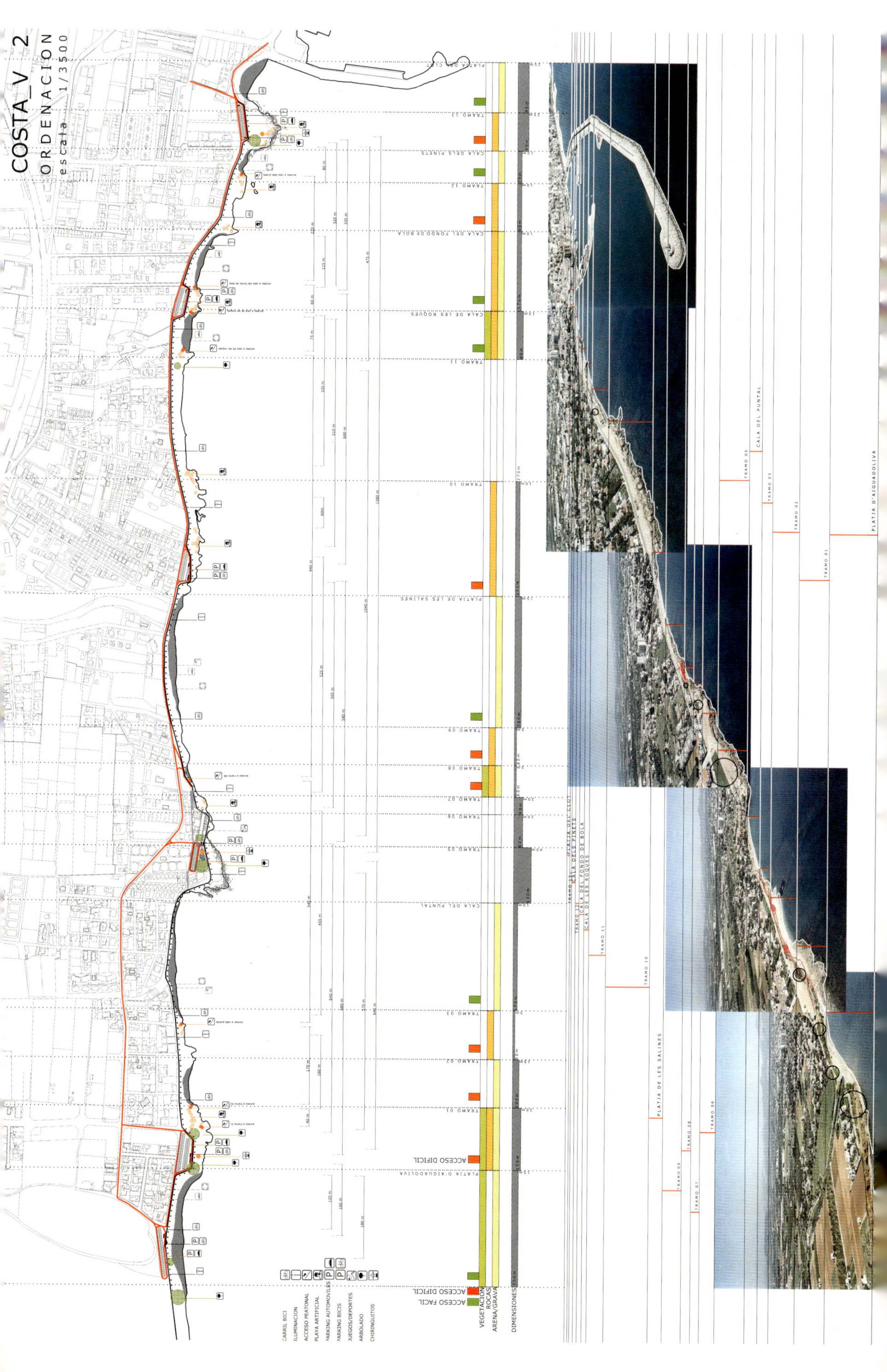
COSTA_V_2
ORDENACION
escala 1/3500
CARRIL BICI
ILUMINACION
ACCESO PEATONAL
PLAYA ARTIFICIAL
PARKING AUTOMOVILES
PARKING BICIS
JUEGOS/DEPORTES
ARBOLADO
CHIRINGUITOS
ACCESO FACIL
ACCESO DIFICIL
VEGETACION
ROCAS
ARENA/GRAVA
DIMENSIONES
PLATJA D'AIGUADOLIVA
CALA DEL PUNTAL
PLATJA DE LES SALINES
CALA DE LES ROQUES
CALA DEL FONDO DE BOLA
CALA DELS PINETS
PLATJA DEL CLOT

needed to regulate and organize the accesses to the gravel beaches and the rocky cliffs with wood stairs that can be easily demountable.

Artificial beach
Based in an hexagonal patron we defined a series of wood platforms that allow the organized enjoyment of the rocky beaches. These platforms encounter themselves in a very flexible way, what permits to adapt themselves to the roughness of the terrain.

Lightning
All along the coast you'll find a lightning system that will illuminate all the traffic zones. We defined a system of variable heights that allow different intensities in the ambience so that it doesn't "contaminate" all the coast with a constant light.

Trees
In the "relax areas" a different variety of native trees will be placed: from one side, palm trees (phoenix dactilifera or phoenix whasingtomnia) and in the other side low native bushes like the crithmo limonietea or the hinojo marino.

Playground areas
In the limit between the city and the sea, we can dispose certain zones of games and sports in order to donate another important content to the coast. Certain ESI (Elemental sports installations) will be installed, where the urban impact will be minimum.

"Chiringuitos"
Certain "Chiringuitos" will be located near the sports facilities, the green zones or the beach accesses; certain with bars and others with snack places (certain with services, certain without). They will be wooden and they will have a second textile cover. They will be mounted in a wooden platform.

Vinaroz南海岸的Ribera游憩大道
瓦伦西亚社区的海岸线长达474公里。正如我们所提出的那样，这一综合性项目将会为从Vinaroz到Pilar de la Horda-da滨海空间中的行人与自行车使用者提供一个不间断的连接道。这条连接道能够根据当地政府所需要的变形原则来将自身调整至不同的状态。
由于Vinaroz的位置朝向北部，它必须调整出一种中庸的态势，以一种非常具有创造力，创新性且富有挑战力的方式来启动这个区域性项目，对该区域的居住者、到访者以及旅游者所处的滨海空间的质量进行评估。
海岸附近的增长活动使得组织紧邻都市中心的传统滨海大道以及海滨沙滩成为必要，并能够利用到附近的峭壁地区和难以进入的地区。
同样，除了这些建筑性的都市活动以外，为了防止海岸被腐蚀，一系列具有生态特性的措施必须被采取。我们可以通过建造人工防波堤或是石墙来保护自然的地形不受到波浪的侵蚀。

蓝色步道保证整个海岸沿线的行人和自行车移动的连续性。这是一条提倡当地地区丰富性的道路，这也就是为什么它极为谨慎地处理与其紧邻环境之间的关系，创造了一种富有休憩区域和观景点的连续道路，并为使用者提供那些最具有魅力的都市景观和滨海景观。
在Vinaroz的南岸，我们发现了石滩、岩石与达到2.892米的小型峭壁。这种特性化的识别正是决定该对以下类别所包含的元素采取何种措施的前提：

行人散步道
沿着整条海岸，必须保证人们具有在一个连续的路径上缓慢地环绕、散步、慢跑或是骑行的能力。自行车道将会用玄武岩石来建造以保证雨水的顺利吸收。

汽车和自行车停车场
为了组织起一个环绕海滩的循环交通系统，在我们能够分解空间的地方为来自高速公路的交通流组织停车区域是非常必要的。停车平台将会使用砂粒和沙来建造以避免沥青对海滩的侵蚀。

接触
得知人们对于海岸边娱乐活动以及海水浴的兴趣，通过建造易于拆卸的木制台阶来规范和组织人们对于砂粒海滩和岩石峭壁的接触是非常必要的。

人工海滩
以六角形平台为基础，我们设定了一系列木制平台来保证石滩上有组织的娱乐活动。这些平台以一种可变的方式组合起来，这样的特点使得它们能够很好地适应地形的起伏。

照明
沿着整个海岸线人们都可以感受到一个照亮整个交通区域的照明系统。我们设定了一个包含了多种高度的系统，以保证不同密度的氛围，让整个海岸免受一种恒定灯光的污染。

树木
多种完全不同的本地植被将被设置在"放松区"内：一边是palm树（拉丁名phoenic dactilifera或phoenic whasingtomnia），另一边则是诸如crithmo limonietea或是hinojo marino之类的当地生矮灌木。

游乐场区
在城市与海之间的有限空间内，我们能够分散出某些空间来为游戏和体育活动提供场地，这一举措同样为海岸活动贡献了一个重要的内容。
在这些都市影响最小的地方，会被安装相当数量的ESI(基础体育装置)。

Chiringuitos
体育设施、绿化区域或海滩的附近将会设置一些"chiringuitos"；其中一些将会配备酒吧或是休闲餐厅（一些备有餐饮服务，一些没有）。它们将会由木制材料建造而成并有第二层松覆盖，被融入木制平台当中。

THE WOODEN BEACH

In those parts of the coast where rocky beaches are at the sea level, we propose a series of modular wooden platforms that allow to enjoy this artificial beach and to create a new organized bathing area. This concept is very popular in certain Mediterranean countries like Italy and the countries from the old Yugoslavia, and done with the accurate sensibility it doesn't bother the littoral landscape and it improves the functional qualities of this difficult coast accidents.
This hexagonal platforms incorporate different relieves that permit the users to lay down or sit on top of them.

THE ORGANISED MOBILITY : A BICYCLE LANE AND A PEDESTRIAN WALK

Next to the traffic lanes, we consider necessary to establish new lanes for the bicycles and for the pedestrians that will permit a more human scale, danger free for the children and young users. This route need to be continuous all along the coast which is basically the whole idea of the project. In addition to the different observation points and the relax areas that will be equipped with the appropriate furnish for each. These lanes will answer as well to a series of vehicles and bicycles parking.
The adequate combination of these spaces will organize the vehicle access to the coast and will generate a powerful and continuous image that can become over time a new territorial identity symbol. This is why we propose a signalization program unique to Vinaroz.

THE RECONSTRUCTION OF THE EROSIONED MARINE FRONT

Vinaroz has a prototype coast of low cliffs, formed by a conglomerate of rocks and compacted earth that can suffer from erosion very easily. The construction of buildings near the coast and the erosion of the same has produced as an effect that certain parts of the coast are completely inaccessible.
In the critic parts where access is very difficult we propose a wooden passage of 3 meters wide, with which we could solve the access problem with a very reduce and ecological cost.

PLAYGROUNDS IN FORNT OF THE SEA

In the superior platforms of the steep ends, we propose certain playgrounds, of a very low maintenance cost that will serve as well as point of encounter for the surrounding low density urbanizations. This zones will have as well palm trees and native vegetation like the crithmo limonietea or the hinojo marino.

木制海滩

在那些岩石海滩位于海平面的海岸，我们提出了一系列模数木制平台的设想，以满足在这片人造海滩进行娱乐活动的求，同时创造一个新的秩序化的海水浴区域。这一概念在诸意大利这类的地中海国家和旧Yugoslavia这类国家当中非流行，并且能够以精确的探知力实施，令其在不干扰滨海景的同时提升这类难以处理的海岸空间的功能利用。
这种六角形平台融入了不同程度的放松空间来允许使用者或坐在上面。

秩序化的移动性：自行车道以及行人步道

我们认为在紧邻机动车道的地方有必要为自行车单独建立的行驶车道，同时也为儿童和年轻的使用者提供了一个更加有宜人尺度的安全空间。这条路线需要沿海岸连续起来，这好是该方案的主旨。除了不同的观景点以及休闲区域将会通适度装饰的形式予以呈现，这些车道也将对一系列的机动车自行车的停车问题予以回应。
适度地将这些空间结合起来组织机动车靠近海岸并生成一有力量的连续意向，这样的意向经过时间的雕琢将会形成一新地域特性的标志。这就是为什么我们要建议一个Vinaroz有的符号化过程。

重建被侵蚀的滨海地区

Vinaroz特有的低悬崖海岸是由非常容易遭受侵蚀的岩石与的聚合体所构成。海岸附近的房屋建造活动和房屋被腐蚀的程共同造就了海岸的某些部分完全不可接近这一事实。
在环境条件严峻以至于人类的活动变得十分困难的地方，我提议建造一种宽3米的木质国道，用一种非常集约的生态型源来解决人们的进入问题。

面朝大海的游乐场

在平台最高层的末端，我们设计了一些日常维护费用很低的乐场，它们同时也为海岸周围的低密度社区提供服务。这些域将会种植桉树和一些诸如Crithmo Limonietea或Hinc Marino这类的本地生植物。

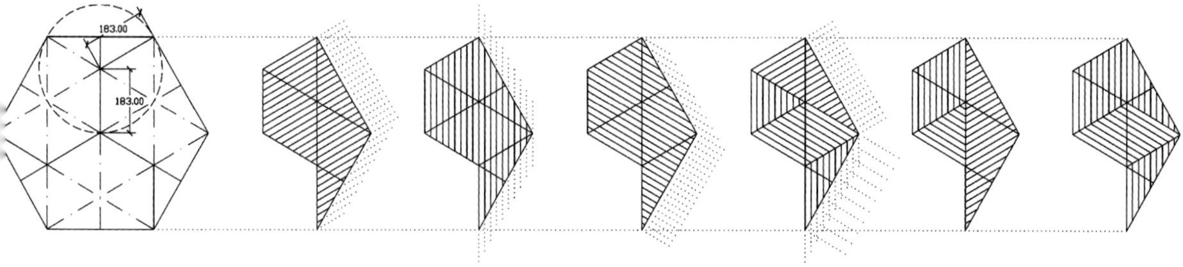
183.00
183.00

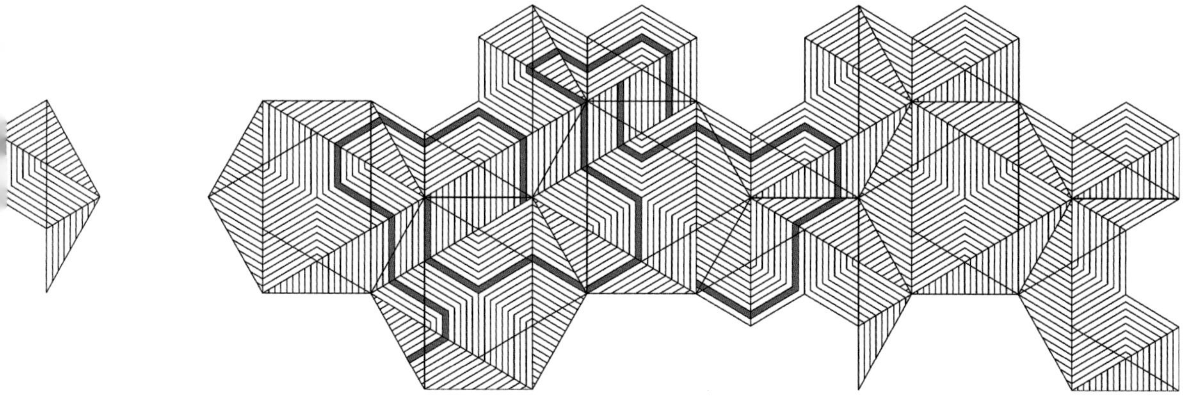

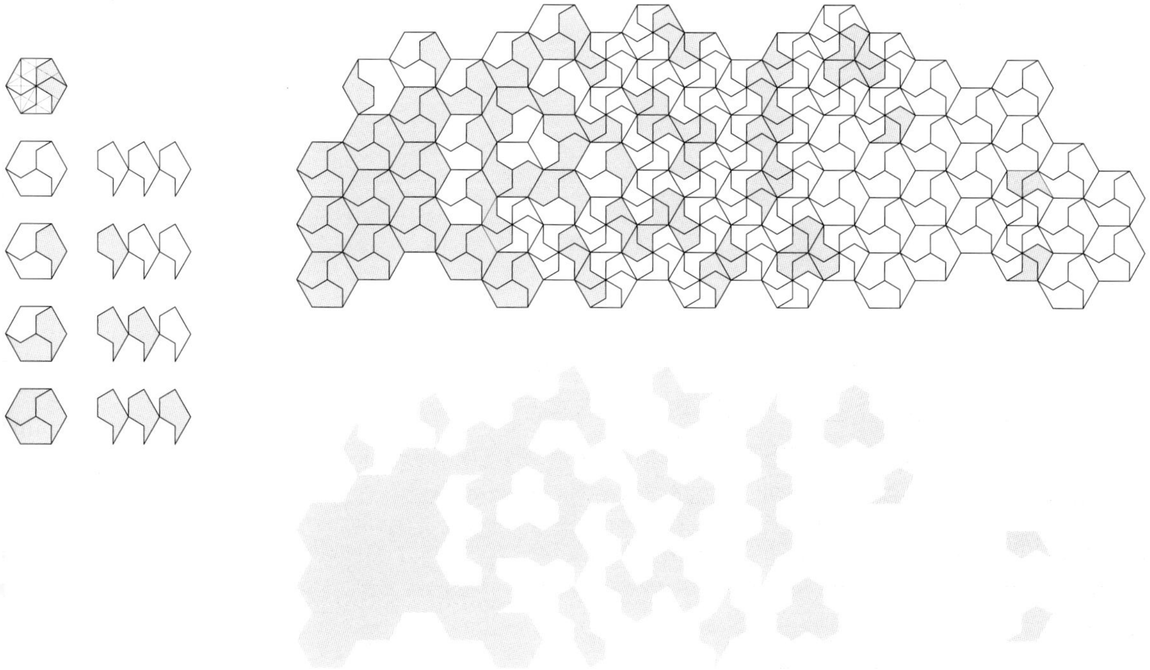

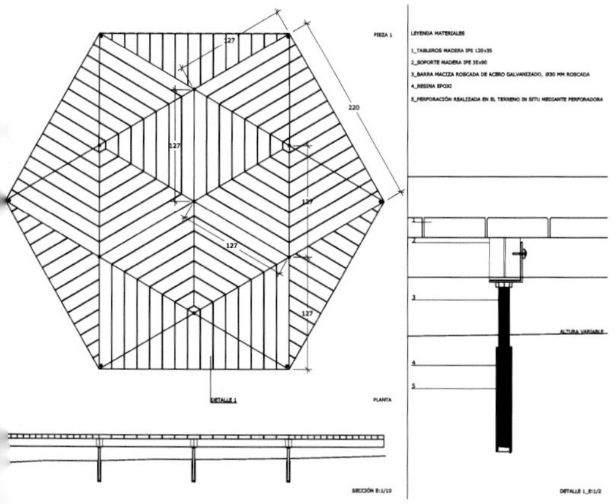
PIEZA 1
LEYENDA MATERIALES
1_TABLEROS MADERA IPE 120x35
2_SOPORTE MADERA IPE 35x90
3_BARRA MACIZA ROSCADA DE ACERO GALVANIZADO, Ø30 MM ROSCADA
4_RESINA EPOXI
5_PERFORACIÓN REALIZADA EN EL TERRENO IN SITU MEDIANTE PERFORADORA
ALTURA VARIABLE
PLANTA
DETALLE 1
SECCIÓN E:1/10
DETALLE 1_E:1/2

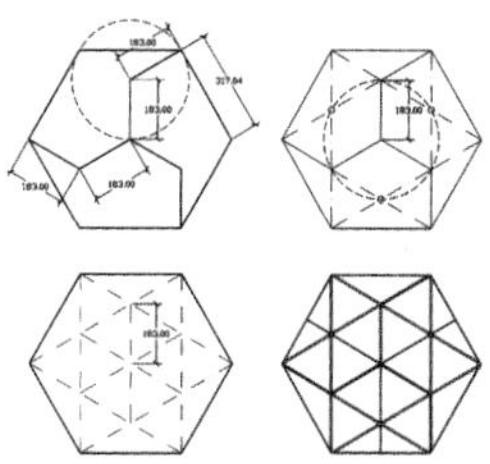

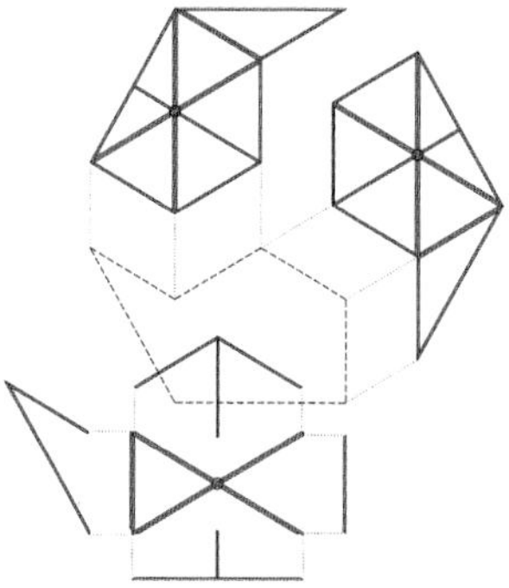

KEELUNG MARITIME PLAZA PHASE 1

基隆海洋广场一期工程

SITE

Keelung is the traditional port of Taipei, a key location for the export of goods and technologies from Taiwan to the rest of the world.
The city's urban centre is thus oriented towards the port. At the same time, however, the activity of the port has now moved towards areas closer to the sea
As is the case in so many port cities at the start of the 21st century, the principal point of interaction between city and port requires a major conversion. Nevertheless, in Keelung the bulk of the city's traffi still circulates in the centre of the city.
In the light of this, until this situation changes, it is necessary to develop strategies that permit the coexistence of the port's activities and intense traffic, with an urban development of real quality.
The presence of the bus and train station on the we side of the port can be used as a significant opportunity to transform the zone.
With this in mind, the actions we propose to undertake here are the following:

1. GATEWAY — Tranformation of the bus and train station to make this the authentic GATEWAY to the city, building a business centre over the new railwa station and a complex for shops and restaurants ov the bus station that will act as a public transport transfer zone, enhancing the urban quality of the square in front of the station. **2. CIVIC AXIS** — Creation of a civic axis that will connect the two sides c the port, as a continuation of the cultural and commercial axis of the internal canal. This axis will crea two tourist zones on the edge of the port. **3. TOURIST PORT** — Creation of a pier oriented towards tou ist use in the maritime plaza to be created in front of the station. **4. SPORTS CENTRE** — Creation of ar indoor sports centre, for the use of the public, on tc of the present car park.

URBAN PROPOSALS

OPTION 1

The seafront plaza is located in front of the port. Th generates a square next to a zone of intense traffic, and is not connected to the civic axis of the canal.

New surface area: The seafront plaza is a major expansion of the harbour. Creation of a large squar between the station and the canal, creating a spacious seafront plaza.
This solution is the most radical, and perhaps also the most expensive.

A maritime plaza associated with a leisure port: Creation of a leisure and tourist port at the end of the commercial port, in much the same way as in histor European cities.

场地

基隆是台北一个传统的港口，是中国台湾地区向世界各地输送他们的产品和技术的重要口岸。

这个城市的都会中心因而朝向港口而建。但是同时港口的活动现在已经迁移到离海很近的地区。正像20世纪初的众多港口城市一样，城市与港口之间的相互反应的重点在于一种主流的转换。

无论如何，基隆市内的全盘交通仍然在市中心不断循环着。

从这点上来讲，我们需要发展一系列的对策来保证港口活动，高强度交通以及都市开发的共存是必要的，直到现有的状况得以改善。

港口西面现存的公共汽车与火车站可以被用来作为区域转化的一个意义重大的良机。

带着这样的想法，我们所计划采取的实际行为如下。

1. 门户

将公共汽车和火车站进行转化成为城市本身的门户，在新的轨道车站上建造一个商业中心，在公交汽车站上建造一个购物餐饮综合体，这两者将作为一处公共的交通工具转换区域，加强站前广场的城市特性。

2. 城市轴线

一条城市轴线的建立将港口的两侧连接起来，成为一个文化和商业的连续界面。这条轴线将会在港口创造出两块不同的游乐场区。

3. 游客港口

我们将在车站前的休闲广场创造一个基于pier面向游客的国内港口。

4. 体育中心

在现存的停车场上方的室内设计建造一个面向公众体育中心。

都市预案

选择1

面海广场被设置在港口的对面。这就在密集交通区域的旁生成了一个广场，并且没有与通道的文化轴线相连接。

新地表区

面海广场是海港的一次大规模的扩张。在车站与通道之间造出一个巨大的，具有空间感的面海广场。

这一解决方案是最具有放射性的，同时也可能是最为昂贵的。

一个与休闲港相连接的海事广场

像很多欧洲的历史性城市一样，在商业港口的尽端设置休闲游港口。

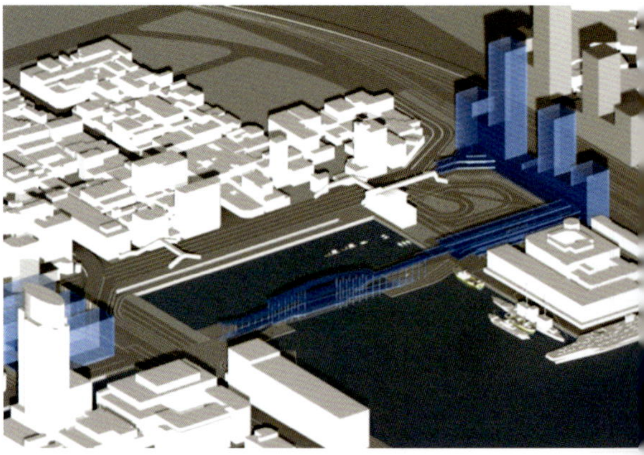

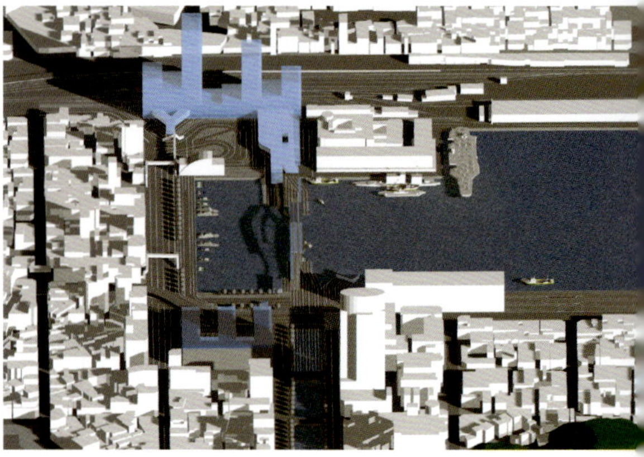

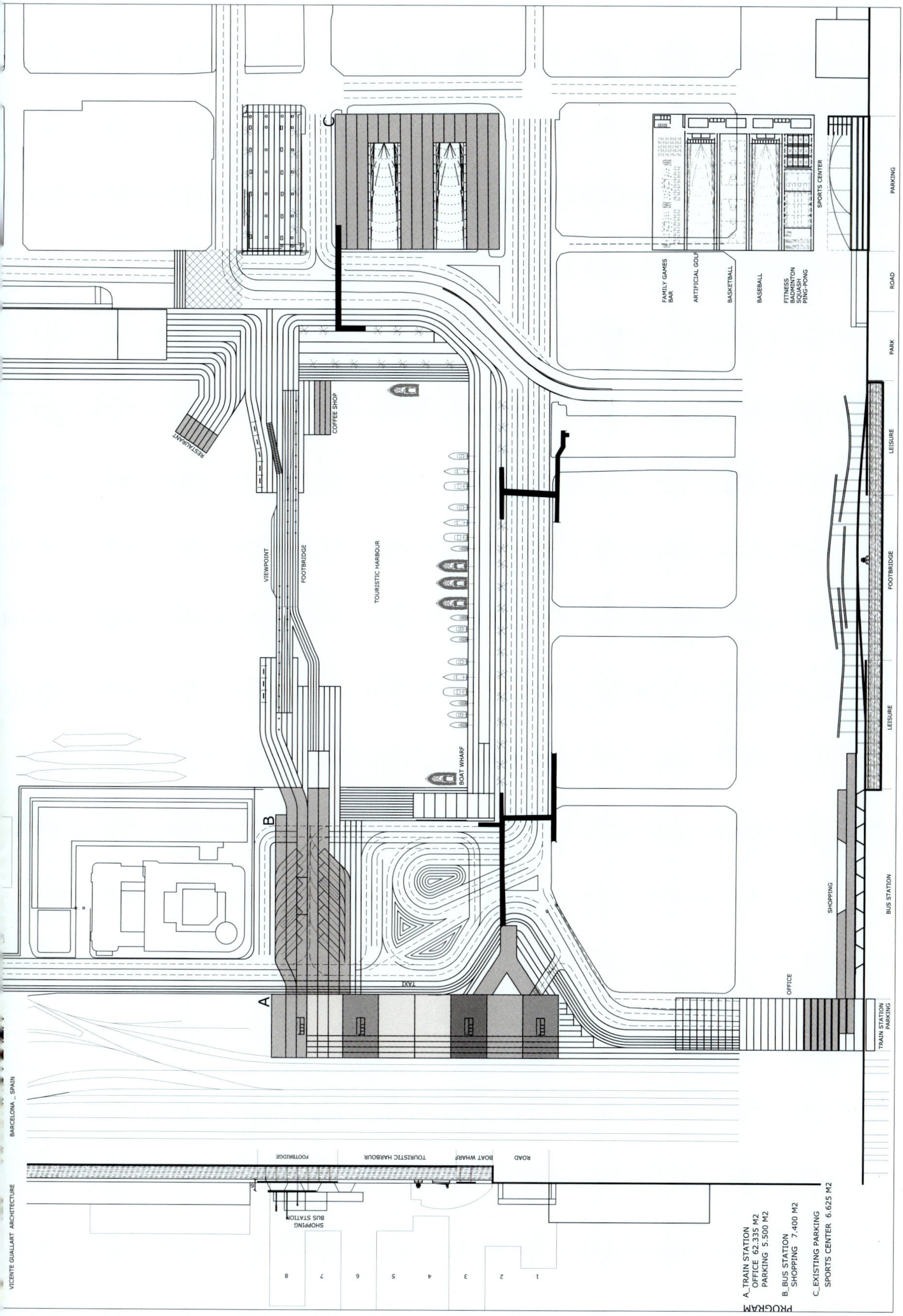
C
FAMILY GAMES BAR
ARTIFICIAL GOLF
BASKETBALL
BASEBALL
FITNESS BADMINTON SQUASH PING-PONG
SPORTS CENTER
PARKING
ROAD
PARK
RESTAURANT
COFFEE SHOP
LEISURE
VIEWPOINT
FOOTBRIDGE
TOURISTIC HARBOUR
FOOTBRIDGE
LEISURE
BOAT WHARF
B
SHOPPING
BUS STATION
TAXI
A
OFFICE
TRAIN STATION PARKING
BARCELONA _ SPAIN
VICENTE GUALLART ARCHITECTURE
FOOTBRIDGE
TOURISTIC HARBOUR
BOAT WHARF
ROAD
SHOPPING
BUS STATION
8
7
6
5
4
3
2
1
PROGRAM
A_TRAIN STATION
OFFICE 62.335 M2
PARKING 5.500 M2
B_BUS STATION
SHOPPING 7.400 M2
C_EXISTING PARKING
SPORTS CENTER 6.625 M2

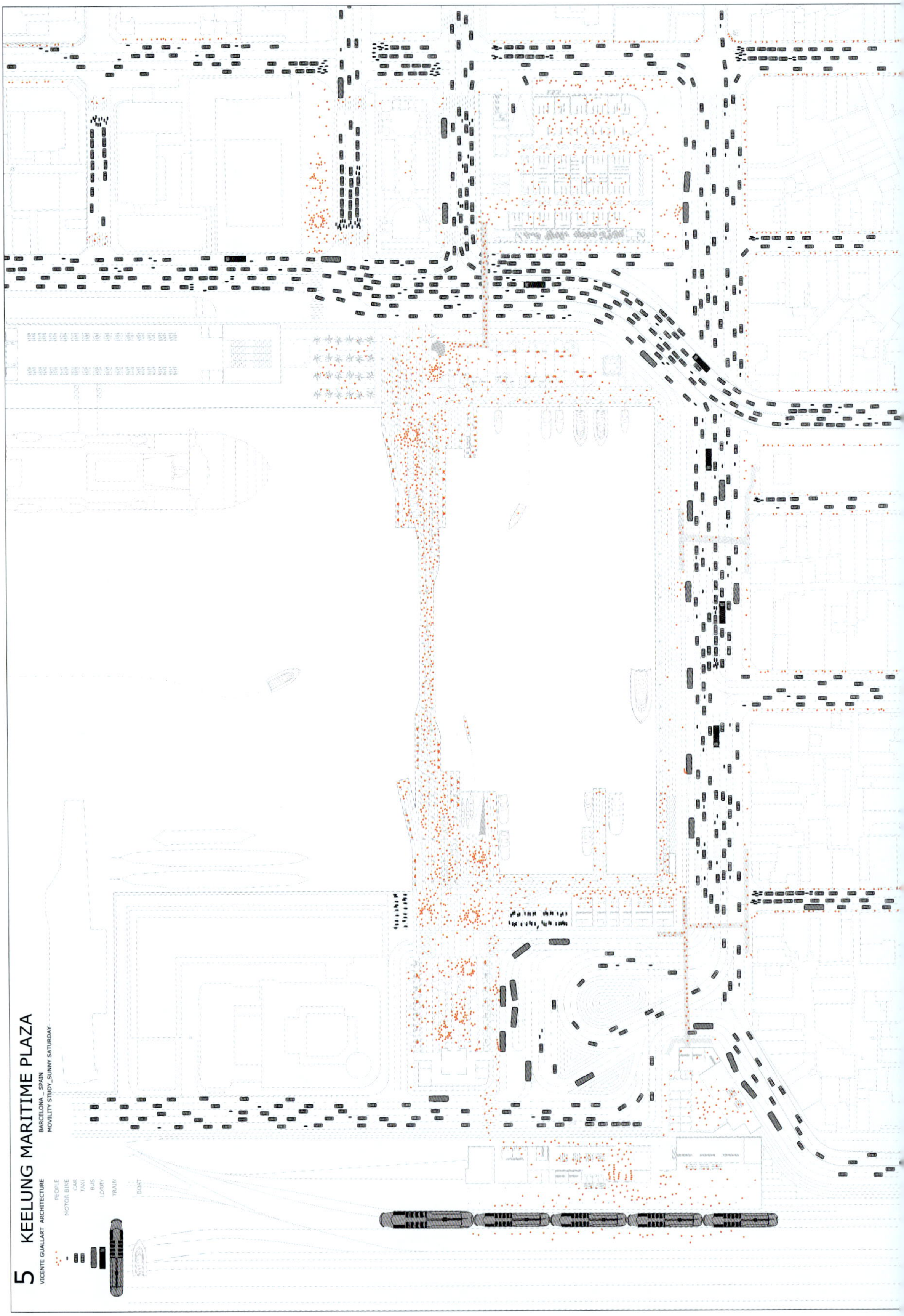
5 KEELUNG MARITIME PLAZA
VICENTE GUALLART ARCHITECTURE
BARCELONA _ SPAIN
MOVILITY STUDY_SUNNY SATURDAY
PEOPLE
MOTOR BIKE
CAR
TAXI
BUS
LORRY
TRAIN
BOAT

KEELUNG MARITIME PLAZA PHASE 2
基隆海洋广场二期工程

AN OPPORTUNITY FOR QUALITY

GENERAL OBJECTIVE
A PROGRAMME TO IMPROVE THE URBAN QUALIT OF KEELUNG

Keelung has emerged with notable success from th quantitative phase of its urban development. The strength of its economy, the youth of its populatio and the sustained growth experienced in recent decades have given rise to a vital and dynamic city But this is not enough. It is time now to move on to the qualitative phase, of improving the quality of life of the citizens, of reordering the most important spaces and of creating an innovative, cutting-edge urban image. To achieve this purpose, architecture alone is not sufficient: this is the moment for urbanism, for concerted action that will establish relationships between the different pieces that make up the central space of the city.

提升质量的良机
总体目标
一个提升基隆城市质量的方案

基隆近年来的城市开发活动取得了引人注目的成就。这座市经济的坚韧性，年轻的居民以及近年来持续增长的经验，令其成长为一个具有活力和生命力的城市。但是，只有这些不够的。现在我们应当开始着重质量，提高居民的生活质量重新组织那些重要的空间，并创造一种具有创新性的，尖端城市意象。想要成就这一目标，只有建筑是远远不够的：这一个都市主义的时代，是一个在组成城市核心空间的不同断之间建立联系的协同行为的时代。

PECIFIC STRATEGIES

TO CREATE A SYMBOLIC AND FUNCTIONAL ENTRALITY.

city is defined internally and externally by the trength and quality of its central space. The citi- ens interpret the city through a reading that is not nly functional but also symbolic, so that it is nec- ssary to design a space through which the people f Keelung can identify with the city. Our proposal ffers the opportunity of creating a more powerful nd complex urban centre than will come to consti- ute Keelung's symbolic space and serve to project n image of the city to the exterior.

TO POSIT A NEW RELATIONSHIP WITH THE ORT AND THE SEA.

eelung as it is today would not exist without the conomic power it derives from its port, but nor ould the port have achieved its present level of ctivity without the explicit support of the city. owever, the city's relationship with the sea is bstructed and limited by the port's activity. It is ssential to incorporate the sea into Keelung as omething more than a mere economic reference, iving a leading role to its space as a vital part of e urban scene.

TO REINFORCE THE URBAN STRUCTURE CON- ECTING THE EAST AND WEST OF THE CITY WITH CIVIC AXIS.

he port has been until now, for all its social and conomic usefulness, a disruptive element in terms f the urban organization of Keelung. The differ- nt parts of the city are functionally segregated y the impact of infrastructures and traffic. A new nd more effective link between the East and West ones of the city will assist in integrating the space f production and relation (East) with the access frastructures (West), promoting the functional nd economic complementarity of both, giving con- nuity to the canal and generating the axis of force at the city has historically lacked.

TO OBTAIN NEW SPACES OF PUBLIC AMENITIES.

city assumes its real significance for its citizens hen it has a network of amenities that satisfies ot only the basic needs but also the demand for ports, culture, contact and other analogous needs. e image of modernity is closely associated with ese functions, which in Keelung are still only an cipient presence. Our proposal links the access frastructures with the zone of amenities around e Cultural Centre, creating new facilities that will rve to reinforce the centrality of the space.

TO REINFORCE RELATIONAL USES.

e structural base of a city is, much more than its reets and buildings, the set of human relation- ips that are created and developed within it. The ty is an accumulation of complex functions and lationships between people, so that limiting it to mmercial spaces implies a reductionist percep- tion of the urban phenomenon. The new amenities and leisure spaces, the effective pedestrian link between the different urban sectors and the bridge itself constitute the bases on which centrality will be reinforced by the intensity and dynamism of the human relationships it accommodates.

• TO REORGANIZE THE TRANSPORT AREA.

In the West zone of the area of intervention the three main transport infrastructures of the city converge. The solution that is envisaged for this space is not confined to embracing their existence as such, but seeks to facilitate the uses that these infrastructures induce and require, spatially orga- nizing their satisfactory distribution and functional integration in the ways that are most convenient for the commuters and proposing cultural and leisure uses that will extend beyond the usual commuter timetable to avoid the risk of it losing its vitality and becoming marginalized.

• TO RESTRUCTURE THE AREA OF SOCIO-CUL- TURAL AMENITIES.

In the West zone there are a variety of cultural, administrative and commercial facilities whose integration is problematic, because of the difficul- ties in the way of pedestrian movement. At the same time there is a need to generate new uses and functions that will reinforce the relational contents of the public spaces, in order to provide a qualita- tively richer urban framework for the citizens. This entails the restructuring of the area of amenities to the East of the port and the incorporation of new functions.

• TO EXTEND THE AREA OF PEDESTRIAN CON- NECTION.

In spite of living from the sea and with the sea, the people of Keelung are confronted by major ob- stacles when it comes to enjoying it, in view of the impact of vehicular traffic on the seafront and the very limited space that has been reserved for pe- destrian circulation. To overcome this obstacle we propose a solution that not only improves the flow of pedestrian circulation but also generates a space that, as in all of the best operations for the restruc- turing of city-port interfaces, permits the creation of a rest space where people can enjoy contemplat- ing the sea and the bustling activity of the port.

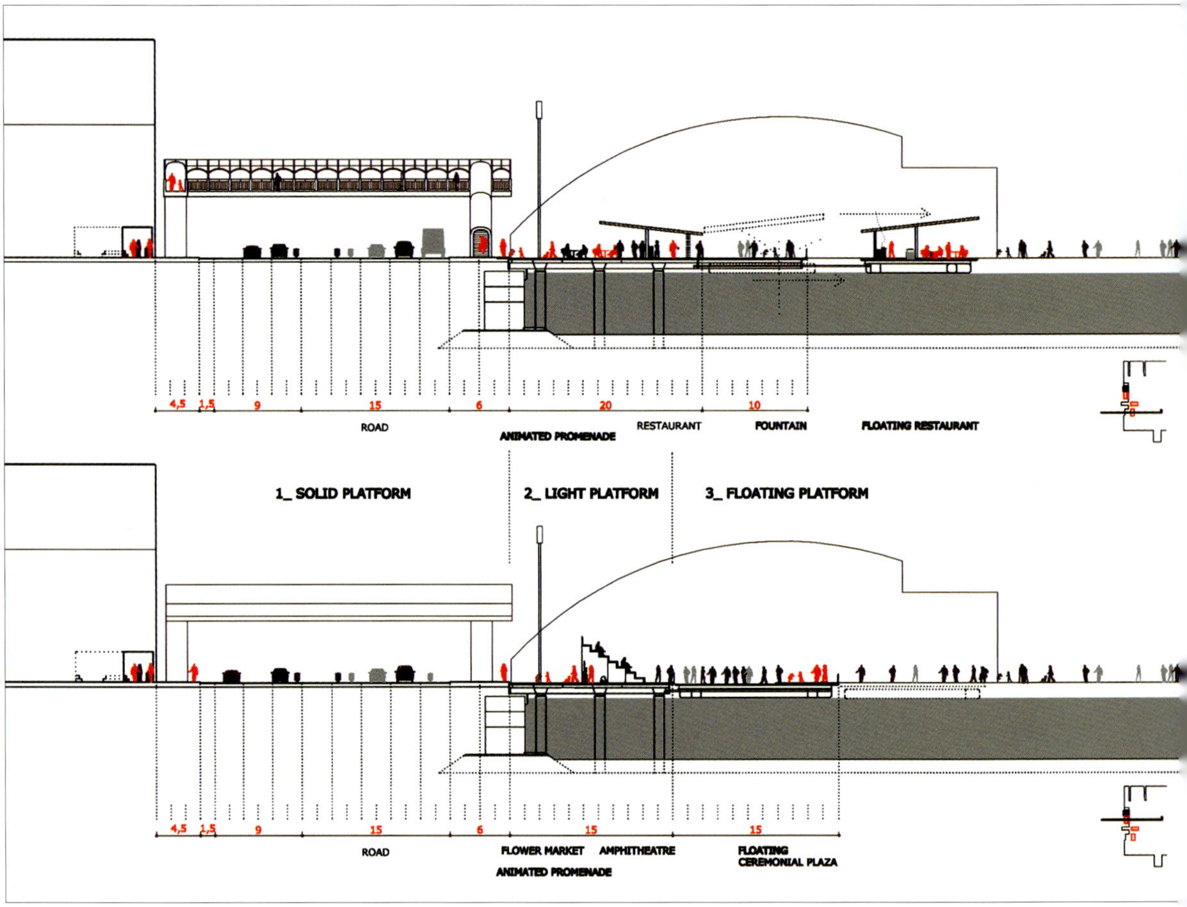

1.COMPACT

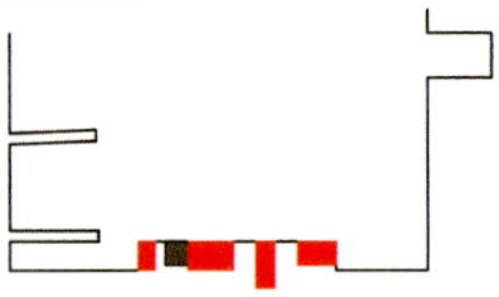

2.OPENING

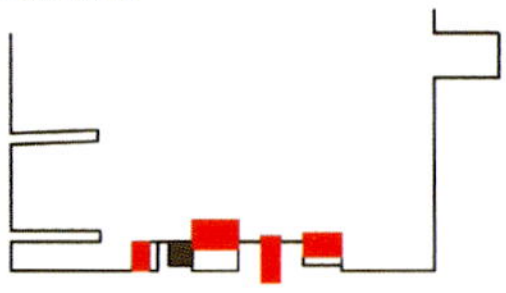

3.OPEN

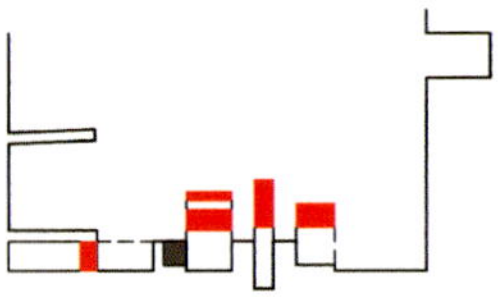

4.MAXIMUM DISPLACEMENT

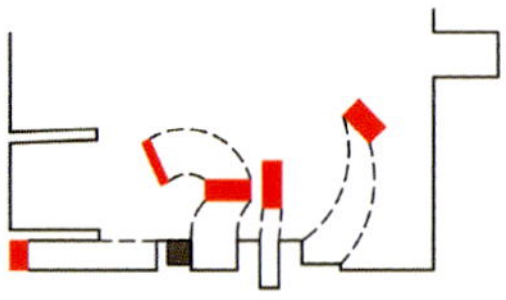

具体策略

创造一种具有标志性和功能性的核心

城市从内部和外部被其核心空间的耐受力和质量所决定。居民通过一种不仅只具有功能性，同时还带有标志性的解读方式来理解城市，因而设计一种基隆居民能够识别的空间是非常必要的。我们的方案提供了创造一种更加具有力量和复杂性的都市重生的良机，这一方案将建立基隆的标志性空间，并同时为城市建立其外在形象。

在港口与海之间建立一种新的关系

如果没有从港口所获得的经济力量，基隆不会发展到今天所具有的生机，但是没有城市作为其坚实的后盾，港口本身也不可能成就其今天的运行水平。但是，城市与海之间的关系也被港口所阻断和限制。将海在空间上作为城市景象充满活力的一个领导性角色融入基隆是非常重要的，而不单单是经济依托。

加强城市结构，将城市的东西两部分用一条文化轴线连接起来

直到今天，港口以其在社会和经济方面的作用在基隆的城市组织结构当中担当着不可替代的角色。城市的不同部分在基础设施和交通的影响之下被划分成不同的功能块。城市的东西不同区域之间，一种新的更加有效的联系方式将会通过基础设施的介入来协助制造空间与联系空间的融合，并推广两者的功能和经济完整性，实现通道的可持续性并生成城市所一直缺少的力量轴线。

获得新的公共休闲空间

当一个城市能够提供一个不仅能满足基本需求，而且能够满足体育、文化、交流以及其他相应要求的休闲空间体系的时候，它才真正在居民的生活中建立起了其本身的重要性。上述功能作为与现代化的城市形象紧密关联的元素，在基隆仅仅是初具雏形而已。我们的方案将基础设施的改造与围绕文化中心的休闲区域联系在一起，创造了加强空间中心性的新设施。

加强关联性的功能

一座城市的结构性基础实际上并不仅仅是它所包含的建筑和街道，而是在这些硬件网络下所生长起来的人际关系网。城市是人与人之间复杂的功能和关系的累积物，因而将城市空间限制在商业空间的层面上意味着对于都市产物过于简化的认知。新的娱乐休闲空间，不同的都市向量之间有效的行人连接道，以及桥本身共同构成的基础上，都市空间的核心性被人类自身的密集和活力不断加强。

重新组织交通区域

在进行干预地区的西部，三条城市主要的交通干道交汇在一起。针对这一空间已成形的解决方案并没有局限于现有的形式，而是寻求囊括这些基础设施所衍生或需要的功能，以一种将便利充分给予通勤者的方式从空间上组织分散它们的最佳状态和功能性交互，同时将文化休闲功能拓展到日常通勤时间以外以避免其失去活力而边缘化的风险。

重新构筑社交文化休闲区域

西区存在着多种多样的文化性、行政性以及商业性设施，而由于行人移动的困难性，这些设施的交汇带来了很多问题。同时，为了给居民提供一个内容更为丰富的都市框架，这里也存在着建设能够加强公共空间相关功能的结构新用途和功能的需要。

拓展行人连接区域

除了以海为生，与海共生以外，当基隆居民想要享受海的时候则遭受了相当的阻碍，不仅因为机动车交通给海岸带来的压力，同时也因为海边及其有限的空间已经被预留给了步行环线系统。为了克服这一障碍，我们提出了一种不仅能够提高行人移动性，同时能够生成一种将城市/港口交互关系进行最优化重构的空间，它允许我们创造一种人们能够同时享受与海共存并灵活参与港口活动的休闲空间。

KEELUNG MARITIME PLAZA
PHASE 3
In association with J.M. Lin/ The Observer Design Group

基隆海洋广场三期工程

协同[illegible]问有限公司联合设计

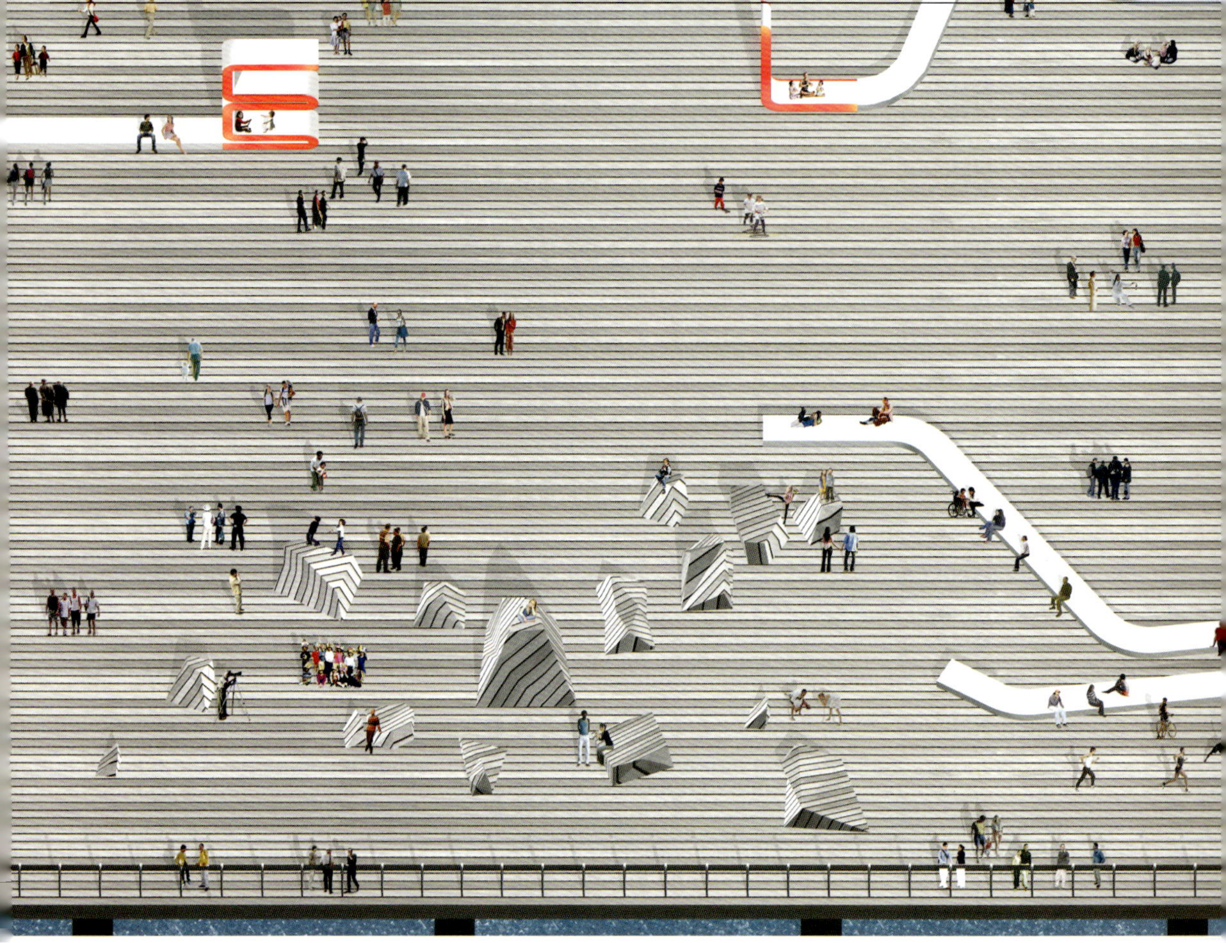

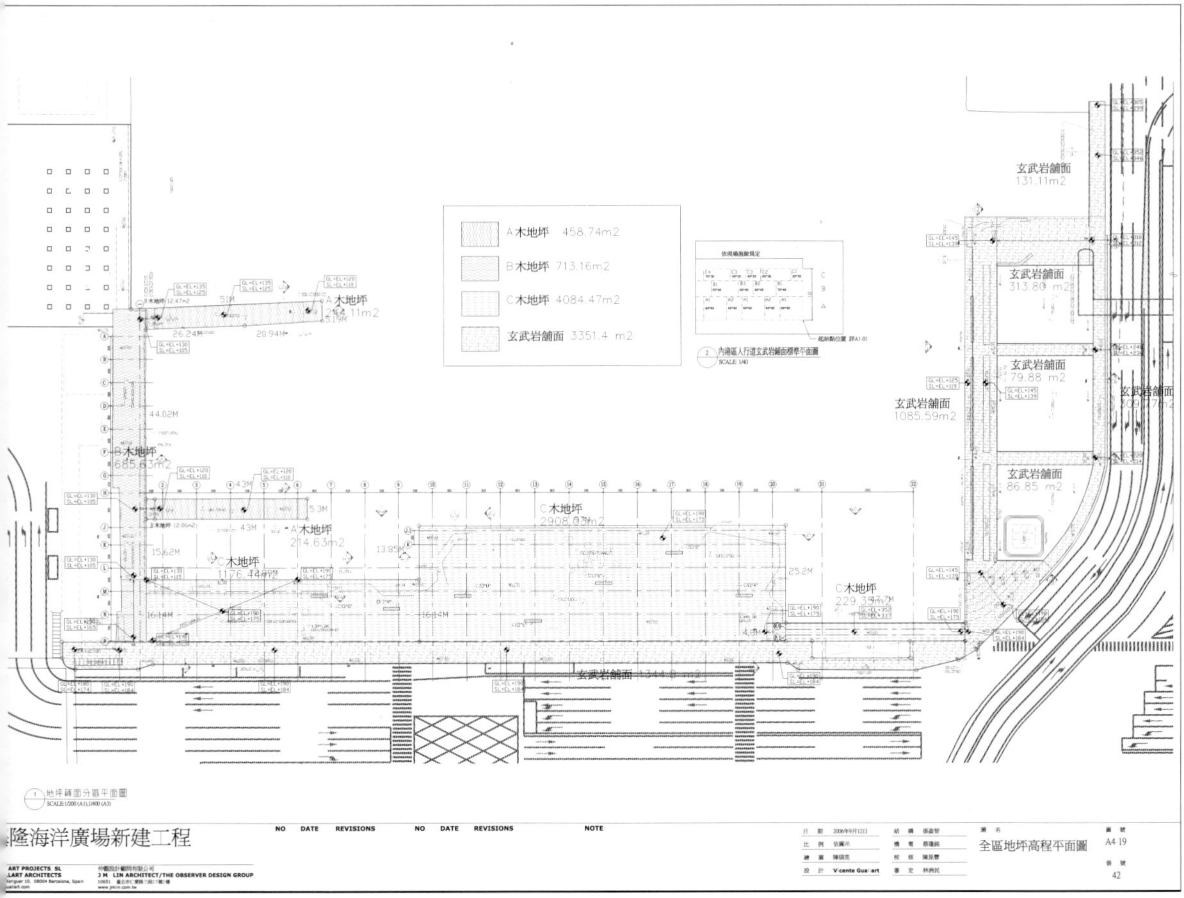
A木地坪 458.74m2
B木地坪 713.16m2
C木地坪 4084.47m2
玄武岩舖面 3351.4 m2
玄武岩舖面 131.11m2
玄武岩舖面 313.80 m2
玄武岩舖面 79.88 m2
玄武岩舖面 1085.59m2
玄武岩舖面 86.85 m2
C木地坪 2908.83m2
C木地坪 1176.44m2
B木地坪 685.63m2
A木地坪 214.63m2
C木地坪 229.35m2
內港區人行道玄武岩舖面標準平面圖
SCALE: 1/40
地坪鋪面分區平面圖
SCALE:1/200 (A1),1/400 (A3)
隆海洋廣場新建工程
NO DATE REVISIONS NO DATE REVISIONS NOTE
ART PROJECTS SL
LART ARCHITECTS
J M LIN ARCHITECT/THE OBSERVER DESIGN GROUP
www.jmlin.com.tw
2006年9月12日
Vicente Guallart
全區地坪高程平面圖
A4-19
42

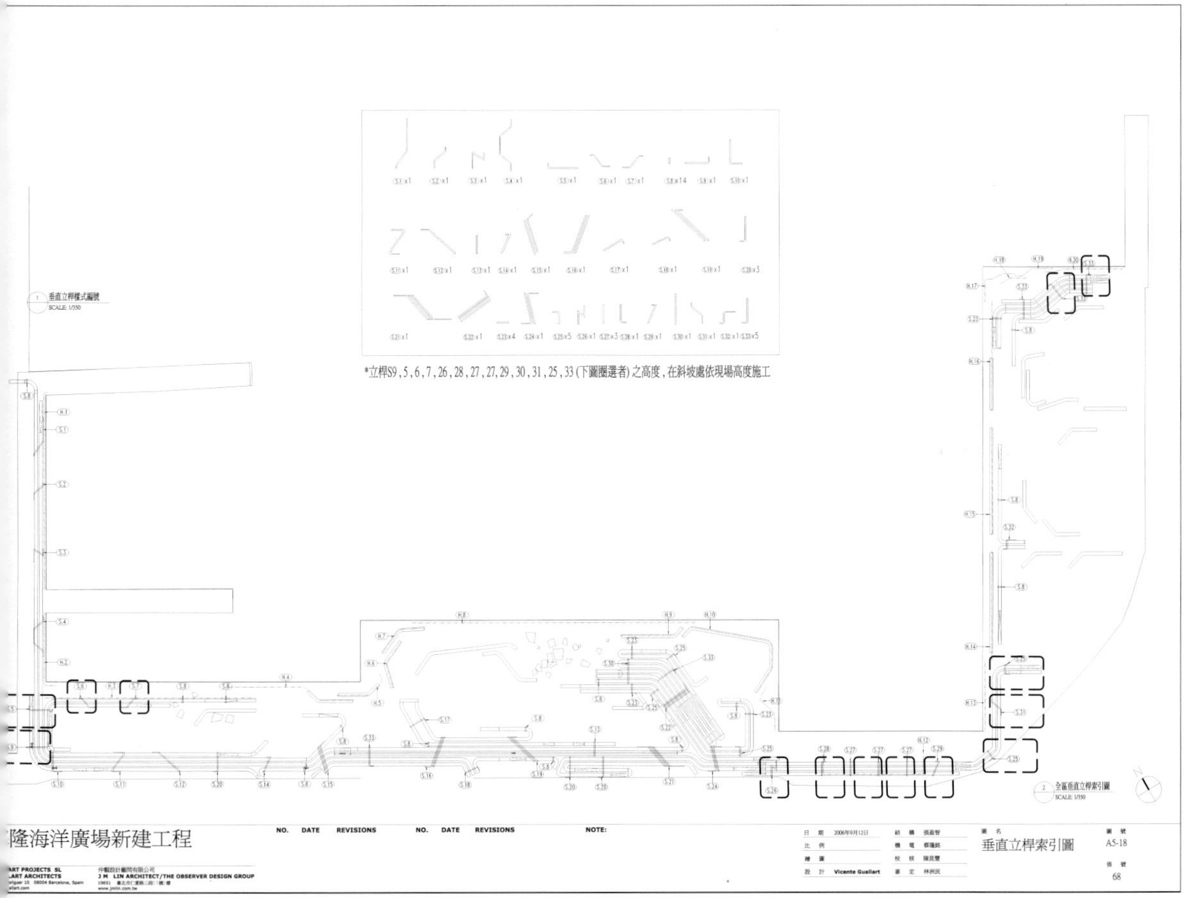
垂直立桿樣式編號
SCALE: 1/150
*立桿S9,5,6,7,26,28,27,27,29,30,31,25,33(下圖圈選者)之高度,在斜坡處依現場高度施工
全區垂直立桿索引圖
SCALE: 1/150
隆海洋廣場新建工程
NO. DATE REVISIONS NO. DATE REVISIONS NOTE:
ART PROJECTS SL
LART ARCHITECTS
J M LIN ARCHITECT/THE OBSERVER DESIGN GROUP
www.jmlin.com.tw
2006年9月12日
Vicente Guallart
垂直立桿索引圖
A5-18
68

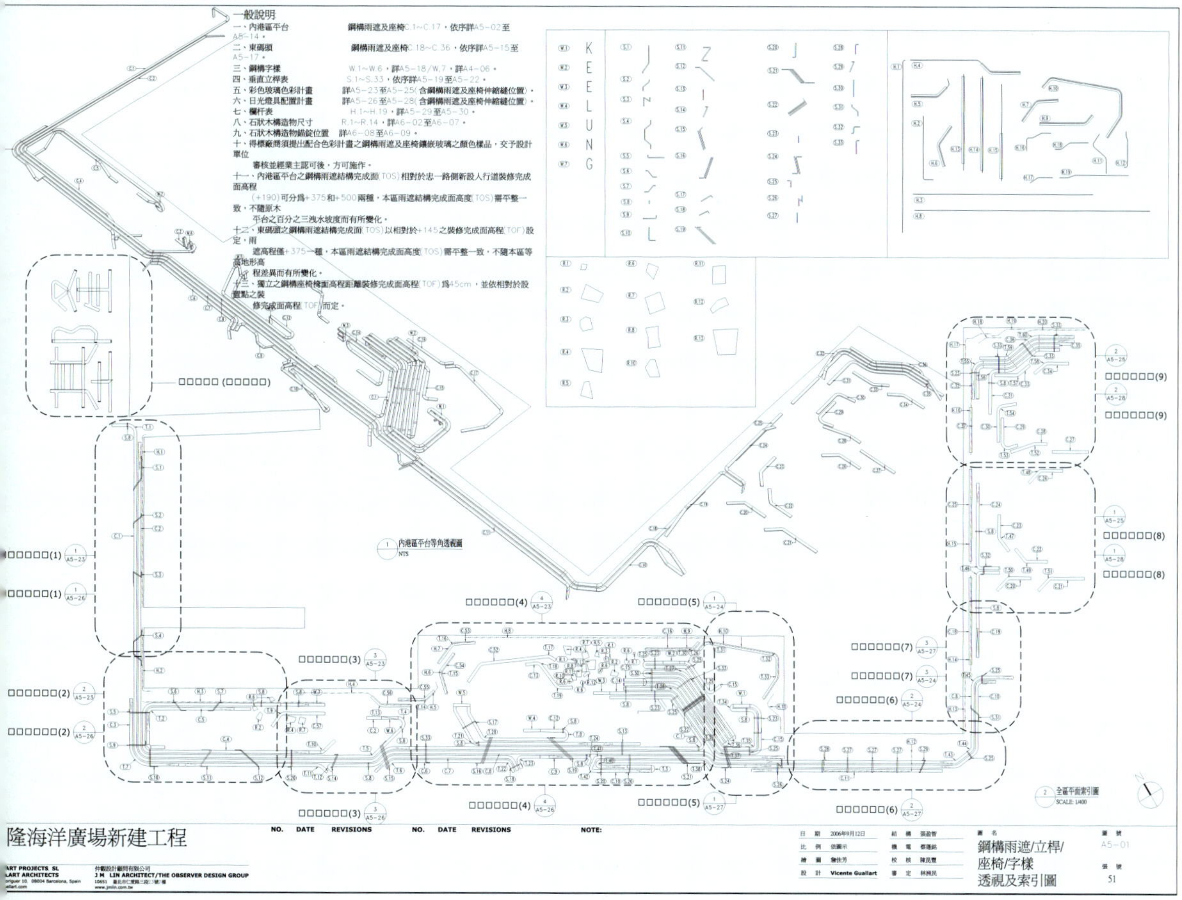
一般說明
一、內港區平台 鋼構雨遮及座椅C.1~C.17，依序詳A5-02至A5-14。
二、東碼頭 鋼構雨遮及座椅C.18~C.36，依序詳A5-15至A5-17。
三、鋼構字樣 W.1~W.6，詳A5-18/W.7，詳A4-06。
四、垂直立桿表 S.1~S.33，依序詳A5-19至A5-22。
五、彩色玻璃色彩計畫 詳A5-23至A5-25(含鋼構雨遮及座椅伸縮縫位置)。
六、日光燈具配置計畫 詳A5-26至A5-28(含鋼構雨遮及座椅伸縮縫位置)。
七、欄杆表 H.1~H.19，詳A5-29至A5-30。
八、石狀木構造物尺寸 R.1~R.14，詳A6-02至A6-07。
九、石狀木構造物錨錠位置 詳A6-08至A6-09。
十、得標廠商須提出配合色彩計畫之鋼構雨遮及座椅鑲嵌玻璃之顏色樣品，交予設計單位審核並經業主認可後，方可施作。
十一、內港區平台之鋼構雨遮結構完成面(TOS)相對於忠一路側新設人行道裝修完成面高程(+190)可分爲+375和+500兩種，本區雨遮結構完成面高度(TOS)需平整一致，不隨原木平台之百分之三洩水坡度而有所變化。
十二、東碼頭之鋼構雨遮結構完成面(TOS)以相對於+145之裝修完成面高程(TOF)設定，雨遮高程僅+375一種，本區雨遮結構完成面高度(TOS)需平整一致，不隨本區等高地形高程差異而有所變化。
十三、獨立之鋼構座椅椅面高程距離裝修完成面高程(TOF)爲45cm，並依相對於設置點之裝修完成面高程(TOF)而定。
KEELUNG
內港區平台等角透視圖
NTS
全區平面索引圖
SCALE: 1/400
隆海洋廣場新建工程
NO. DATE REVISIONS NO. DATE REVISIONS NOTE:
日期 2006年9月12日
比例 依圖示
繪圖 詹佳芳
設計 Vicente Guallart
結構 張盈智
機電 蔡蓮銘
校核 陳昆豐
審定 林洲民
圖名 鋼構雨遮/立桿/座椅/字樣透視及索引圖
圖號 A5-01
張號 51

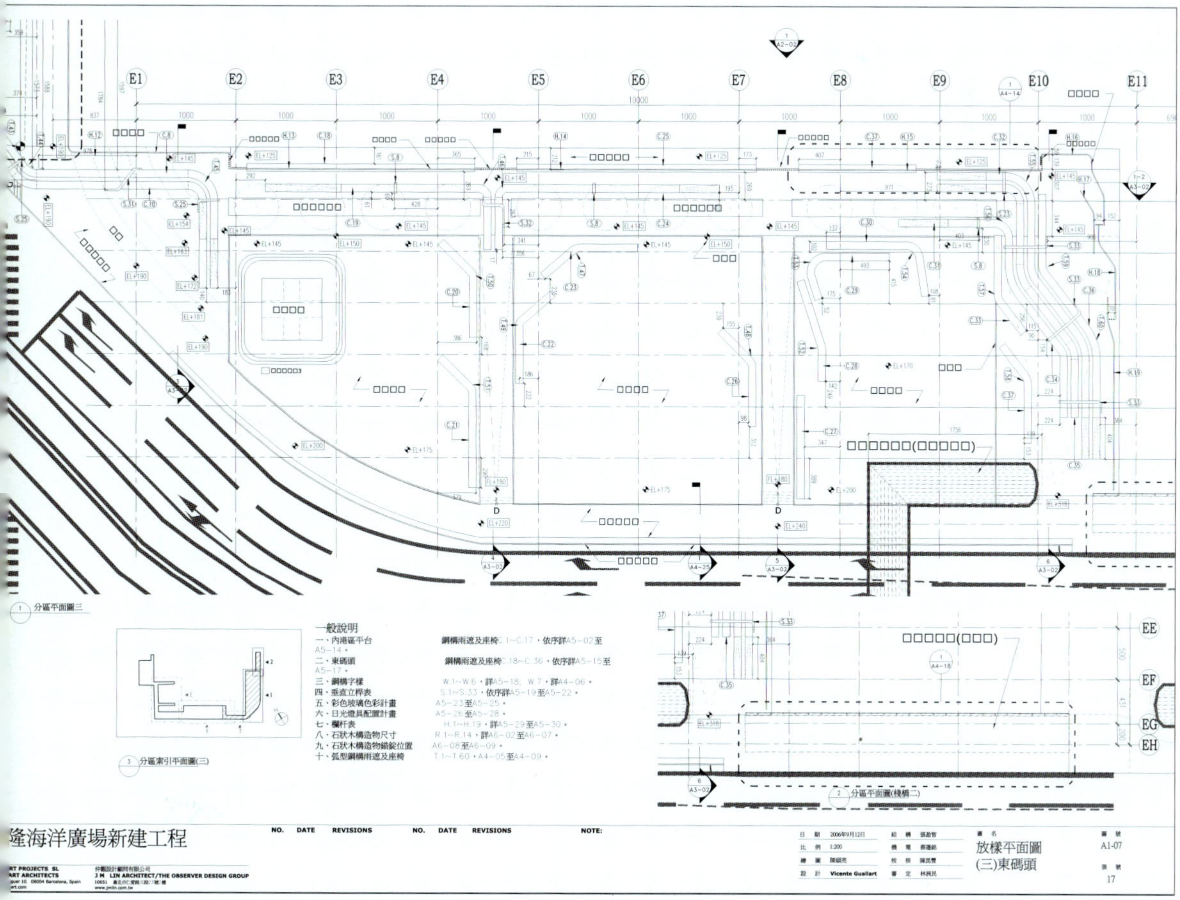
E1 E2 E3 E4 E5 E6 E7 E8 E9 E10 E11
分區平面圖三
分區索引平面圖(三)
一般說明
一、內港區平台 鋼構雨遮及座椅C.1~C.17，依序詳A5-02至A5-14。
二、東碼頭 鋼構雨遮及座椅C.18~C.36，依序詳A5-15至A5-17。
三、鋼構字樣 W.1~W.6，詳A5-18；W.7，詳A4-06。
四、垂直立桿表 S.1~S.33，依序詳A5-19至A5-22。
五、彩色玻璃色彩計畫 A5-23至A5-25。
六、日光燈具配置計畫 A5-26至A5-28。
七、欄杆表 H.1~H.19，詳A5-29至A5-30。
八、石狀木構造物尺寸 R.1~R.14，詳A6-02至A6-07。
九、石狀木構造物錨錠位置 A6-08至A6-09。
十、弧型鋼構雨遮及座椅 T.1~T.60，A4-05至A4-09。
分區平面圖(機橋二)
EE EF EG EH
隆海洋廣場新建工程
NO. DATE REVISIONS NO. DATE REVISIONS NOTE:
日期 2006年9月12日
比例 1:200
繪圖 陳國亮
設計 Vicente Guallart
結構 張盈智
機電 蔡蓮銘
校核 陳昆豐
審定 林洲民
圖名 放樣平面圖(三)東碼頭
圖號 A1-07
張號 17

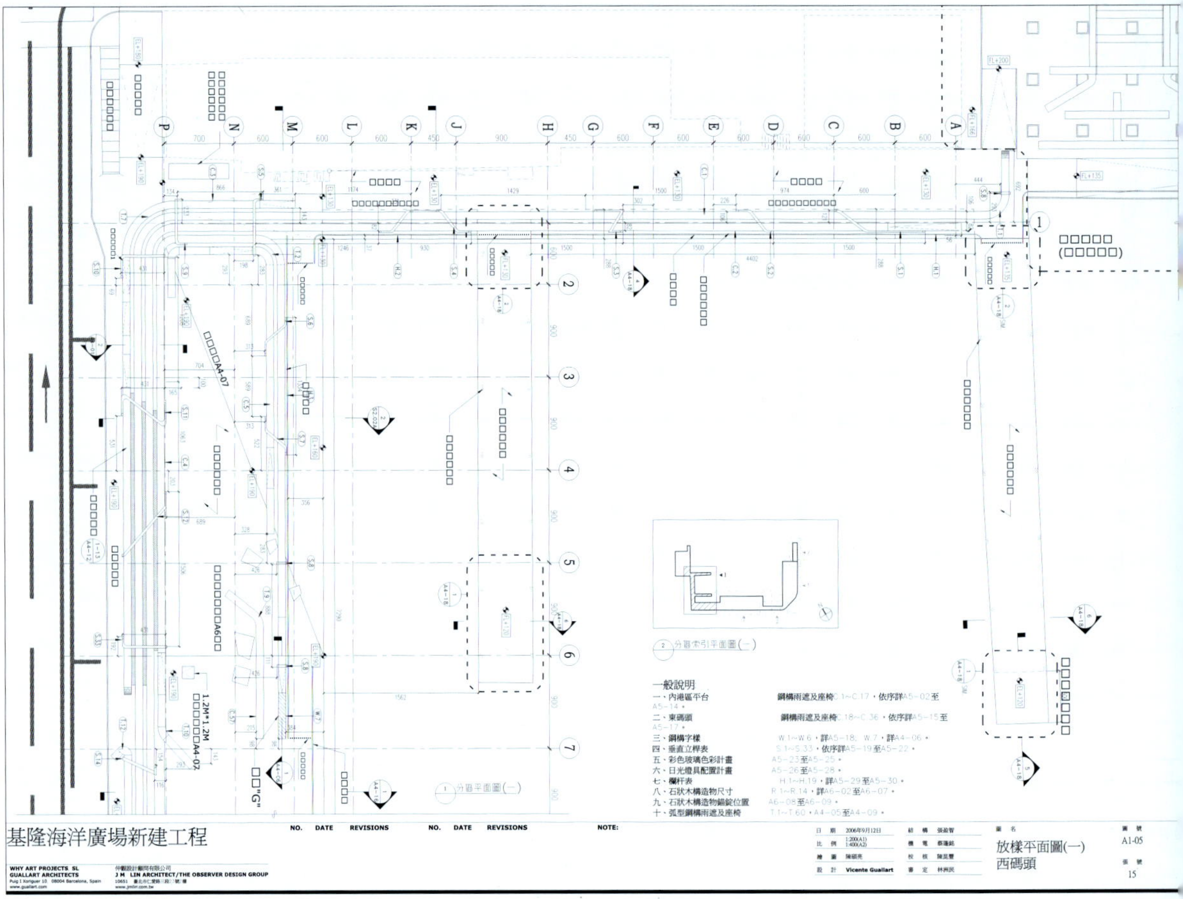
一般說明
一、內港區平台
二、東碼頭
三、鋼構字樣
四、垂直立桿表
五、彩色玻璃色彩計畫
六、日光燈具配置計畫
七、欄杆表
八、石狀木構造物尺寸
九、石狀木構造物錨錠位置
十、弧型鋼構雨遮及座椅
基隆海洋廣場新建工程
NO. DATE REVISIONS
NOTE:
WHY ART PROJECTS SL
GUALLART ARCHITECTS
J M LIN ARCHITECT/THE OBSERVER DESIGN GROUP
Vicente Guallart
放樣平面圖(一)
西碼頭
A1-05
15

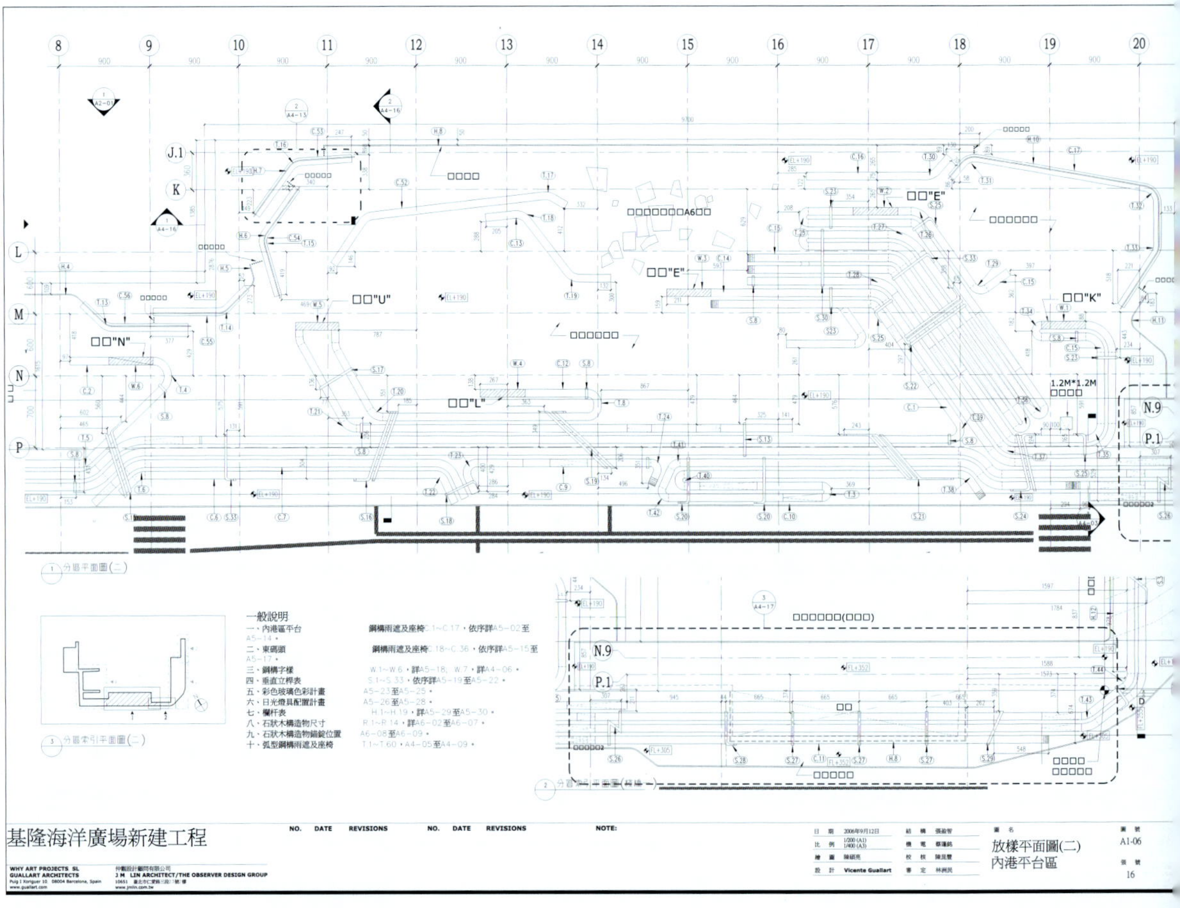
一般說明
一、內港區平台
二、東碼頭
三、鋼構字樣
四、垂直立桿表
五、彩色玻璃色彩計畫
六、日光燈具配置計畫
七、欄杆表
八、石狀木構造物尺寸
九、石狀木構造物錨錠位置
十、弧型鋼構雨遮及座椅
基隆海洋廣場新建工程
NO. DATE REVISIONS
NOTE:
WHY ART PROJECTS SL
GUALLART ARCHITECTS
J M LIN ARCHITECT/THE OBSERVER DESIGN GROUP
Vicente Guallart
放樣平面圖(二)
內港平台區
A1-06
16

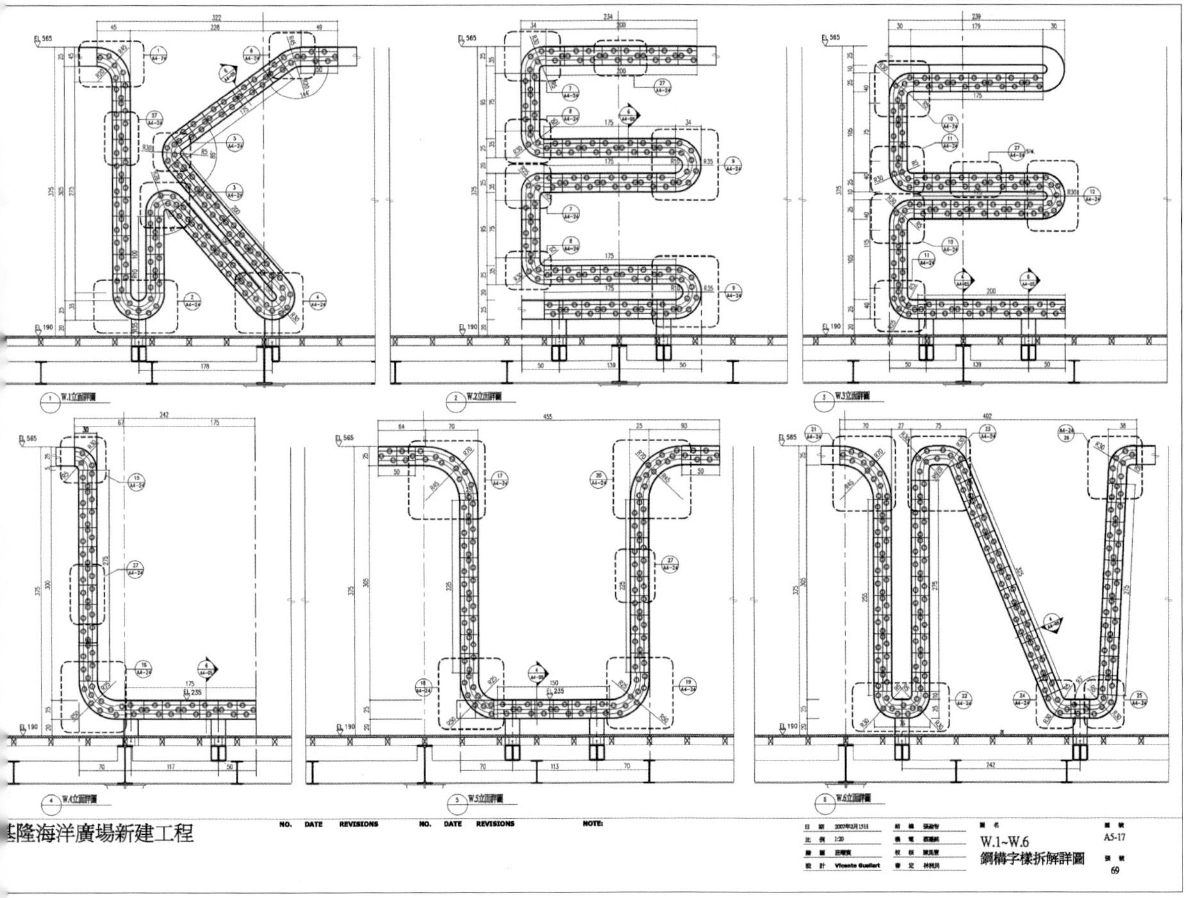
W.1立面詳圖
W.2立面詳圖
W.3立面詳圖
W.4立面詳圖
W.5立面詳圖
W.6立面詳圖
隆海洋廣場新建工程
NO. DATE REVISIONS
NO. DATE REVISIONS
NOTE:
Vicente Guallart
W.1~W.6
鋼構字樣拆解詳圖
A5-17
69

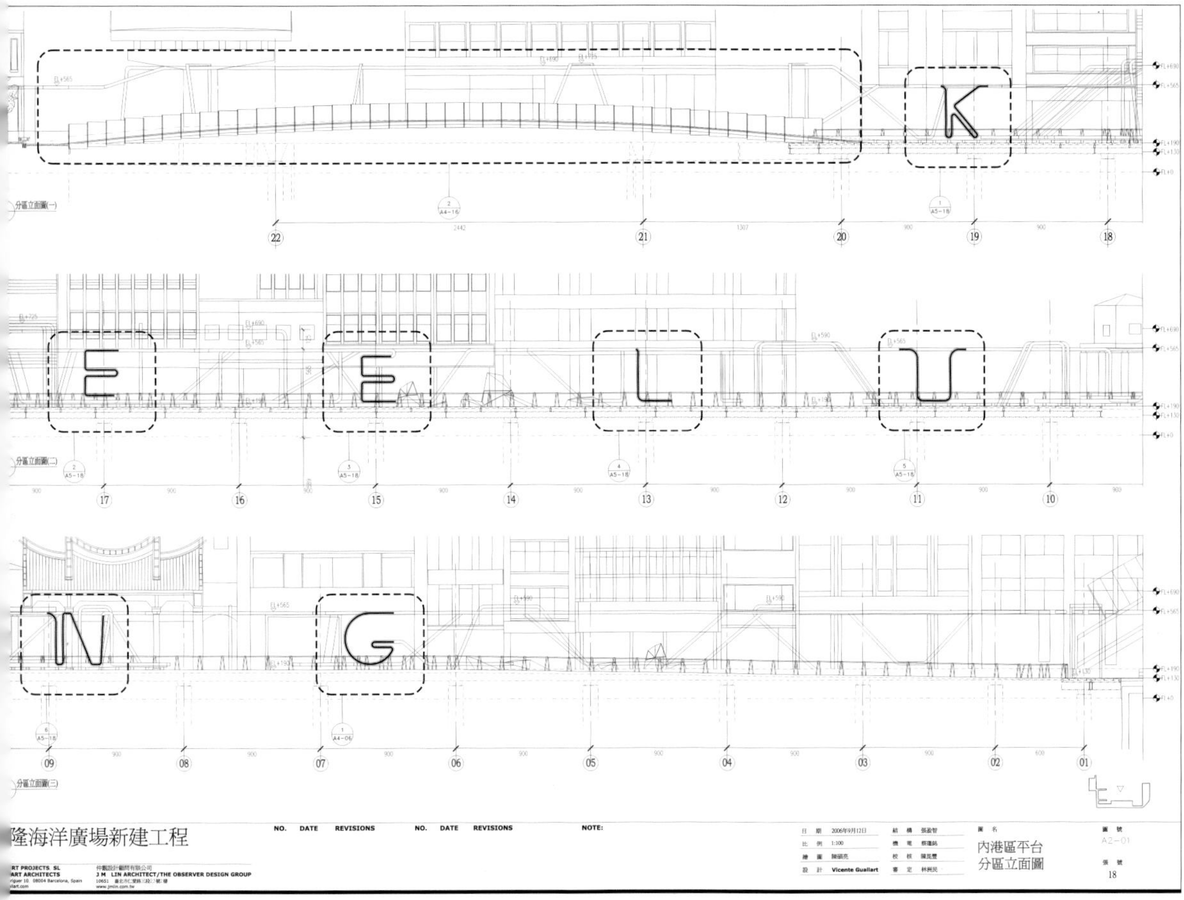
分區立面圖(一)
分區立面圖(二)
分區立面圖(三)
隆海洋廣場新建工程
NO. DATE REVISIONS
NO. DATE REVISIONS
NOTE:
J M LIN ARCHITECT/THE OBSERVER DESIGN GROUP
Vicente Guallart
內港區平台
分區立面圖
A2-01
18

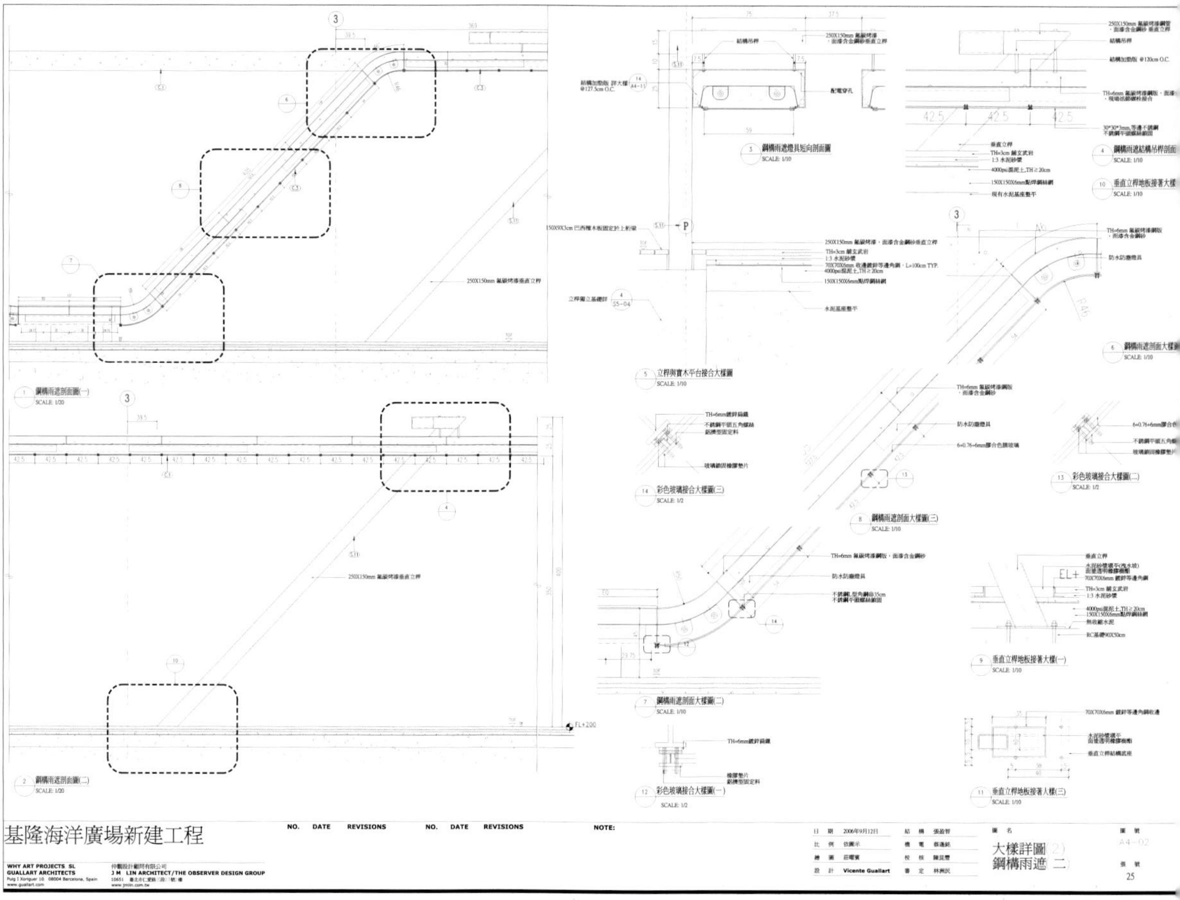
鋼構雨遮剖面圖(一)
SCALE: 1/20
鋼構雨遮剖面圖(二)
SCALE: 1/20
鋼構雨遮燈具短向剖面圖
SCALE: 1/10
鋼構雨遮結構吊桿剖面
SCALE: 1/10
垂直立桿地板接著大樣
SCALE: 1/10
立桿與實木平台接合大樣圖
SCALE: 1/10
鋼構雨遮剖面大樣圖
SCALE: 1/10
彩色玻璃接合大樣圖(二)
SCALE: 1/2
鋼構雨遮剖面大樣圖(三)
SCALE: 1/10
彩色玻璃接合大樣圖(二)
SCALE: 1/2
垂直立桿地板接著大樣(一)
SCALE: 1/10
鋼構雨遮剖面大樣圖(二)
SCALE: 1/10
彩色玻璃接合大樣圖(一)
SCALE: 1/2
垂直立桿地板接著大樣(三)
SCALE: 1/10
基隆海洋廣場新建工程
WHY ART PROJECTS SL
GUALLART ARCHITECTS
仲觀設計顧問有限公司
J M LIN ARCHITECT/THE OBSERVER DESIGN GROUP
NO. DATE REVISIONS
NO. DATE REVISIONS
NOTE:
2006年9月12日
Vicente Guallart
大樣詳圖(2)
鋼構雨遮(二)
A4-02
25

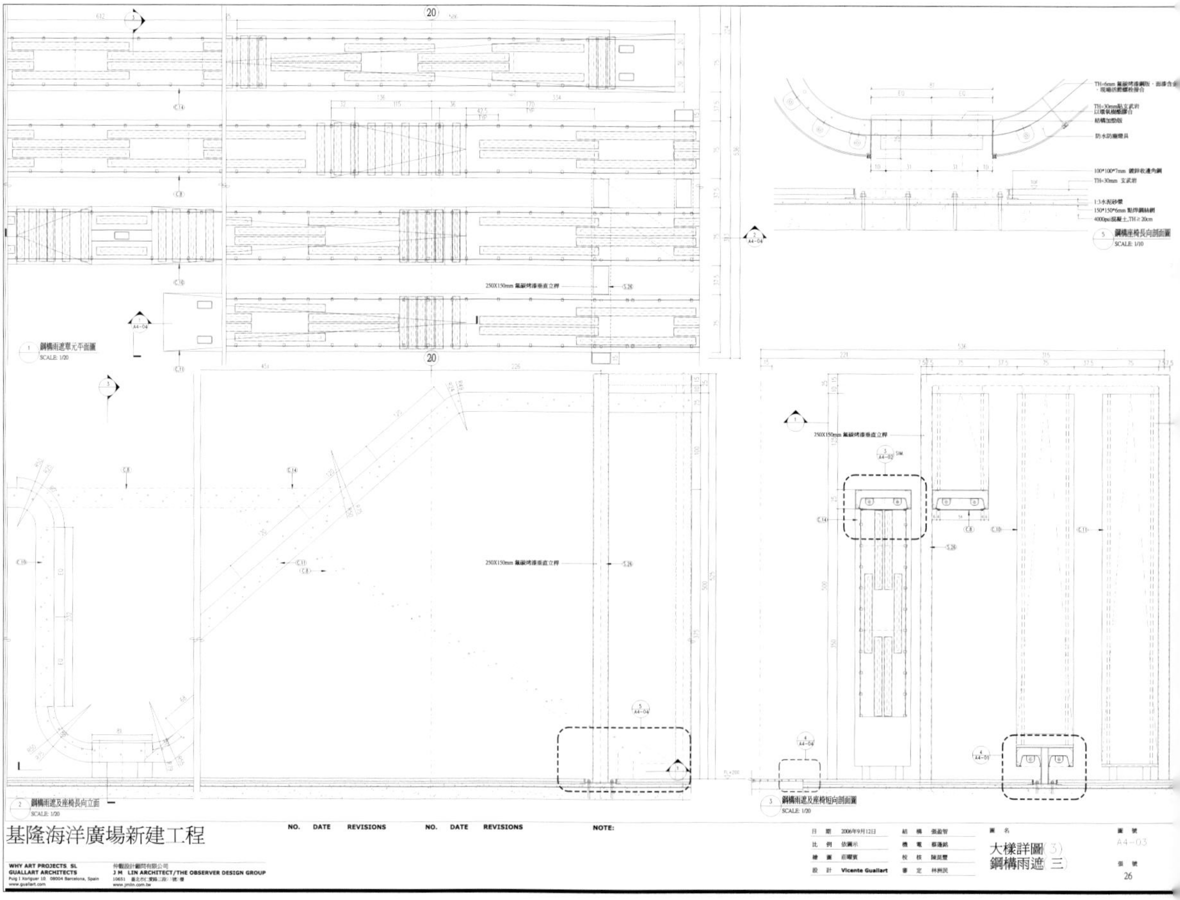
鋼構雨遮單元平面圖
SCALE: 1/20
鋼構雨遮及座椅長向立面
SCALE: 1/20
鋼構座椅長向剖面圖
SCALE: 1/10
鋼構雨遮及座椅短向剖面圖
SCALE: 1/20
250X150mm 氟碳烤漆垂直立桿
基隆海洋廣場新建工程
WHY ART PROJECTS SL
GUALLART ARCHITECTS
仲觀設計顧問有限公司
J M LIN ARCHITECT/THE OBSERVER DESIGN GROUP
NO. DATE REVISIONS
NO. DATE REVISIONS
NOTE:
2006年9月12日
Vicente Guallart
大樣詳圖(3)
鋼構雨遮(三)
A4-03
26

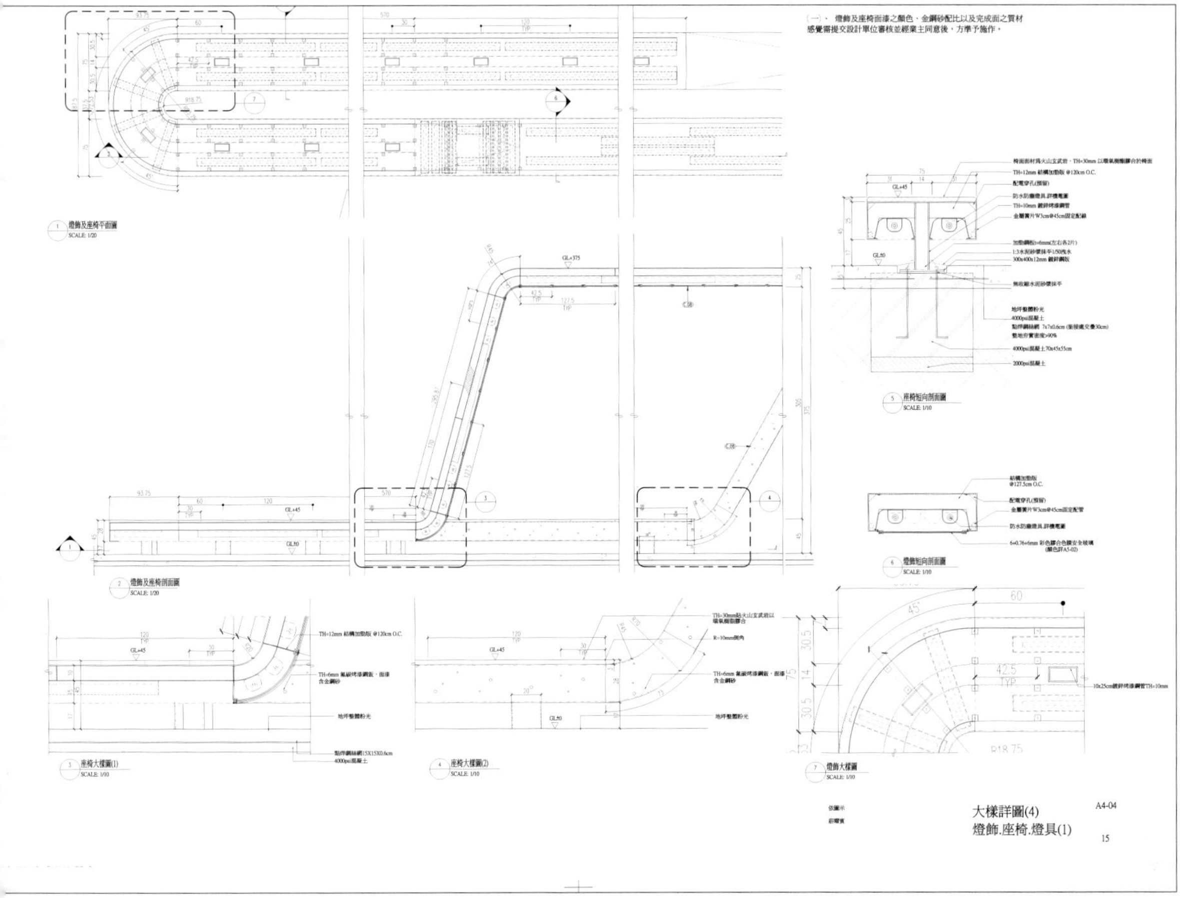
(一)． 燈飾及座椅面漆之顏色．金鋼砂配比以及完成面之質材感覺需提交設計單位審核並經業主同意後，方準予施作。
燈飾及座椅平面圖
SCALE: 1/20
燈飾及座椅剖面圖
SCALE: 1/20
座椅短向剖面圖
SCALE: 1/10
座椅大樣圖(1)
SCALE: 1/10
座椅大樣圖(2)
SCALE: 1/10
燈飾大樣圖
SCALE: 1/10
大樣詳圖(4)
燈飾.座椅.燈具(1)
A4-04
15

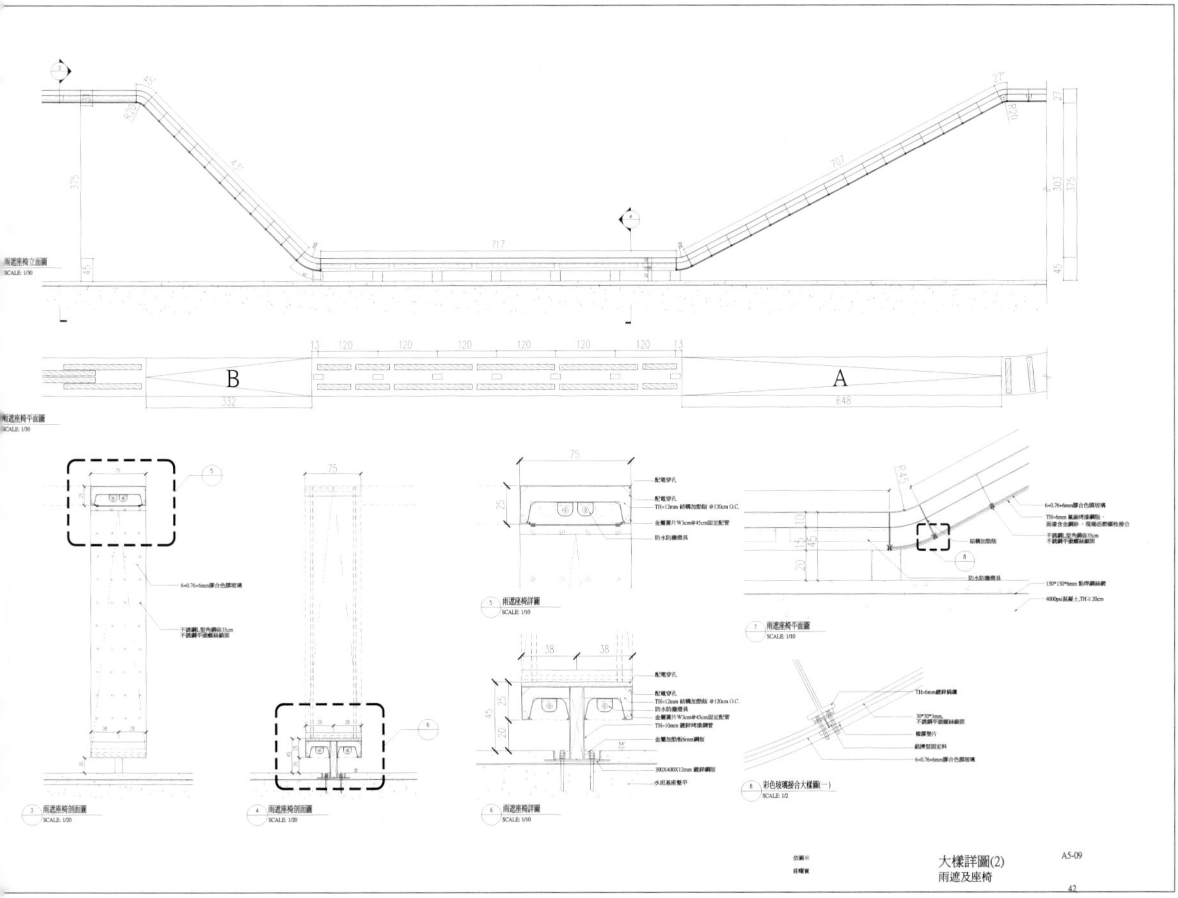
雨遮座椅立面圖
SCALE: 1/30
B
A
雨遮座椅平面圖
SCALE: 1/30
雨遮座椅剖面圖
SCALE: 1/20
雨遮座椅詳圖
SCALE: 1/10
雨遮座椅平面圖
SCALE: 1/10
彩色玻璃接合大樣圖(一)
SCALE: 1/2
大樣詳圖(2)
雨遮及座椅
A5-09
42

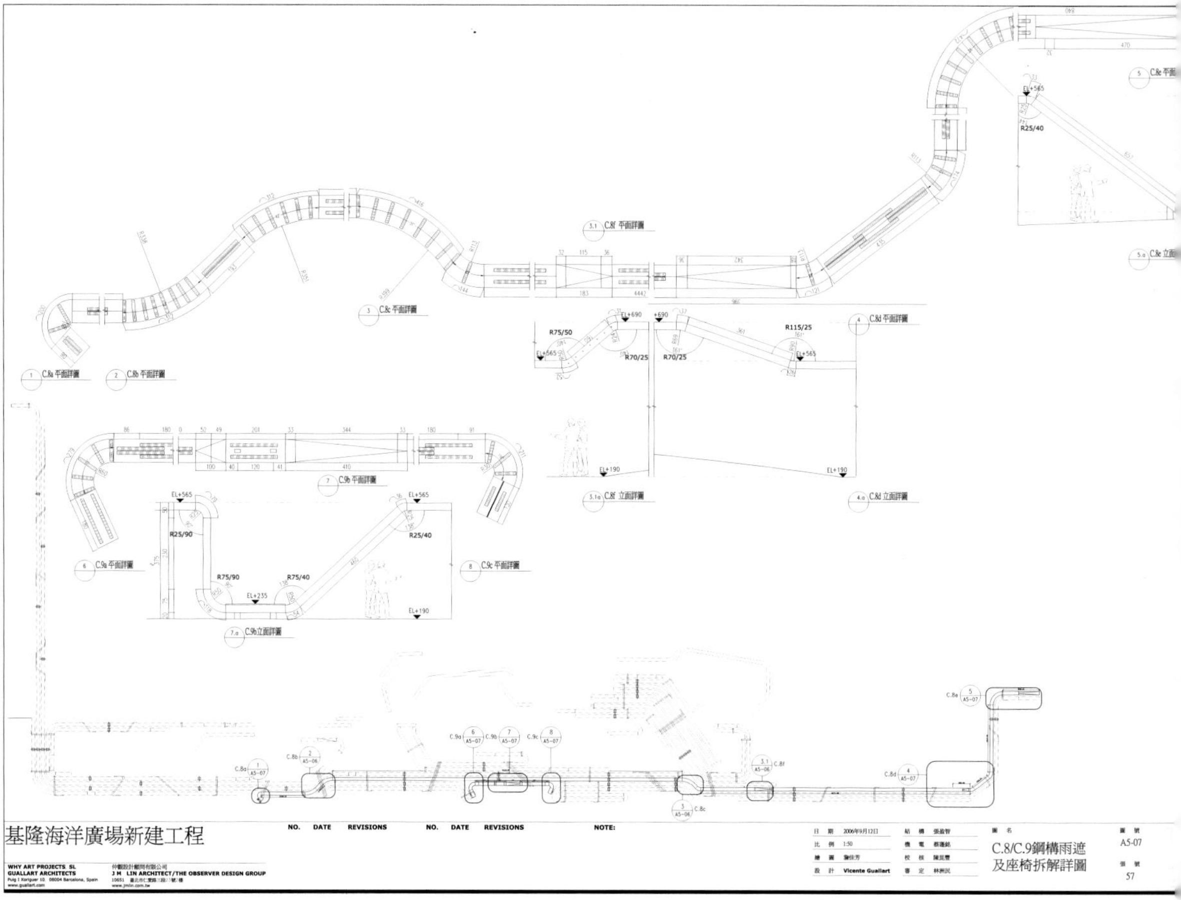
C.8a 平面詳圖
C.8b 平面詳圖
C.8c 平面詳圖
C.8f 平面詳圖
C.8d 平面詳圖
C.8f 立面詳圖
C.8d 立面詳圖
C.9b 平面詳圖
C.9a 平面詳圖
C.9c 平面詳圖
C.9b立面詳圖
基隆海洋廣場新建工程
NO. DATE REVISIONS NO. DATE REVISIONS NOTE:
WHY ART PROJECTS SL
GUALLART ARCHITECTS
J M LIN ARCHITECT/THE OBSERVER DESIGN GROUP
2006年9月12日
1:50
Vicente Guallart
C.8/C.9鋼構雨遮
及座椅拆解詳圖
A5-07
57

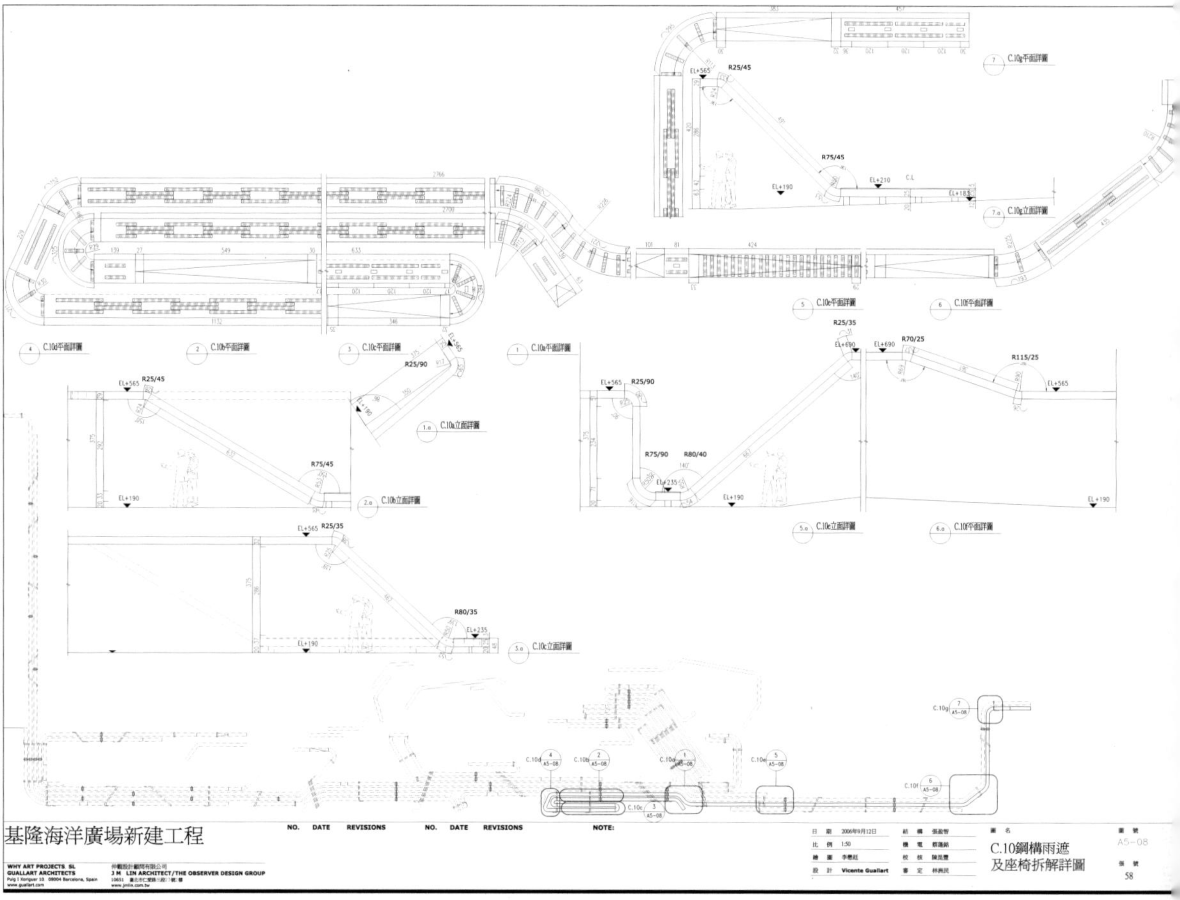
C.10g平面詳圖
C.10g立面詳圖
C.10d平面詳圖
C.10b平面詳圖
C.10c平面詳圖
C.10a平面詳圖
C.10e平面詳圖
C.10f平面詳圖
C.10a立面詳圖
C.10b立面詳圖
C.10e立面詳圖
C.10f立面詳圖
C.10c立面詳圖
基隆海洋廣場新建工程
NO. DATE REVISIONS NO. DATE REVISIONS NOTE:
WHY ART PROJECTS SL
GUALLART ARCHITECTS
J M LIN ARCHITECT/THE OBSERVER DESIGN GROUP
2006年9月12日
1:50
Vicente Guallart
C.10鋼構雨遮
及座椅拆解詳圖
A5-08
58

隆海洋廣場新建工程

NO.	DATE	REVISIONS	NO.	DATE	REVISIONS	NOTE:

ART PROJECTS SL
LART ARCHITECTS
origuer 10, 08004 Barcelona, Spain
ualart.com

仲觀設計顧問有限公司
J M LIN ARCHITECT/THE OBSERVER DESIGN GROUP
10651 臺北市仁愛路三段□3號□樓
www.jmlin.com.tw

日期	2006年9月12日	結構	張盈智
比例	依圖示	機電	蔡瀛銘
繪圖	李懋廷	校核	陳昆豐
設計	Vicente Guallart	審定	林洲民

圖名 R.6/R.7/R.9 石狀木構造物 尺寸詳圖
圖號 A6-04
張號 87

隆海洋廣場新建工程

NO.	DATE	REVISIONS	NO.	DATE	REVISIONS	NOTE:

ART PROJECTS SL
LART ARCHITECTS
origuer 10, 08004 Barcelona, Spain
ualart.com

仲觀設計顧問有限公司
J M LIN ARCHITECT/THE OBSERVER DESIGN GROUP
10651 臺北市仁愛路三段□3號□樓
www.jmlin.com.tw

日期	2006年9月12日	結構	張盈智
比例	1/20	機電	蔡瀛銘
繪圖	李懋廷	校核	陳昆豐
設計	Vicente Guallart	審定	林洲民

圖名 R.1~R.3石狀木構造物 尺寸詳圖
圖號 A6-02
張號 85

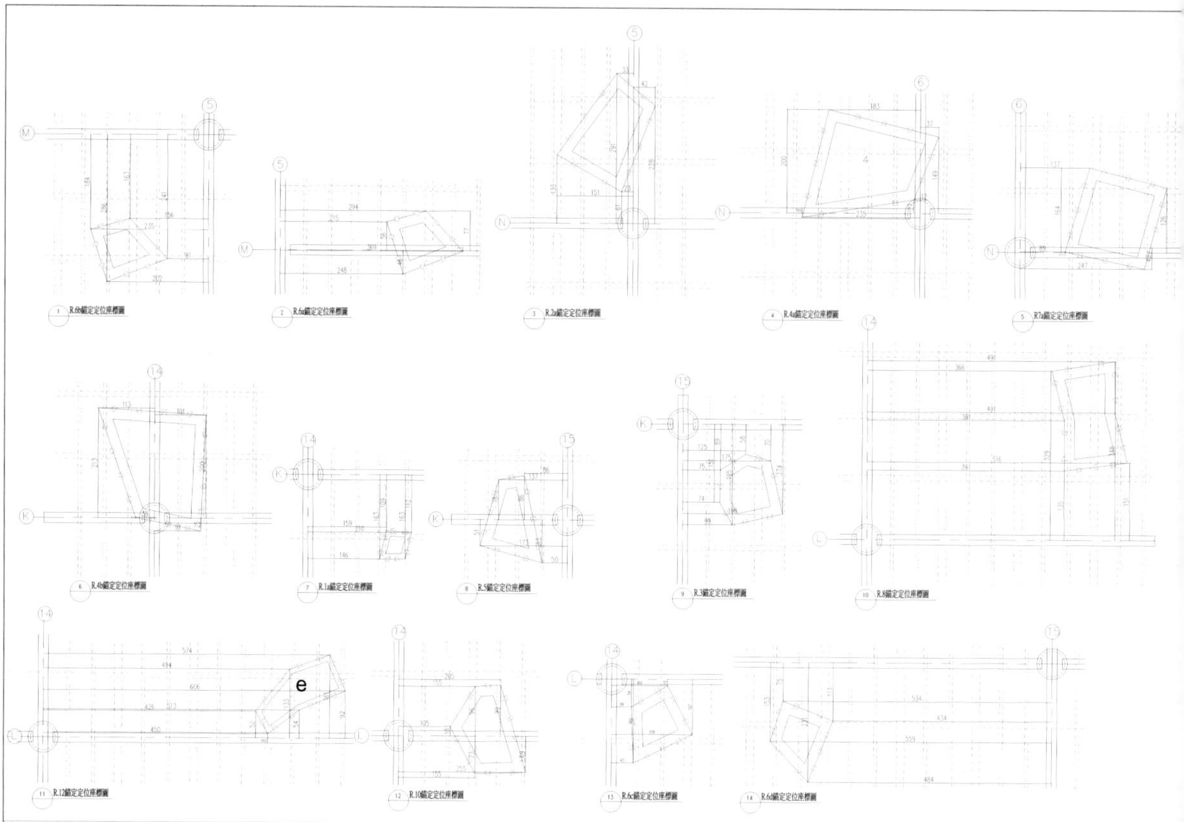

基隆海洋廣場新建工程

NO. DATE REVISIONS NO. DATE REVISIONS NOTE:

WHY ART PROJECTS SL
GUALLART ARCHITECTS
Puig I Xoriguer 10, 08004 Barcelona, Spain
www.guallart.com

仲觀設計顧問有限公司
J M LIN ARCHITECT/THE OBSERVER DESIGN GROUP
10651 臺北市仁愛路三段 號 樓
www.jmlin.com.tw

日期 2006年9月12日
比例 1:30
繪圖 林靜雯
設計 Vicente Guallart
結構 張盈智
機電 蔡蓮銘
校核 陳昆豐
審定 林洲民

圖名 石狀木構造物錨錠位置詳圖(1)

圖號 A6-08

張號 91

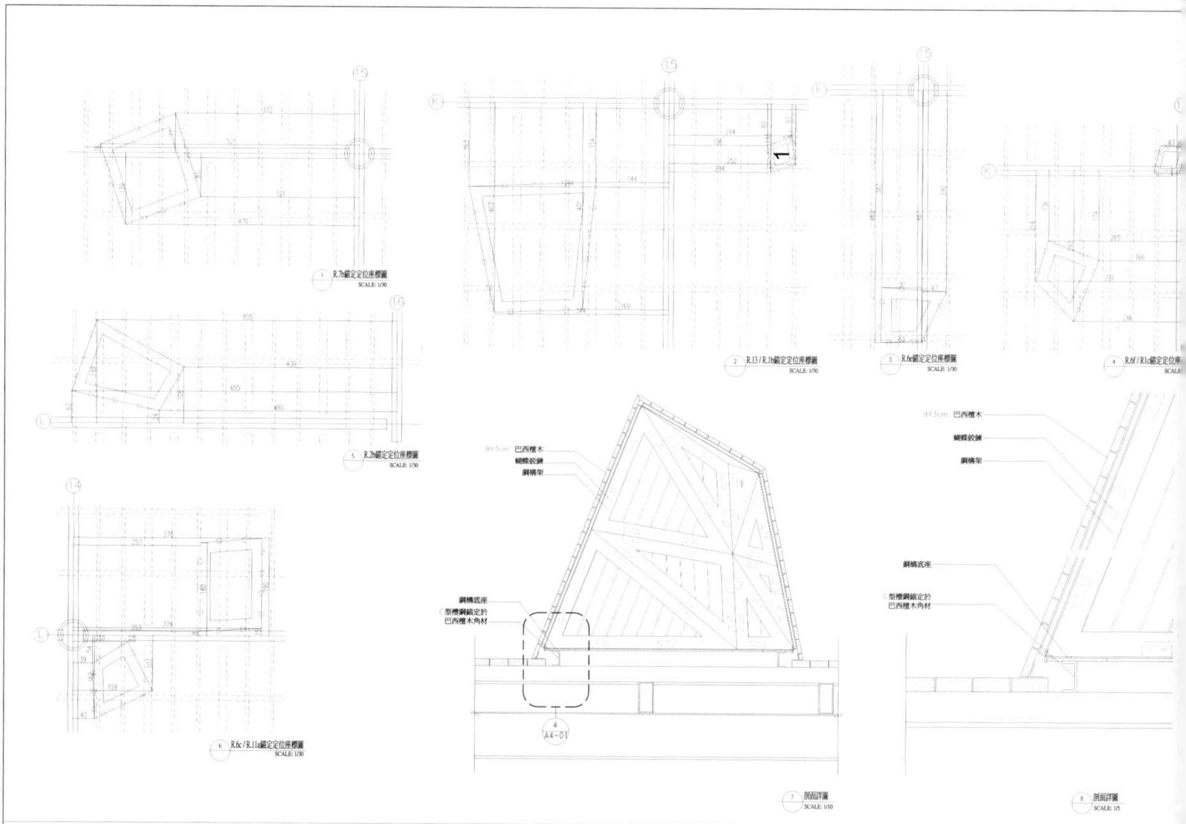

基隆海洋廣場新建工程

NO. DATE REVISIONS NO. DATE REVISIONS NOTE:

WHY ART PROJECTS SL
GUALLART ARCHITECTS
Puig I Xoriguer 10, 08004 Barcelona, Spain
www.guallart.com

仲觀設計顧問有限公司
J M LIN ARCHITECT/THE OBSERVER DESIGN GROUP
10651 臺北市仁愛路三段 號 樓
www.jmlin.com.tw

日期 2006年9月12日
比例 依圖示
繪圖 林靜雯
設計 Vicente Guallart
結構 張盈智
機電 蔡蓮銘
校核 陳昆豐
審定 林洲民

圖名 石狀木構造物錨錠位置詳圖(2)及剖面詳圖

圖號 A6-09

張號 92

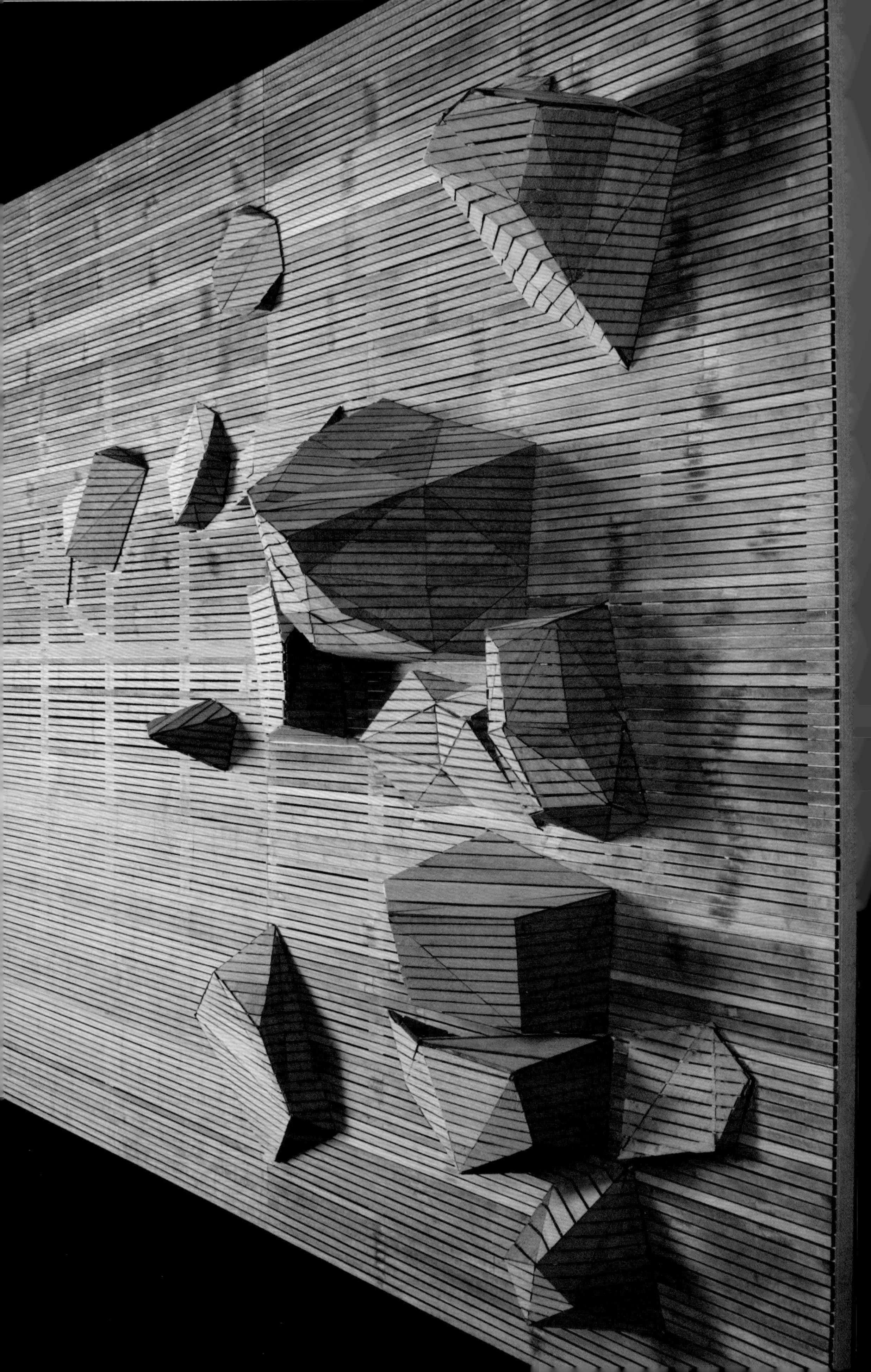

BISHA
In association with J.M. Lin/ The Observer Design Group

八斗子渔港
协同仲观设计顾问有限公司联合设计

Batoutz is situated 50 minutes by car from Taipei, nea Keelung, the city´s traditional port. The size of ther port, its proximity to the capital and its lanscape all facillitate it conversion into an international tourist destination while conserving its fishing activity.
For this to be achieved, what is needed in addition to quality tourist provision based on cuisine, are facilities geared towards a Conventions and Trade, Culture and shopping tourism that will provide the impetus for the desired shift in scale.
We propose:
A. Public facilities required:
Fish market, seafood restaurant and traditional shops
B. Proposed private facilities:
1. A hotel with panoramic views at the entrance of the port. 2. Hotel-Spa-Convention Centre, reconstructin the mountain overlooking the port. 3. A high-tech shopping mall and multi-screen cinemas. 4. A traditional open-air market. 5. An artificial sports complex
C. Public facilities in develpment:
The Ocean Museum and the park

LINEAR PATTERNS

We propose to establish a vibration between the land and the sea. To establish dozens of possible coastlines.
To define linear patterns that order the growth anc the implantation of functions in the port on the basis of a principle of mobility.
Systems scaled to the dimensions of the user (ship buses, automobiles, crowds, people, bicycles)that will order the implantation of buildings, the positionintg of materials, the definition of the lighting. A system open to the design of these that respects the force fields of the place.

PHOTOVOLTAIC PERGOLA

The space between buildings is traversed by a pergola convered with photovoltaic panel that wil serve to generate all of the energy consumed by t public space of the port.
The pergola acts like a tree whose leaves (the panels) collect sunligth, send energy down to the bas (the roots) and produce a fruit that is electric ligh The latest advanced systems allow electricity to b sent directly to the network, from which the energ needed at any given moment is taken, thus avoidir the use of environmentally harmful batteries.

QUARRY

In the central part of Batoutz, opposite the entran to the harbour, there is a quarry from which stone was extrated to build the port. We propose to reconstruct this and use it as the site for a Hotel, Sp and Convention Centre, to create an attractor for professional-sector tourism.
The interior volume will be reconstructed in keepi with the logic of the mountain´s own strata.
The facade will have glazed zones following the p tern of the rock formation to which it belongs.

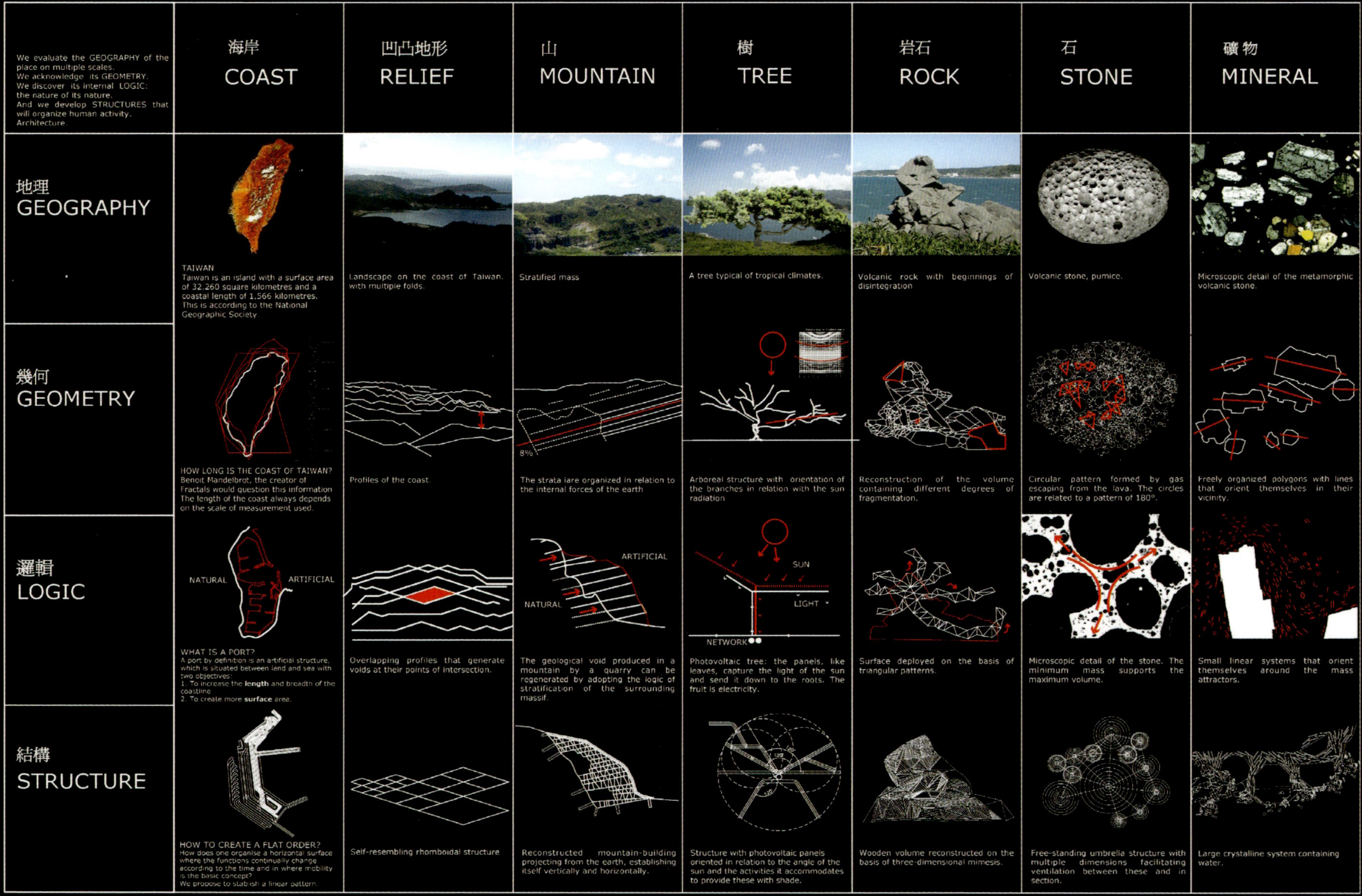
We evaluate the GEOGRAPHY of the place on multiple scales.
We acknowledge its GEOMETRY.
We discover its internal LOGIC: the nature of its nature.
And we develop STRUCTURES that will organize human activity. Architecture.
海岸 COAST
凹凸地形 RELIEF
山 MOUNTAIN
樹 TREE
岩石 ROCK
石 STONE
礦物 MINERAL
地理 GEOGRAPHY
TAIWAN
Taiwan is an island with a surface area of 32,260 square kilometres and a coastal length of 1,566 kilometres. This is according to the National Geographic Society.
Landscape on the coast of Taiwan. with multiple folds.
Stratified mass
A tree typical of tropical climates.
Volcanic rock with beginnings of disintegration
Volcanic stone, pumice.
Microscopic detail of the metamorphic volcanic stone.
幾何 GEOMETRY
HOW LONG IS THE COAST OF TAIWAN?
Benoit Mandelbrot, the creator of Fractals would question this information The length of the coast always depends on the scale of measurement used.
Profiles of the coast.
8%
The strata iare organized in relation to the internal forces of the earth
Arboreal structure with orientation of the branches in relation with the sun radiation
Reconstruction of the volume containing different degrees of fragmentation.
Circular pattern formed by gas escaping from the lava. The circles are related to a pattern of 180°.
Freely organized polygons with lines that orient themselves in their vicinity.
邏輯 LOGIC
NATURAL
ARTIFICIAL
WHAT IS A PORT?
A port by definition is an artificial structure, which is situated between land and sea with two objectives:
1. To increase the length and breadth of the coastline
2. To create more surface area.
Overlapping profiles that generate voids at their points of intersection.
ARTIFICIAL
NATURAL
The geological void produced in a mountain by a quarry can be regenerated by adopting the logic of stratification of the surrounding massif.
SUN
LIGHT
NETWORK
Photovoltaic tree: the panels, like leaves, capture the light of the sun and send it down to the roots. The fruit is electricity.
Surface deployed on the basis of triangular patterns.
Microscopic detail of the stone. The minimum mass supports the maximum volume.
Small linear systems that orient themselves around the mass attractors.
結構 STRUCTURE
HOW TO CREATE A FLAT ORDER?
How does one organise a horizontal surface where the functions continually change according to the time and in where mobility is the basic concept?
We propose to stablish a linear pattern.
Self-resembling rhomboidal structure
Reconstructed mountain-building projecting from the earth, establishing itself vertically and horizontally.
Structure with photovoltaic panels oriented in relation to the angle of the sun and the activities it accommodates to provide these with shade.
Wooden volume reconstructed on the basis of three-dimensional mimesis.
Free-standing umbrella structure with multiple dimensions facilitating ventilation between these and in section.
Large crystalline system containing water.

3 BATOUTZ MASTERPLAN

VICENTE GUALLART ARCHITECTURE BARCELONA _ SPAIN

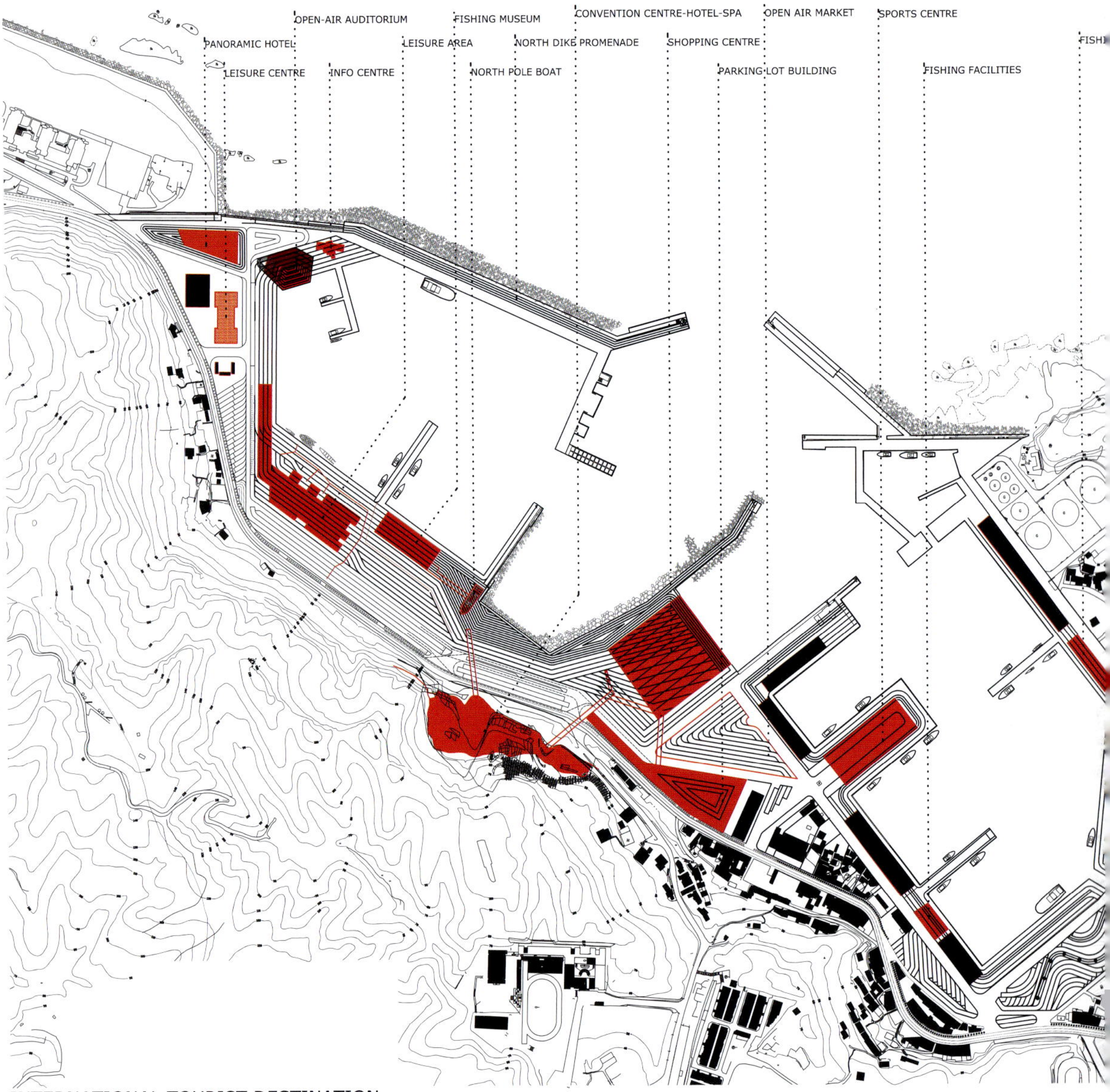

INTERNATIONAL TOURIST DESTINATION

Batouz is situated 50 minutes by car from Taipei, near Keelung, the city's traditional port. The size of the port, its proximity to the capital and its landscape all facilitate it conversion into an international tourist destination while conserving its fishing activity.
For this to be achieved, what is needed in addition to quality tourist provision based on cuisine, are facilities geared towards a Conventions and Trade, Culture and shopping tourism that will provide the impetus for the desired shift in scale.

We propose:

A. Public facilities required:
Fish market, seafood restaurant and traditional shops

B. Proposed private facilities:
1. A hotel with panoramic views at the entrance of the port
2. Hotel-Spa-Convention Centre, reconstructing the mountain overlooking the port.
3. A high-tech shopping mall and multi-screen cinemas
4. A traditional open-air market
5. An artificial sports complex

C. Public facilities in development:
The Ocean Museum and the park

					DEVELOPER		PHASE				CHRONOGRAM. RHYTHM OF ACTIVITY IN LEISURE PROV														
	USES		AREA	UNITS	PUBLIC	PRIVATE	1	2	3	NATURE	9	10	11	12	13	14	15	16	17	18	19	20	21	22	2
	A1	LEISURE AREA	5.180m2	1	X		X			RELIEF															
		CAFÉS	665m2	15	X		X			RELIEF	X	X	X	X	X	X	X	X	X	X	X	X	X	X	X
		RETAIL FISH MARKET	428m2	16	X		X			RELIEF	X	X	X	X	X			X	X	X	X	X			
		STORES	1.000m2	42	X		X			RELIEF	X	X	X	X	X	X	X	X	X	X	X	X	X		
		FOOD PLAZA RESTAURANTS	889m2	13	X		X			RELIEF	X	X	X	X	X	X	X	X	X	X	X	X	X	X	X
		RESTAURANTS	2.128m2	15	X		X			RELIEF	X	X	X	X	X	X	X	X	X	X	X	X	X	X	X
	A2	PHOTOVOLTAIC·PERGOLA	569m	1	X		X			TREE	X	X	X	X	X	X	X	X	X	X	X	X	X	X	X
	A3	OPEN-AIR AUDITORIUM	2.700m2	1000 P.	X		X			COAST															
	A4	INFO CENTRE	1000m2	1	X		X				X	X	X	X	X	X	X	X	X	X	X	X	X	X	X
	A5	NORTH DIKE PROMENADE	490m	1	X		X			COAST	X	X	X	X	X	X	X	X	X	X	X	X	X	X	X
	A6	OCEAN TOURISM CRUISE	70m2	1	X		X																		
		NORTH COAST FAST FERRY		1	X		X				X	X	X	X	X	X	X	X	X	X					
		SUBMARINE (VISIT TO ISLAND)		1	X		X				X	X	X	X	X	X	X	X	X	X					
		BOAT HIRE		1	X		X				X	X	X	X	X	X	X	X	X	X					
		FISHING RENTAL EXPERIENCE		1	X		X				X	X	X	X	X	X	X	X	X	X					
	A7	PARKING LOT		23BUSES 458CARS 481	X		X			COAST	X	X	X	X	X	X	X	X	X	X	X	X	X	X	X

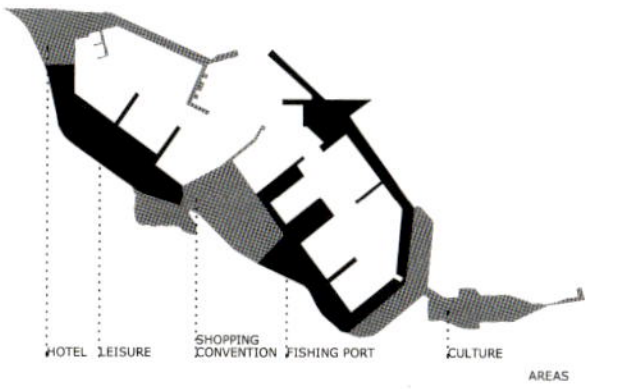

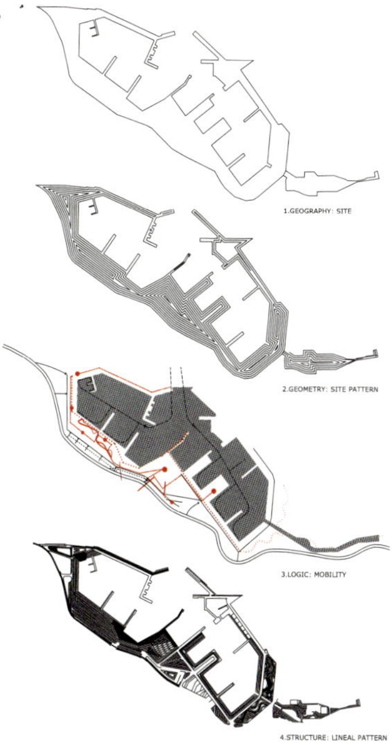

	AREA	UNITS	DEVELOPER PUBLIC	DEVELOPER PRIVATE	PHASE 1	PHASE 2	PHASE 3	NATURE	9	10	11	12	13	14	15	16	17	18	19	20	21	22	23	24	1	2
MS		1	X		X																					
FISHING MUSEUM	2.800m2	1	X		X				X	X	X	X	X		X	X	X	X	X							
OLE BOAT		1	X		X				X	X	X	X	X		X	X	X	X	X							
AMIC HOTEL				X		X																				
	58.884m2	888 ROOMS		X		X		STRATA	X	X	X	X	X	X	X	X	X	X	X	X	X	X	X	X	X	X
SWIMMING POOL	930m2	1		X		X		STRATA	X	X	X	X	X	X	X	X	X	X	X	X	X	X				
RANT	1.500m2	4		X		X		STRATA					X	X	X							X	X	X		
IEQUE	301m2	1		X		X		STRATA															X	X	X	X
E	241m2	1		X		X		STRATA											X	X	X	X	X	X	X	X
MIC VIEW ELEVATOR	1.650m2	1		X		X		STRATA	X	X	X	X			X	X	X	X	X	X	X	X				
CIAL MOUNTAIN		1		X		X		MOUNTAIN																		
TION CENTRE	2.075m2	1500 p.		X		X		MOUNTAIN	X	X	X	X	X	X	X	X	X	X	X	X						
	12.770m2	700 ROOMS		X		X		MOUNTAIN	X	X	X	X	X	X	X	X	X	X	X	X	X	X	X	X	X	X
assotherapy	2.165m2	1		X		X		MOUNTAIN	X	X	X	X				X	X	X	X	X						
RANT	3000m2	4		X		X		MOUNTAIN					X	X	X						X	X	X			
CENTRE	1.073m2	1		X		X		MOUNTAIN	X	X	X	X				X	X	X	X	X						

USES		AREA	UNITS	DEVELOPER PUBLIC	DEVELOPER PRIVATE	PHASE 1	PHASE 2	PHASE 3	NATURE	9	10	11	12	13	14	15	16	17	18	19	20	21	22	23	24	1	2
B3	SHOPPING CENTRE		1		X		X		MINERAL																		
	FASHION SHOPS	2.010m2	30		X		X		MINERAL	X	X	X	X	X	X	X	X	X	X	X	X	X					
	HIGH-TECH MARKET	3.760m2	54		X		X		MINERAL	X	X	X	X	X	X	X	X	X	X	X	X	X					
	CINEMAS	1.848m2	8		X		X		MINERAL							X	X	X	X	X	X	X	X	X	X		
	CAFÉ TERRACES AND MUSIC	2.657m2	15		X		X		MINERAL	X	X	X	X	X	X	X	X	X	X	X	X	X	X	X	X	X	
B4	PARKING LOT BUILDING	14.683m2	238USES 394CARS 413		X		X		RELIEF	X	X	X	X	X	X	X	X	X	X	X	X	X	X	X	X	X	X
B5	SPORTS CENTRE		1		X		X		RELIEF																		
	ARTIFICIAL GOLF/BASEBALL	6.366m2	1		X		X		RELIEF	X	X	X	X	X	X	X	X	X	X	X	X						
	TENNIS-BASKETBALL-MINI FOOTBALL	3.312m2	4		X		X		RELIEF	X	X	X	X	X	X	X	X	X	X	X	X	X	X				
	FAMILY GAMES	6.366m2	1		X		X		RELIEF	X	X	X	X	X	X	X	X	X	X								
B6	FISHING FACILITIES	7.611m2	2	X			X		COAST	X	X	X	X	X	X	X	X	X	X								
B7	OPEN AIR MARKET		1	X			X		COAST	X	X	X	X	X	X	X	X	X	X	X	X	X	X	X	X	X	X
C1	CULTURAL PLAZA		1	X				X	COAST																		
	OCEAN MUSEUM	34.770m2	1	X				X	COAST	X	X	X	X	X			X	X	X	X	X						
	CULTURE PARK	16.000m2	1	X				X	COAST	X	X	X	X	X			X	X	X	X	X						
C2	LEISURE CENTRE	9.760m2	1	X				X		X	X	X	X	X	X	X	X	X	X	X	X	X	X				

SECCIÓN +220
SECCIÓN +180
SECCIÓN +140
SECCIÓN +100
SECCIÓN +80
SECCIÓN +40
SECCIÓN +20

ATOUTZ
TERNATIONAL TOURIST DESTINATION
ne proposal is based on the conviction that the ccessful development of tourist activity consists making available a wide range of leisure uses roughout the course of the day, and these will ave to be complemented by a sufficient level of ac- ommodation to ensure a continuity of demand for e catering establishments and shops. Because the oposal is centred exclusively on the port area, it ntains a limited development of the hotel capac- y, which will need to be increased by means of in- rventions in the immediate periphery of the port. nce both phases have been concluded, the total commodation will be not less than 10,000 m2, of hich at least 70% will be occupied by hotels.

ort Batoutz as a component of the network of astal tourist nodes:
e proposal is conceived from the conviction that atoutz is the hub of a network of tourist destina- ons that extends all along the coast of Taiwan. ese nodes will have to contain amenities that will tisfy the leisure needs of their own residents and the same time complement the others, so that ch of these, including Batoutz, will be able to add its own extensive range of services and activi- es those provided by other destinations and by e capital of Taipei itself, places which tourists are ely to visit in the course of their stay. The provi- on thus configured will be sufficient to satisfy the ectrum of leisure needs for a minimum of 7 days, though the length of stay could be extended or e stay repeated in successive years if comple- ented with visits to other destinations. The loss of nsumer revenue represented by the 'loss' of tour- s making excursions outside of Batoutz would be mpensated by the visit of others tourists staying other resorts in the network.

toutz距离台北50分钟车程，靠近城市的传统港口基隆。港的规模，它到市中心的便捷性，以及其原本所具有的景观都其具备了在继续现有的捕捞活动的同时，转化成为一个国际的旅游胜地的条件。
要成就这样的目标，除了需要提供高质量的旅游环境所必需餐饮条件以外，硬件需求转向了贸易、文化以及购物性旅游所能提供的尺度转化性的条件。
门提出：
公共设施需求：水产品市场、海鲜餐馆以及传统风格的商铺
私有设施需求：
.港口入口处有一个能够看到全景的的旅馆
.酒店、水疗及会议中心，重建能够俯瞰整个港口的山丘。
.高科技购物中心和多荧幕电影院
.一个传统的露天市场
.一座健身中心
开发当中的公共设施：
海洋博物馆及公园

线性模式：
我们提出在土地与海洋之间建立一种弹性关系。进而建立多条可能的海岸线。在一个移动性原则的基础之上对于港口功能的增长和内嵌进行秩序化的线性模式进行定义。
整个系统在尺度上进行了调整以适应使用者（船只、公共汽车、机动车、人群、行人、自行车），而这些使用者则负责对建筑的嵌入、材料的位置、照明的定义进行组织。
这个系统开放了对于那些顺从该处力场的设计。

光电面板
建筑群之间的空间被一条带有光电面板的藤架穿过，这些光电面板负责生成港口公共空间所消耗的所有能源。
藤架的运行方式类似于树木，树叶（面板）负责采集阳光，然后将能量传输到基础（根部），最终所制造的果实就是电力照明。
通过最新技术改进的系统允许电力被直接输送到网络当中，能量则在任何给定的时间从这一网络中提取其所需的电力，进而避免使用对环境有害的电池。

采石场
在Batoutz的中心，与港口的入口相对处有一个为建造港口提供石材的采石场。我们计划对它进行重建，并将它用作一个酒店、水疗及休闲中心的场地，为专业地区的旅游创造一个吸引游人的魅力点。
内部体量将会以保持山丘原本层次的逻辑进行重建。
立面将通过块状玻璃来划分以顺延它所在环境的岩石构造模式。

Batoutz
国际旅游者的目的地
这一方案的提出基于这样的认识：一个成功的旅游活动开发需要在一天的行程内尽可能包含各种各样的休闲项目，同时这样的要求则需要充足的商住空间来保证其与餐饮和购物需求的连续性。该方案将是港口区域的焦点，因而它包含了有限的酒店资源开发措施，这将需要通过在紧邻港口的外围进行干预来拓展。一旦两种措施都被采用，总体的商住面积就可以达到不少于一万平方米的标准，而这其中的70%都将被酒店占用。

Batoutz港口作为海岸旅游系统中的一个部分节点：
该方案将Batoutz作为连接整个台湾海岸旅游目的地网络的路由器。这些节点将必须包含能够在满足当地居民的休闲需求同时补充其他地区需要的休闲设施，那么包括Batoutz在内的这些旅游目的地就都能够获得自身的服务和其他地点以及台北市本身提供的活动，而这些地点都是旅游者们在其逗留期间希望到访的地方。由此确定的前提条件将会足够丰富以满足最少七天的游览需求量，在与游览其他地区相结合的情况下，逗留时间的长度有可能被延长，或是在接下来的几年中会再次到访。通过Batoutz以外的旅游者观览活动所体现出的消费者总收入的损失，将会通过在网络中其他休闲设施中逗留的旅游者的到访来弥补。

4 BATOUTZ USES

VICENTE GUALLART ARCHITECTURE BARCELONA _ SPAIN

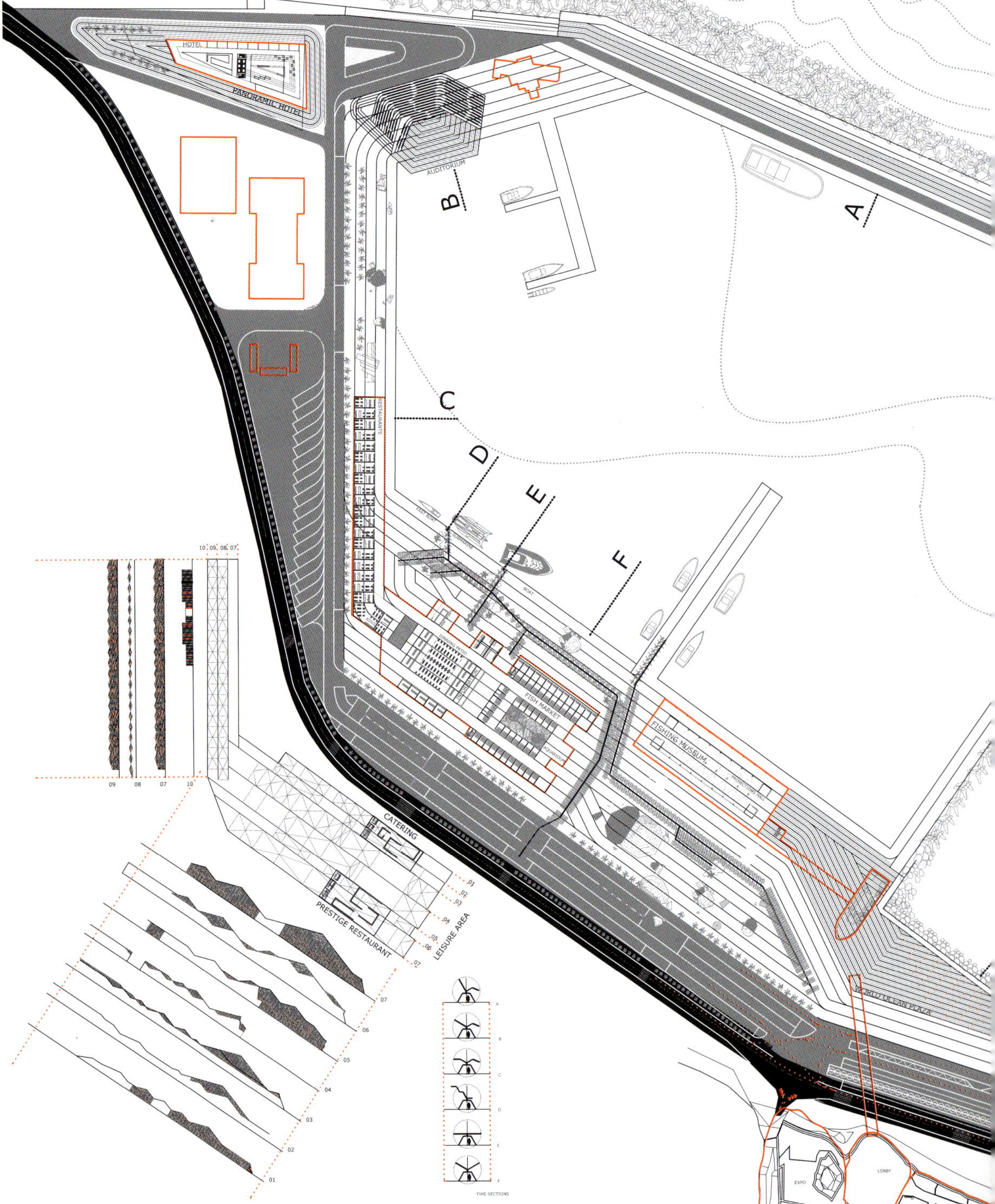

LINEAR PATTERNS

We propose to establish a vibration between the land and the sea. To establish dozens of possible coastlines.
To define linear patterns that order the growth and the implantation of functions in the port on the basis of a principle of mobility.
Systems scaled to the dimensions of the user (ships, buses, automobiles, crowds, people, bicycles) that will order the implantation of buildings, the positioning of materials, the definition of the lighting...
A system open to the design of these that respects the force fields of the place.

PHOTOVOLTAIC PERGOLA

The space between buildings is traversed by a pergola covered with photovoltaic panel that will serve to generate all of the energy consumed by the public space of the port.
The pergola acts like a tree whose leaves (the panels) collect sunlight, send energy down to the base (the roots) and produce a fruit that is electric light.
The latest advanced systems allow electricity to be sent directly to the network, from which the energy needed at any given moment is taken, thus avoiding the use of environmentally harmful batteries.

QUARRY

In the central part of Batoutz, opposite the entrance to the harbour, there is a quarry from which stone was extracted to build the port.
We propose to reconstruct this and use it as the site for a Hotel, Spa and Convention Centre, to create an attractor for professional-sector tourism.
The interior volume will be reconstructed in keeping with the logic of the mountain's own strata.
The facade will have glazed zones following the pattern of the rock formation to which it belongs.

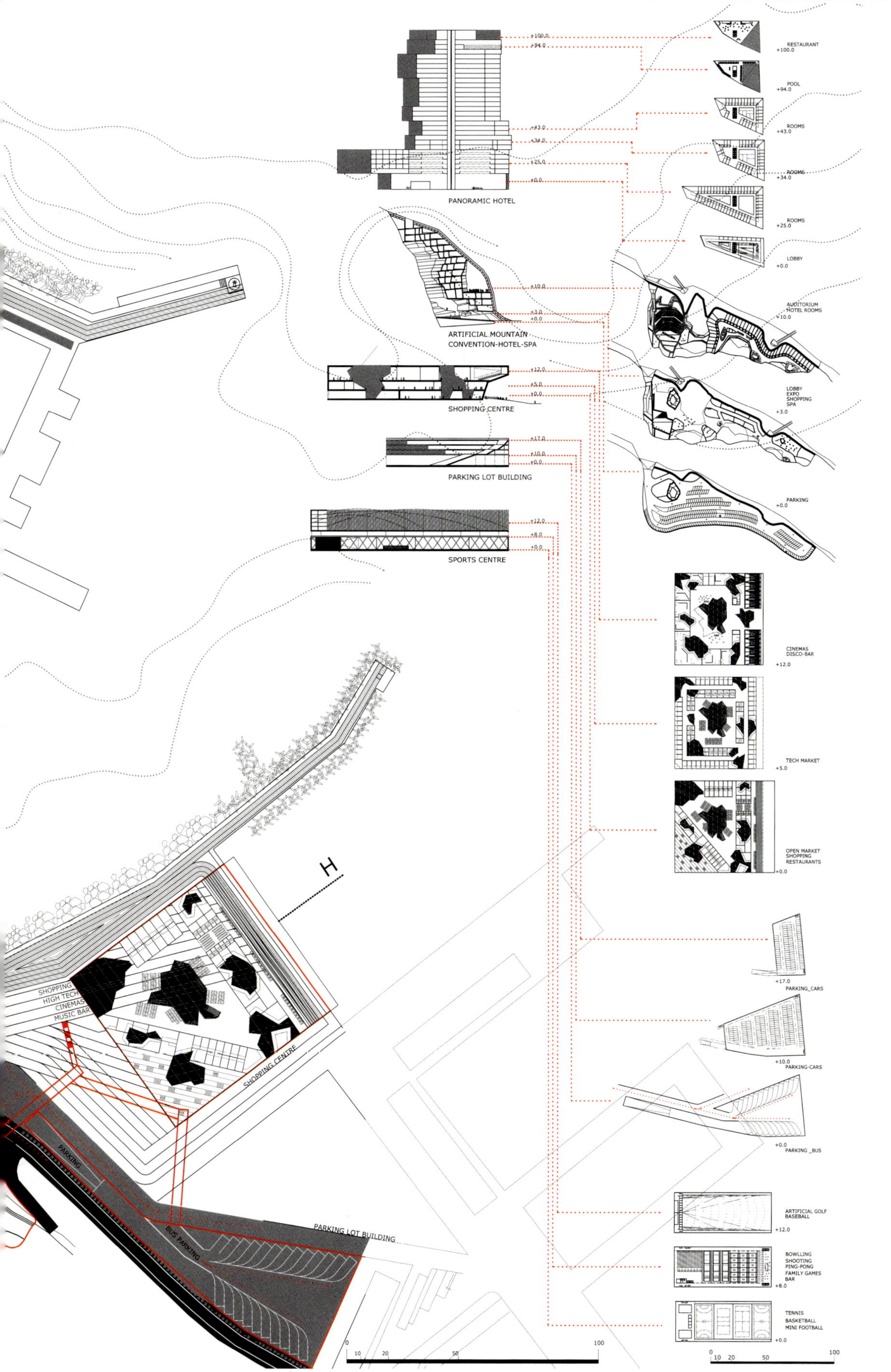
PANORAMIC HOTEL
ARTIFICIAL MOUNTAIN
CONVENTION-HOTEL-SPA
SHOPPING CENTRE
PARKING LOT BUILDING
SPORTS CENTRE
RESTAURANT
+100.0
POOL
+94.0
ROOMS
+43.0
ROOMS
+34.0
ROOMS
+25.0
LOBBY
+0.0
AUDITORIUM
HOTEL ROOMS
+10.0
LOBBY
EXPO
SHOPPING
SPA
+3.0
PARKING
+0.0
CINEMAS
DISCO-BAR
+12.0
TECH MARKET
+5.0
OPEN MARKET
SHOPPING
RESTAURANTS
+0.0
+17.0
PARKING_CARS
+10.0
PARKING-CARS
+0.0
PARKING _BUS
ARTIFICIAL GOLF
BASEBALL
+12.0
BOWLLING
SHOOTING
PING-PONG
FAMILY GAMES
BAR
+8.0
TENNIS
BASKETBALL
MINI FOOTBALL
+0.0
H
SHOPPING
HIGH TECH
CINEMAS
MUSIC BAR
SHOPPING CENTRE
PARKING
BUS PARKING
PARKING LOT BUILDING
0 10 20 50 100
0 10 20 50 100

THE ARCHITECTURE OF TIME: LEISURE SPACES

If success in tourism is related to the quantity, variety and quality of the leisure provision, what is needed is a programme of uses that can satisfy the multiple demands that may be made by tourists, in such a way as to offer a range of functional alternatives all through the day. The organization of these sequences is known as 'the architecture of time'. The considerable size of Port Batoutz and its area of influence allows us to anticipate a wide spectrum of leisure provision that - although constituting no more than the embryo of what will be available once the subsequent phases of the project are implemented - will keep the visitor active, with at least 35 functions from 9 in the morning on, hours and with at least 10 functions during the night. These functions will cover all the needs for the use of time, enabling us to assert with confidence that the project fulfils all the functional requirements necessary for a satisfactory stay.

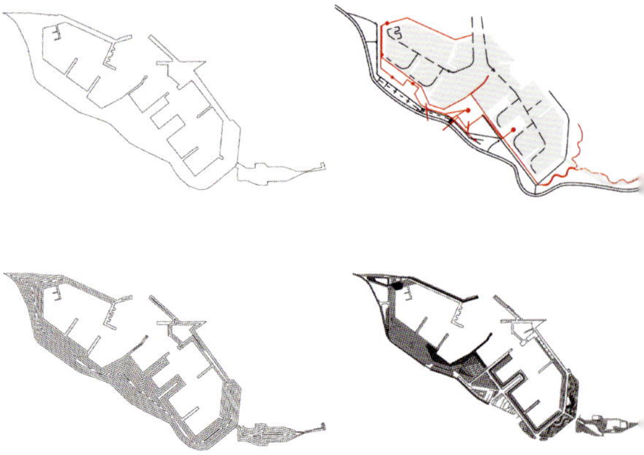

THE ORGANIZATION OF TIME IN SPACE

In order to give complete satisfaction with such a range of provision and meet the objective of satisfying the desires of the tourist, what is needed a spatial organization of uses that guarantee the functional efficiency of the system as a whole. To do this, it is essential to plan a distribution of facilities that takes due account of the potential and the limitations of each space. In the project that we are proposing we have organized 11 different spaces (Cultural Plaza, Fishing Area, Artificial Mountain, Shopping Centre, Traditional Open-air Market, Sports Centre, Museums, Leisure Area, Boat Access, Hotel Tower and Excursions), most of which contain a variety of establishments and options. The set of typical uses is estimated at 40 within the area, complemented by 5 proposed excursions: together these constitute one of the most important concentrations of tourist leisure provision in Taiwan.

时间的建筑：休闲空间

如果说与旅游业的成功息息相关的是休闲设施的数量、种和质量的话，我们所需要的是一个与功能相关的组织程序，便在旅游过程中持续提供更大范围的可供选择功能来满足游者可能提出的多种需要。这些功能的组织被称为“时间的筑”。Batoutz港口的客观体量以及其在区域内的影响让我能够充分调动更大范围内的休闲娱乐措施——尽管这些措仅限于那些一旦方案相关阶段被实施就能够实现的功能—它们仍将保证来访者的活跃性，包括早九点起开始生效的十五个功能，而夜间则至少有十个功能是可供使用的。这些能将会囊括一切的即时需要，令我们对该方案能满足来访的一切功能需求有充分的自信。

时间在空间当中的组织

为了让这样一系列的措施提供令人满意的服务，并满足游客一切客观需求，我们需要将整个系统作为一个整体，将那保证该系统功能有效性的使用方式进行空间上的组织。而到这一点最为重要的就是充分考虑每个空间的潜力与限制通过规划令不同设施的使用时间分散开来。在这个方案当中我们提出组织十一个不同的空间（文化广场、捕捞区、人造丘、购物中心、传统露天市场、体育中心、博物馆、休闲娱区、游船区、酒店大厦和其他观览点），而这其中大部分空都将包括大量不同的设备和选择。区域内一个典型的使用式序列建立在四十个左右，同时受到五个预先设置的观览的补充支持：这些措施共同建立了一个最为重要的台湾旅游闲环境浓缩体。

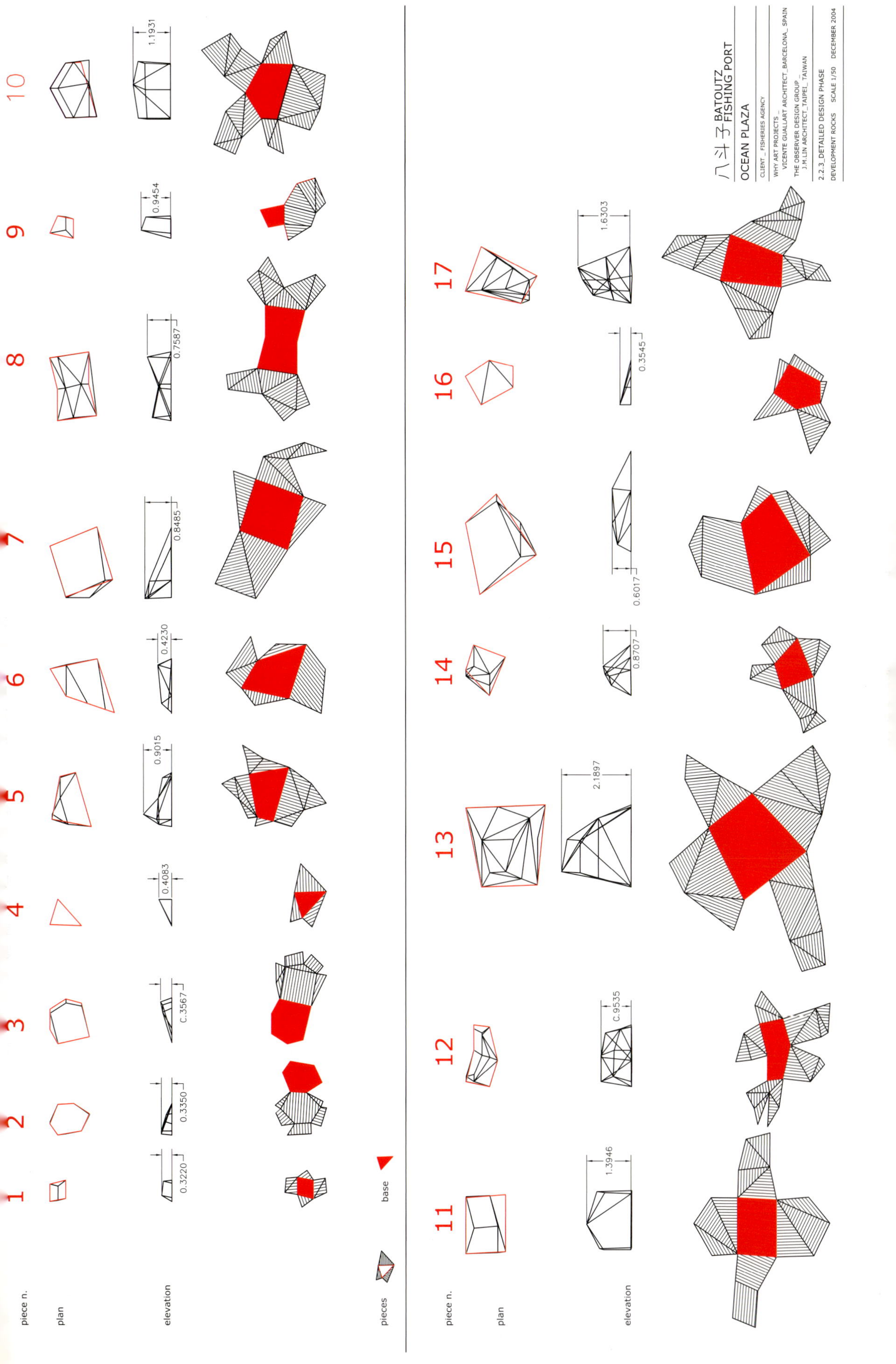

八斗子 BATOUTZ FISHING PORT
OCEAN PLAZA
CLIENT _ FISHERIES AGENCY
WHY ART PROJECTS _
VICENTE GUALLART ARCHITECT_BARCELONA_ SPAIN
THE OBSERVER DESIGN GROUP _
J.M.LIN ARCHITECT_TAIPEI_ TAIWAN
2.2.3_DETAILED DESIGN PHASE
DEVELOPMENT ROCKS SCALE 1/50 DECEMBER 2004
piece n.
plan
elevation
pieces
base
1 2 3 4 5 6 7 8 9 10
0.3220 0.3350 C.3567 0.4083 0.9015 0.4230 0.8485 0.7587 0.9454 1.1931
11 12 13 14 15 16 17
1.3946 C.9535 2.1897 0.8707 0.6017 0.3545 1.6303

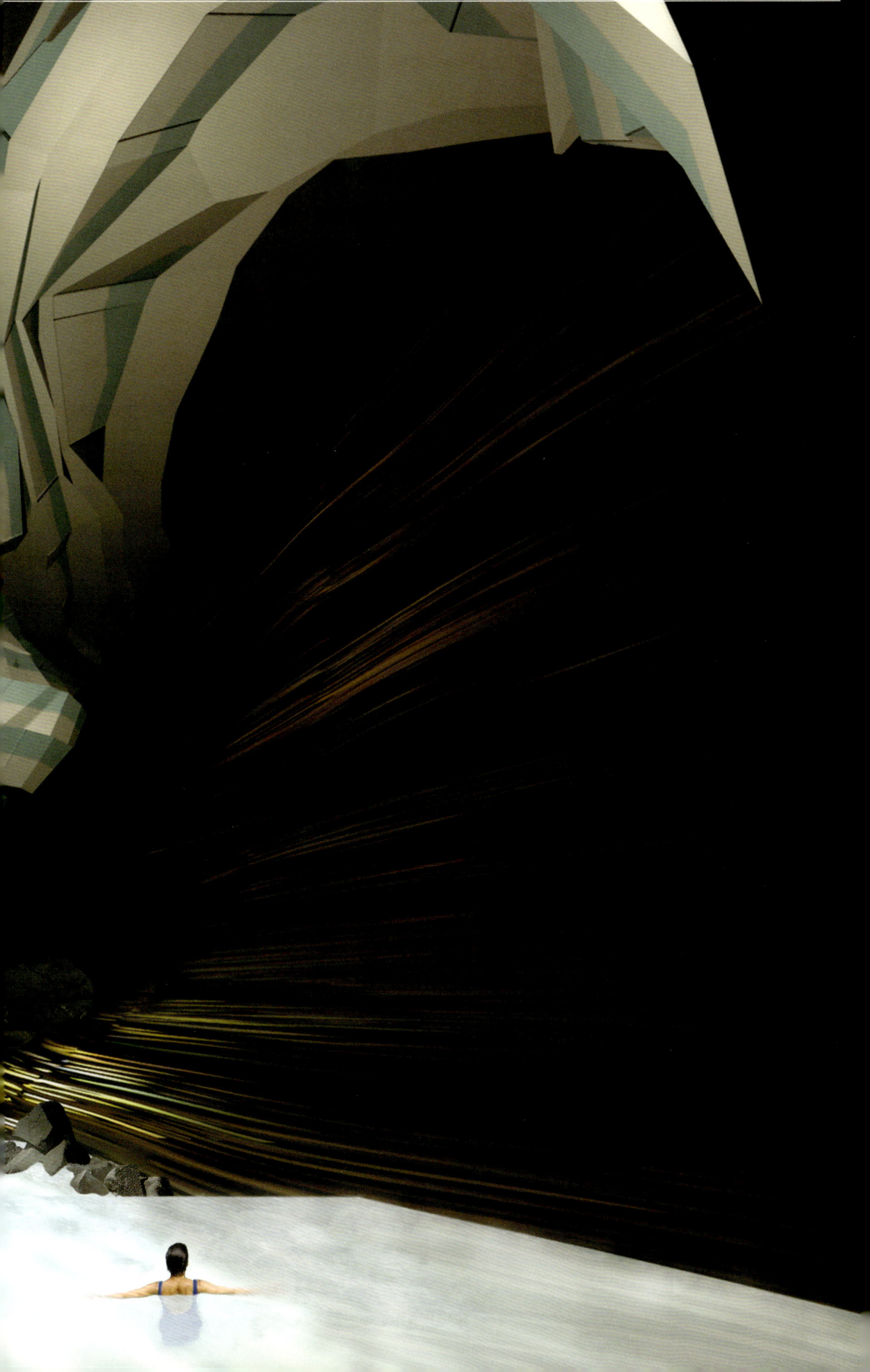

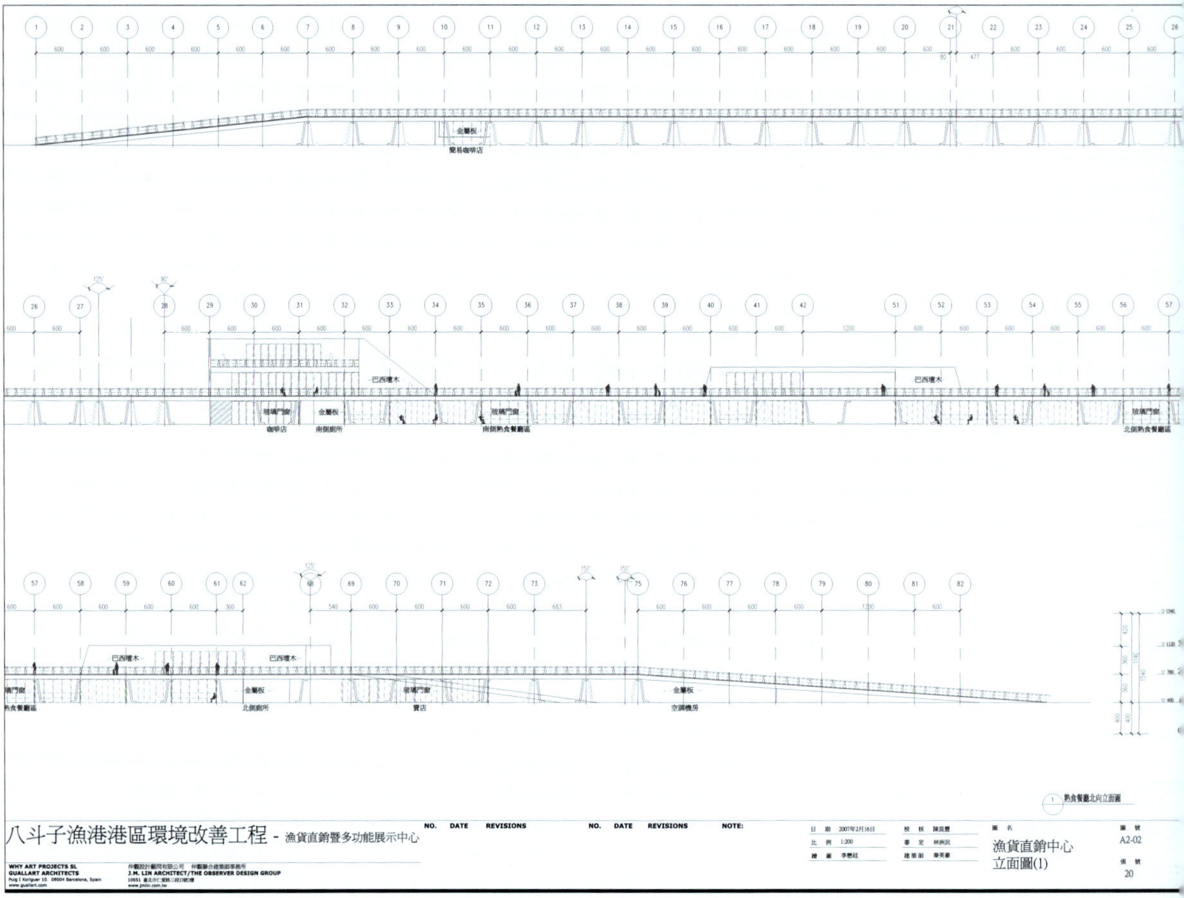
熟食餐廳北向立面圖
八斗子漁港港區環境改善工程 - 漁貨直銷暨多功能展示中心
NO. DATE REVISIONS
NO. DATE REVISIONS
NOTE:
日期 2007年2月16日
比例 1:200
漁貨直銷中心
立面圖(1)
A2-02
20
WHY ART PROJECTS SL
GUALLART ARCHITECTS
Puig I Xoriguer 10. 08004 Barcelona, Spain
www.guallart.com
J.M. LIN ARCHITECT/THE OBSERVER DESIGN GROUP
www.jmlin.com.tw

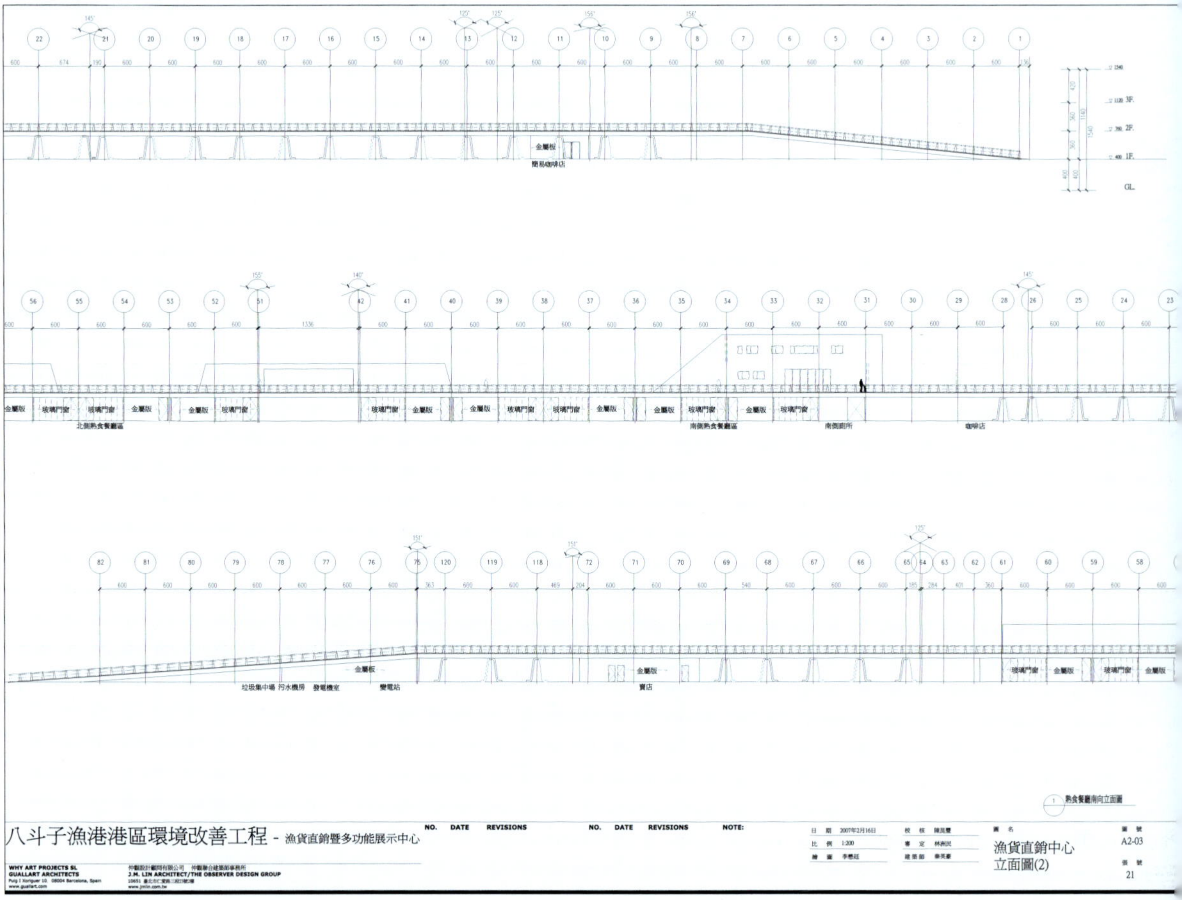
熟食餐廳南向立面圖
八斗子漁港港區環境改善工程 - 漁貨直銷暨多功能展示中心
NO. DATE REVISIONS
NO. DATE REVISIONS
NOTE:
日期 2007年2月16日
比例 1:200
漁貨直銷中心
立面圖(2)
A2-03
21
WHY ART PROJECTS SL
GUALLART ARCHITECTS
Puig I Xoriguer 10. 08004 Barcelona, Spain
www.guallart.com
J.M. LIN ARCHITECT/THE OBSERVER DESIGN GROUP
www.jmlin.com.tw

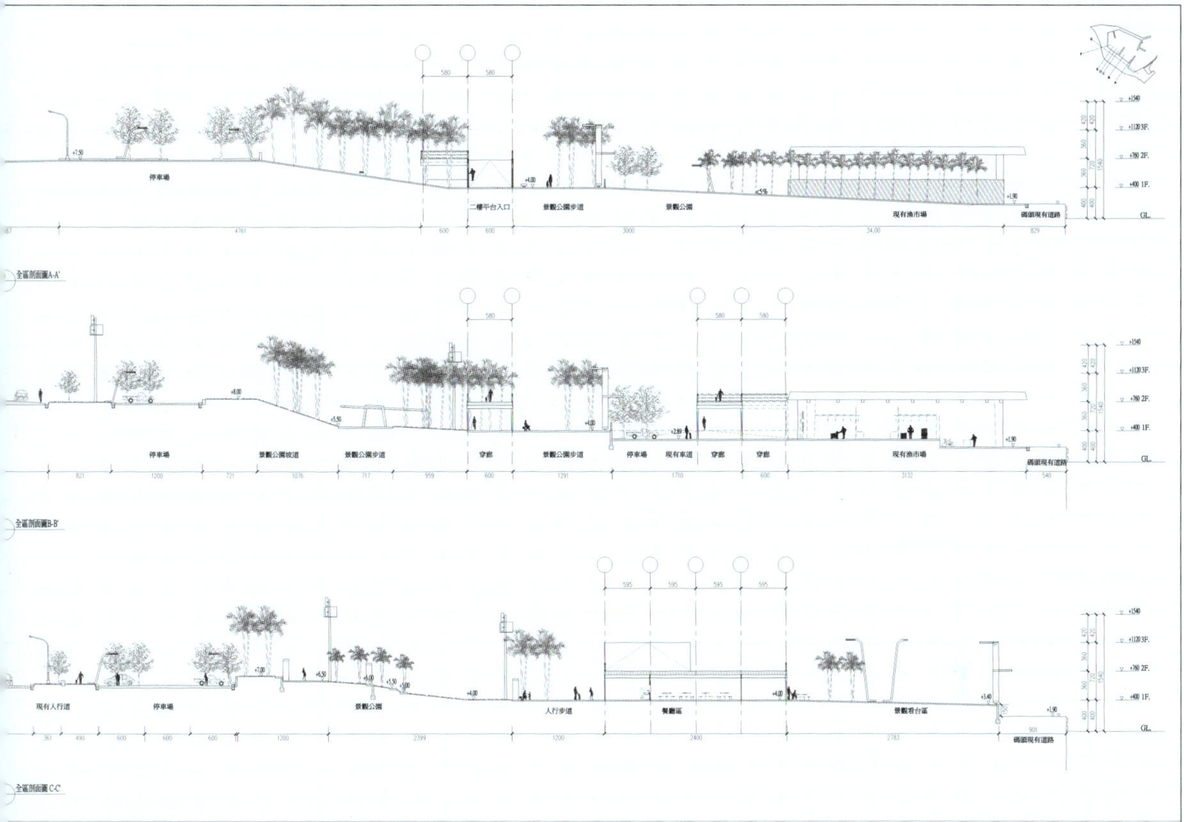

斗子漁港港區環境改善工程 - 海景廣場

NO. DATE REVISIONS NO. DATE REVISIONS NOTE:

日期 2007年1月15日
比例 1:200
繪圖 詹佳芳
校核 陳晁豐
審定 林洲民
建築師 秦英豪

圖名 魚貨直銷中心 全區短向剖面圖(1)

圖號 A3-03
張號 24

ART PROJECTS SL
ART ARCHITECTS
...riguer 10. 08004 Barcelona, Spain
...art.com

仲觀設計顧問有限公司
J.M. LIN ARCHITECT/THE OBSERVER DESIGN GROUP
10651 臺北市仁愛路三段21號2樓
www.jmlin.com.tw

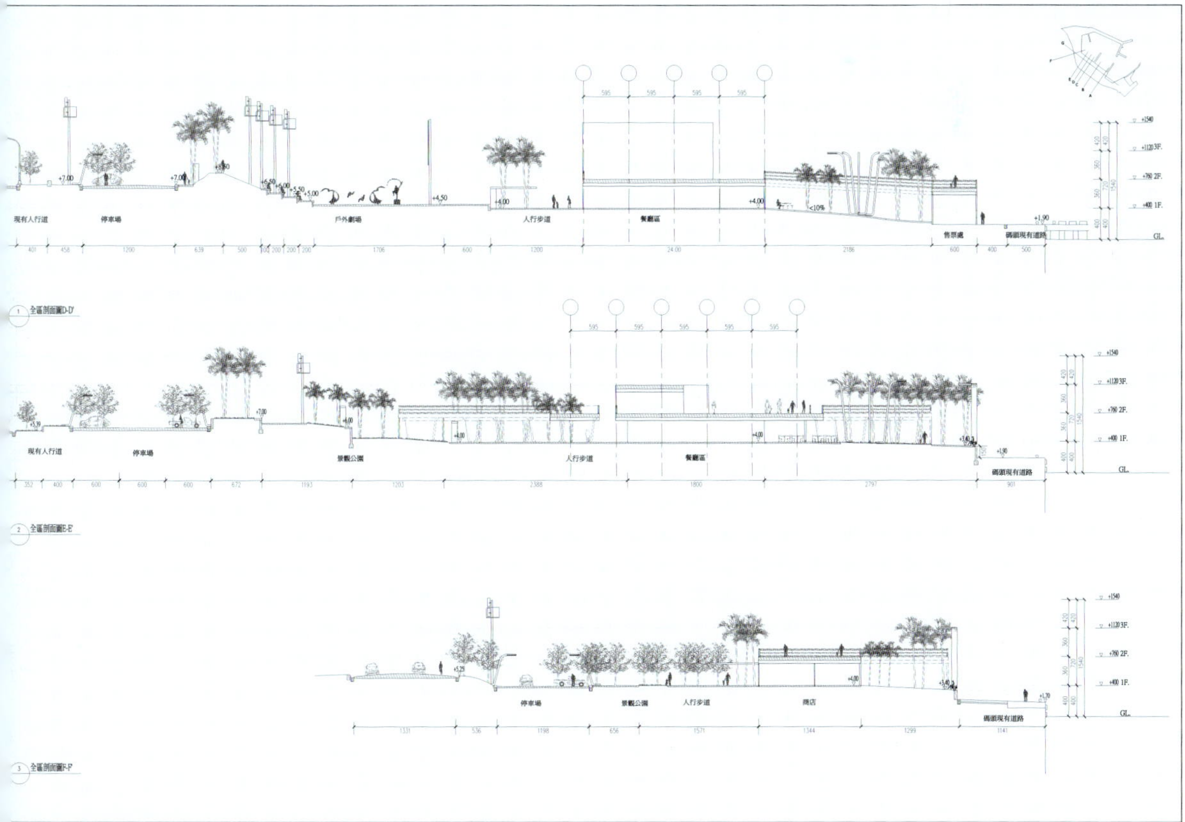

斗子漁港港區環境改善工程 - 漁貨直銷暨多功能展示中心

NO. DATE REVISIONS NO. DATE REVISIONS NOTE:

日期 2007年2月16日
比例 1:200
繪圖 詹佳芳
校核 陳晁豐
審定 林洲民
建築師 秦英豪

圖名 漁貨直銷中心 全區短向剖面圖(2)

圖號 A3-04
張號 25

ART PROJECTS SL
ART ARCHITECTS
...guer 10. 08004 Barcelona, Spain
...art.com

仲觀設計顧問有限公司 仲觀聯合建築師事務所
J.M. LIN ARCHITECT/THE OBSERVER DESIGN GROUP
10651 臺北市仁愛路三段21號2樓
www.jmlin.com.tw

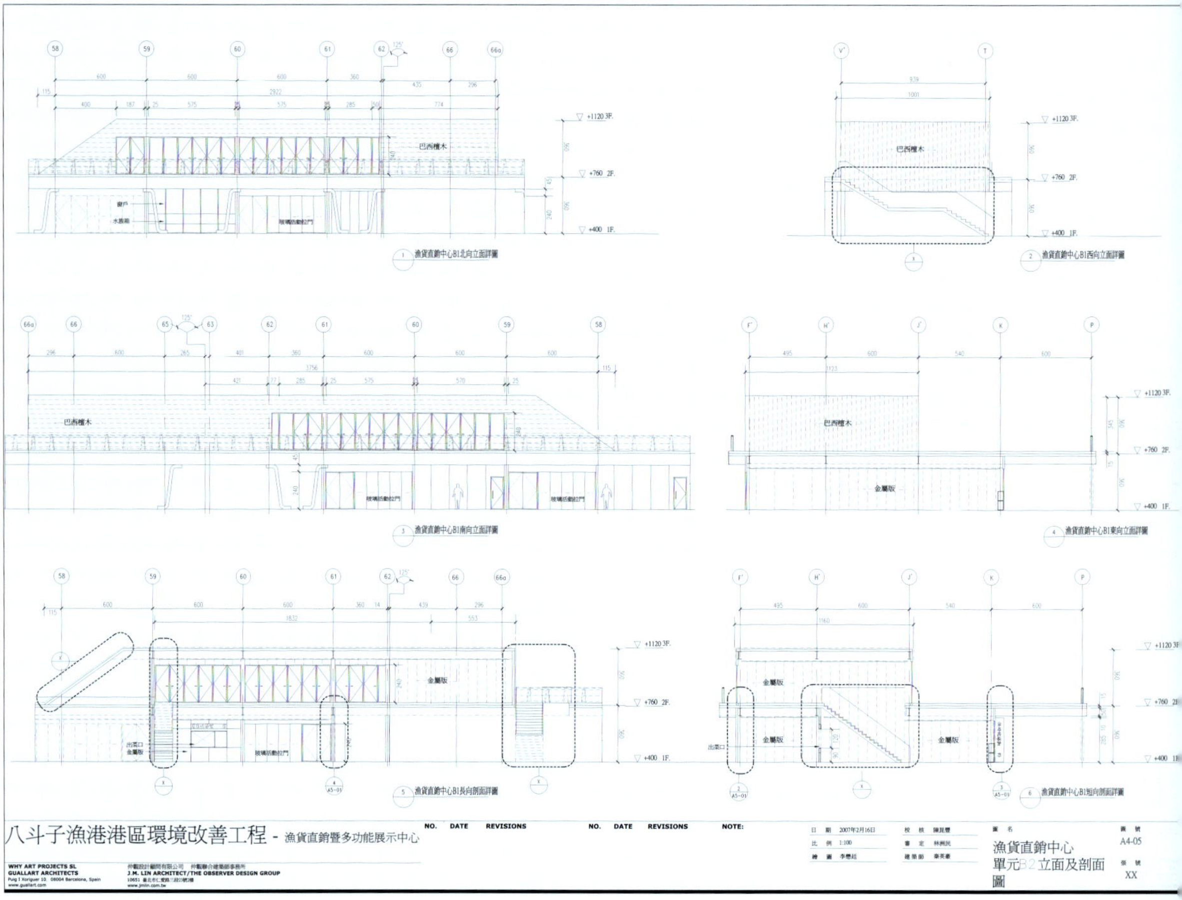
漁貨直銷中心B1北向立面詳圖
漁貨直銷中心B1西向立面詳圖
巴西檀木
漁貨直銷中心B1南向立面詳圖
漁貨直銷中心B1東向立面詳圖
金屬版
漁貨直銷中心B1長向剖面詳圖
漁貨直銷中心B1短向剖面詳圖
八斗子漁港港區環境改善工程 - 漁貨直銷暨多功能展示中心
NO. DATE REVISIONS
NO. DATE REVISIONS
NOTE:
WHY ART PROJECTS SL
GUALLART ARCHITECTS
J.M. LIN ARCHITECT/THE OBSERVER DESIGN GROUP
2007年2月16日
1:100
漁貨直銷中心
單元B2立面及剖面圖
A4-05
XX

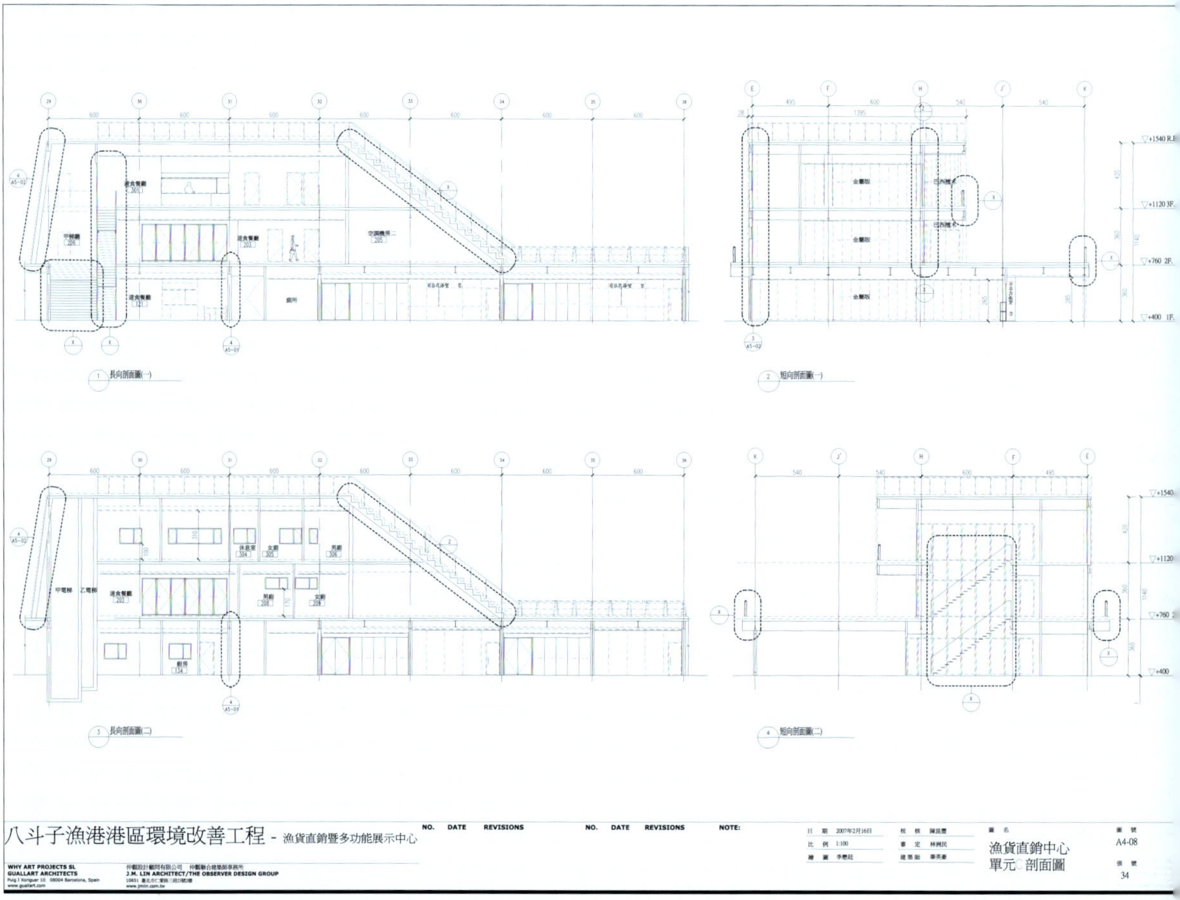
長向剖面圖(一)
短向剖面圖(一)
長向剖面圖(二)
短向剖面圖(二)
八斗子漁港港區環境改善工程 - 漁貨直銷暨多功能展示中心
NO. DATE REVISIONS
NO. DATE REVISIONS
NOTE:
WHY ART PROJECTS SL
GUALLART ARCHITECTS
J.M. LIN ARCHITECT/THE OBSERVER DESIGN GROUP
2007年2月16日
1:100
漁貨直銷中心
單元C剖面圖
A4-08
34

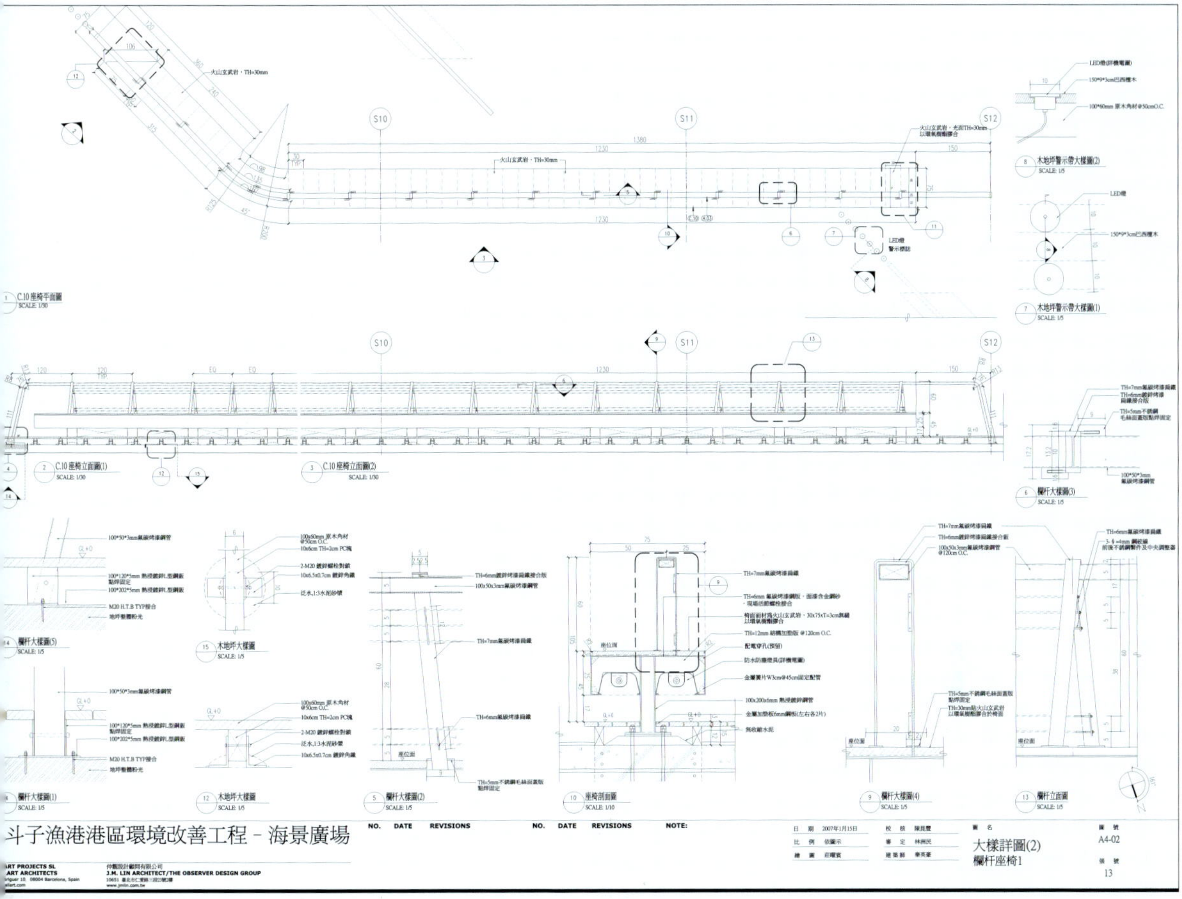
斗子漁港港區環境改善工程－海景廣場
NO. DATE REVISIONS
NO. DATE REVISIONS
NOTE:
大樣詳圖(2)
欄杆座椅1
A4-02
13
仲觀設計顧問有限公司
J.M. LIN ARCHITECT/THE OBSERVER DESIGN GROUP
C.10 座椅平面圖
C.10 座椅立面圖(1)
C.10 座椅立面圖(2)
木地坪警示帶大樣圖(2)
木地坪警示帶大樣圖(1)
欄杆大樣圖(3)
欄杆大樣圖(5)
木地坪大樣圖
欄杆大樣圖(2)
座椅剖面圖
欄杆大樣圖(4)
欄杆立面圖
欄杆大樣圖(1)
木地坪大樣圖

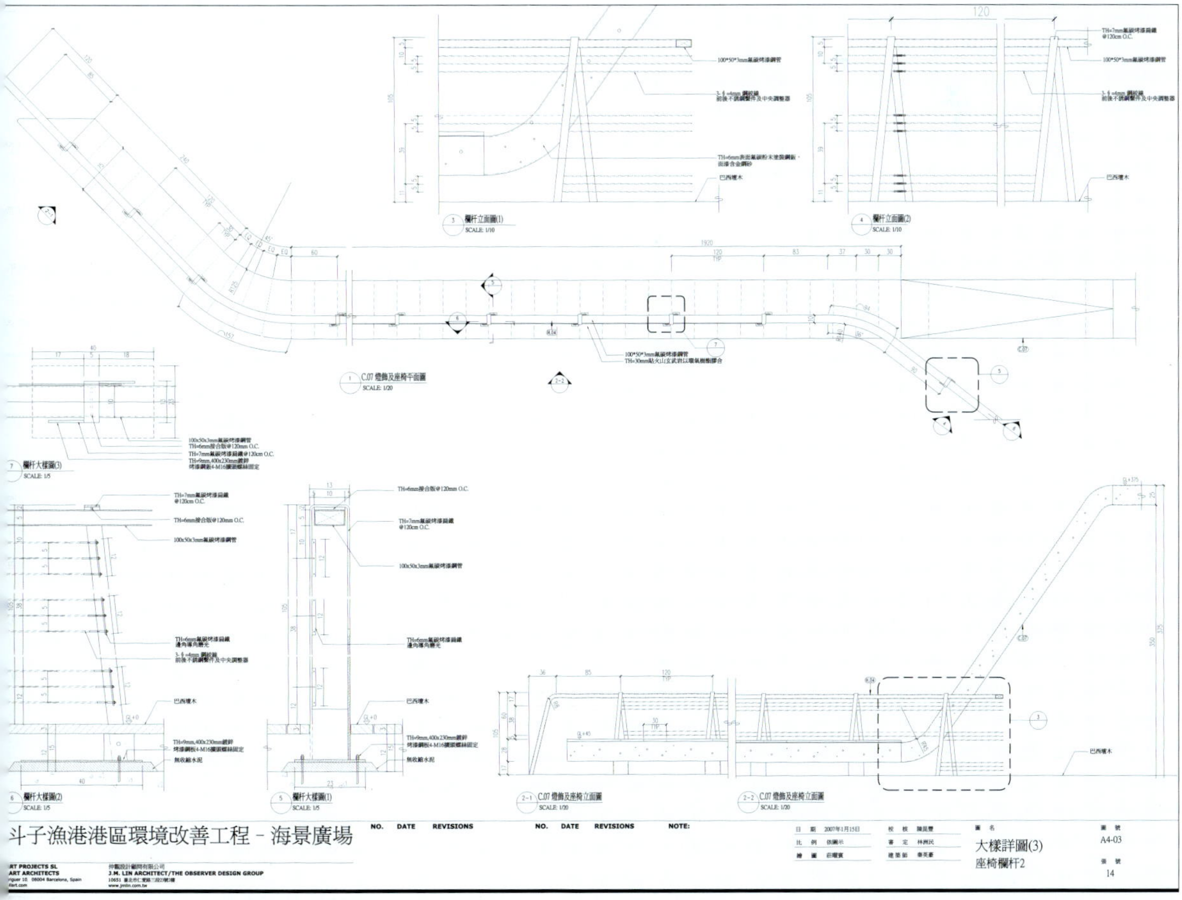
斗子漁港港區環境改善工程－海景廣場
NO. DATE REVISIONS
NO. DATE REVISIONS
NOTE:
大樣詳圖(3)
座椅欄杆2
A4-03
14
仲觀設計顧問有限公司
J.M. LIN ARCHITECT/THE OBSERVER DESIGN GROUP
欄杆立面圖(1)
欄杆立面圖(2)
C.07 燈飾及座椅平面圖
欄杆大樣圖(3)
欄杆大樣圖(2)
欄杆大樣圖(1)
C.07 燈飾及座椅立面圖
C.07 燈飾及座椅立面圖

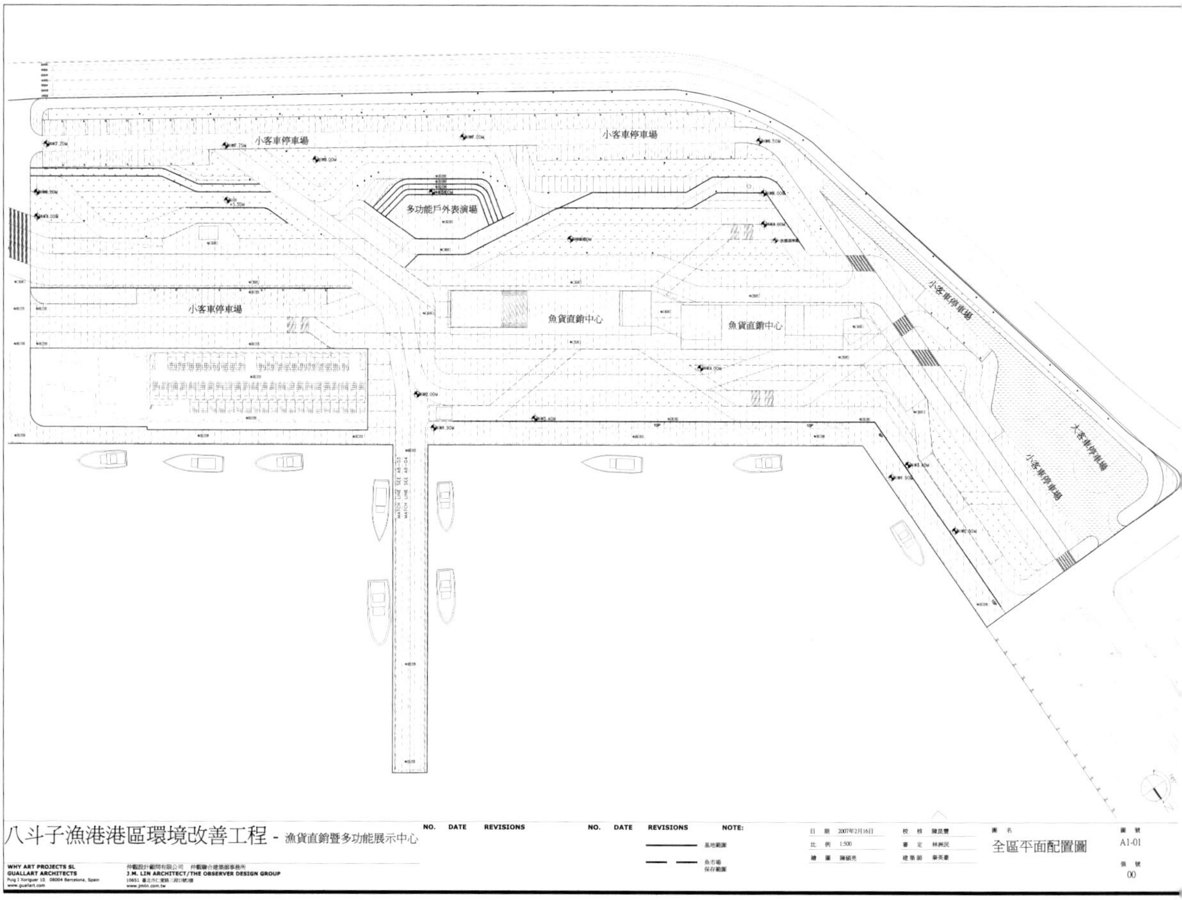

小客車停車場
多功能戶外表演場
魚貨直銷中心
大客車停車場
八斗子漁港港區環境改善工程 - 漁貨直銷暨多功能展示中心
NO. DATE REVISIONS
NO. DATE REVISIONS
NOTE:
基地範圍
魚市場保存範圍
日期 2007年2月16日
比例 1:500
全區平面配置圖
A1-01
00
WHY ART PROJECTS SL
GUALLART ARCHITECTS
Puig I Xoriguer 10. 08004 Barcelona, Spain
www.guallart.com
J.M. LIN ARCHITECT/THE OBSERVER DESIGN GROUP
www.jmlin.com.tw

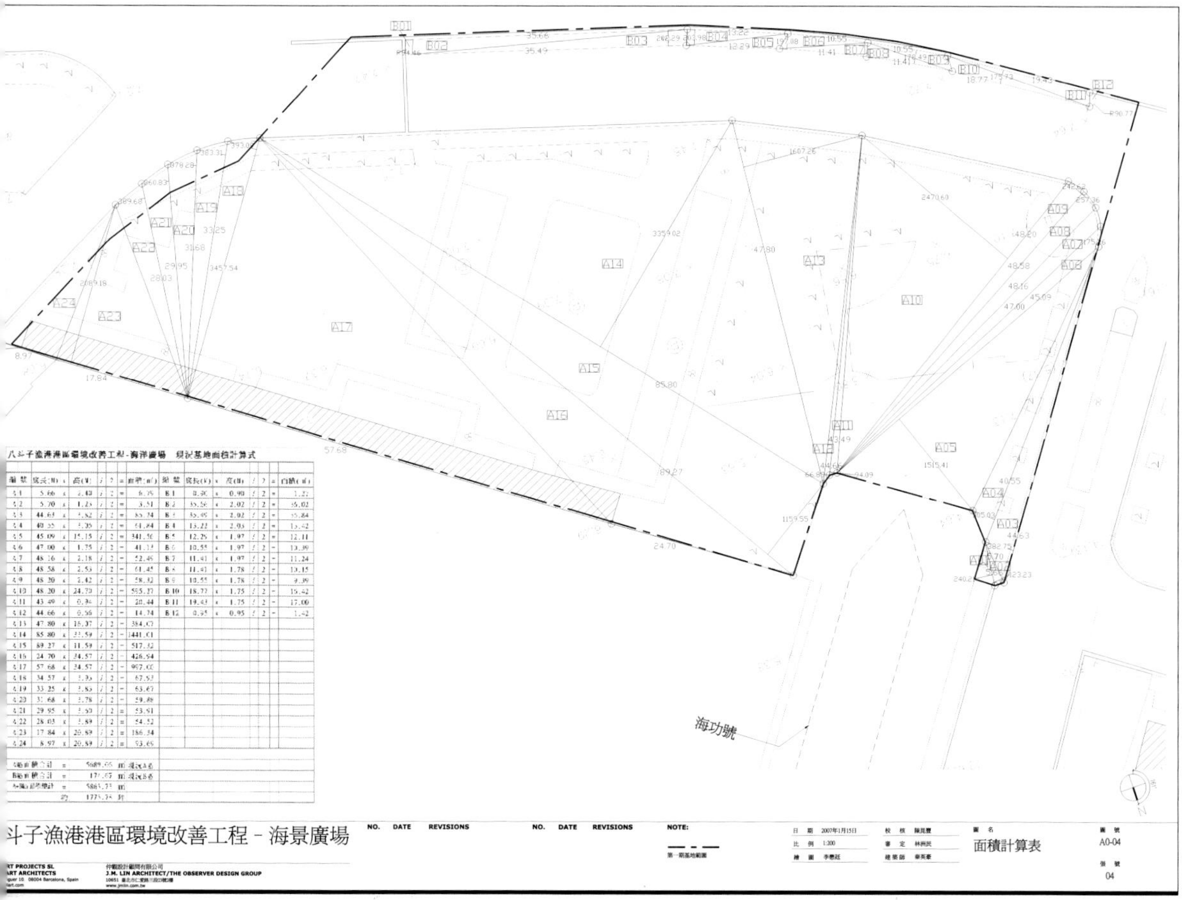

八斗子漁港港區環境改善工程-海洋廣場 現況基地面積計算式

編號	底長(M)	x	高(M)	/	2	=	面積(㎡)	編號	底長(M)	x	高(M)	/	2	=	面積(㎡)
A1	5.66	x	2.40	/	2	=	6.75	B1	0.90	x	0.90	/	2	=	1.27
A2	5.70	x	1.23	/	2	=	3.51	B2	35.56	x	2.02	/	2	=	35.02
A3	44.63	x	3.82	/	2	=	85.24	B3	35.49	x	2.02	/	2	=	35.84
A4	40.55	x	3.05	/	2	=	61.84	B4	13.22	x	2.03	/	2	=	13.42
A5	45.09	x	15.15	/	2	=	341.56	B5	12.25	x	1.97	/	2	=	12.11
A6	47.00	x	1.75	/	2	=	41.13	B6	10.55	x	1.97	/	2	=	10.39
A7	48.16	x	2.18	/	2	=	52.46	B7	11.41	x	1.97	/	2	=	11.24
A8	48.58	x	2.53	/	2	=	61.45	B8	11.41	x	1.78	/	2	=	10.15
A9	48.20	x	2.42	/	2	=	58.32	B9	10.55	x	1.78	/	2	=	9.39
A10	48.20	x	24.70	/	2	=	595.27	B10	18.77	x	1.75	/	2	=	16.42
A11	43.49	x	0.94	/	2	=	20.44	B11	19.43	x	1.75	/	2	=	17.00
A12	44.66	x	0.66	/	2	=	14.74	B12	0.95	x	0.95	/	2	=	1.42
A13	47.80	x	16.07	/	2	=	384.07								
A14	85.80	x	33.59	/	2	=	1441.01								
A15	89.27	x	11.59	/	2	=	517.32								
A16	24.70	x	34.57	/	2	=	426.94								
A17	57.68	x	34.57	/	2	=	997.00								
A18	34.57	x	3.95	/	2	=	67.53								
A19	33.25	x	3.83	/	2	=	63.67								
A20	31.68	x	3.78	/	2	=	59.88								
A21	29.95	x	3.60	/	2	=	53.91								
A22	28.03	x	3.89	/	2	=	54.52								
A23	17.84	x	20.89	/	2	=	186.34								
A24	8.97	x	20.89	/	2	=	93.69								

A塊面積合計 = 5689.65 ㎡ 現況A塊
B塊面積合計 = 176.67 ㎡ 現況B塊
A+B面積總計 = 5865.77 ㎡
約 1775.38 坪

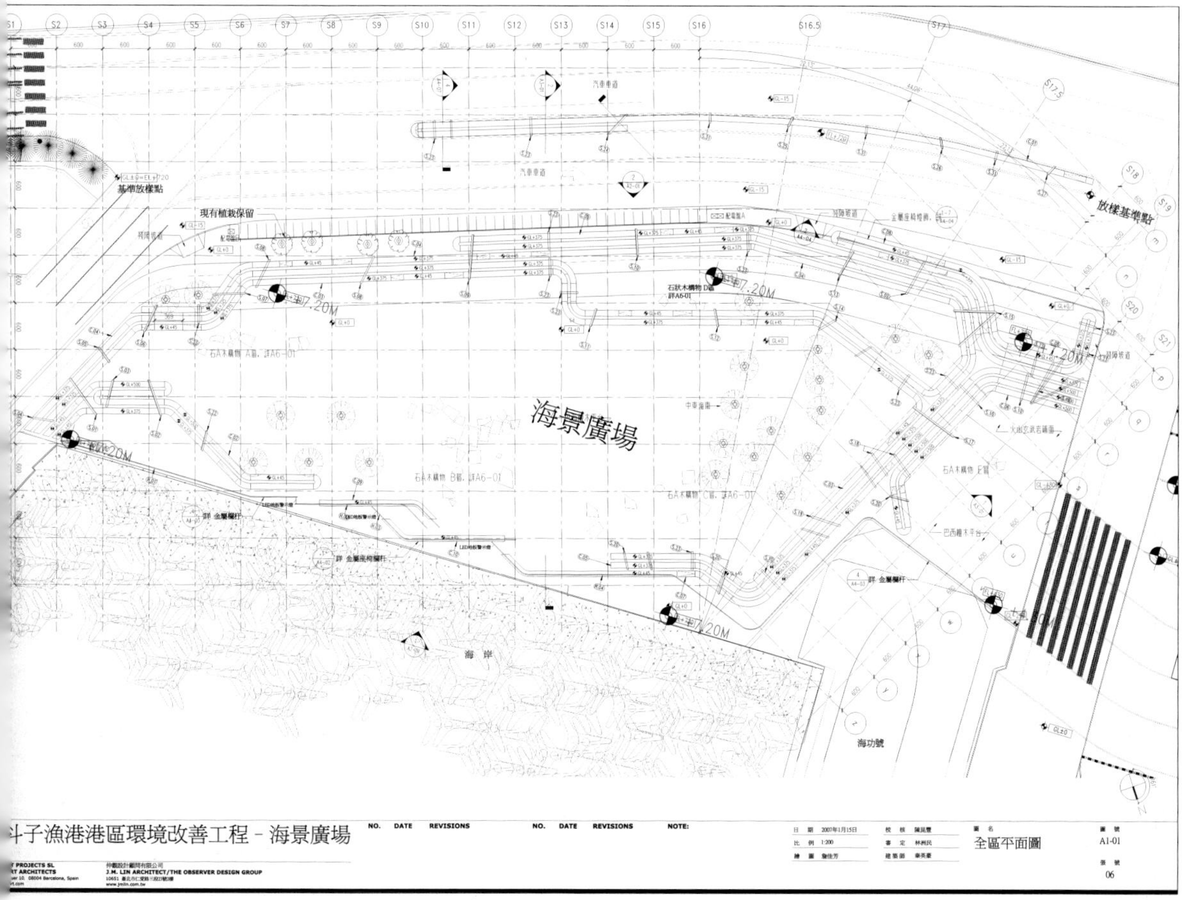

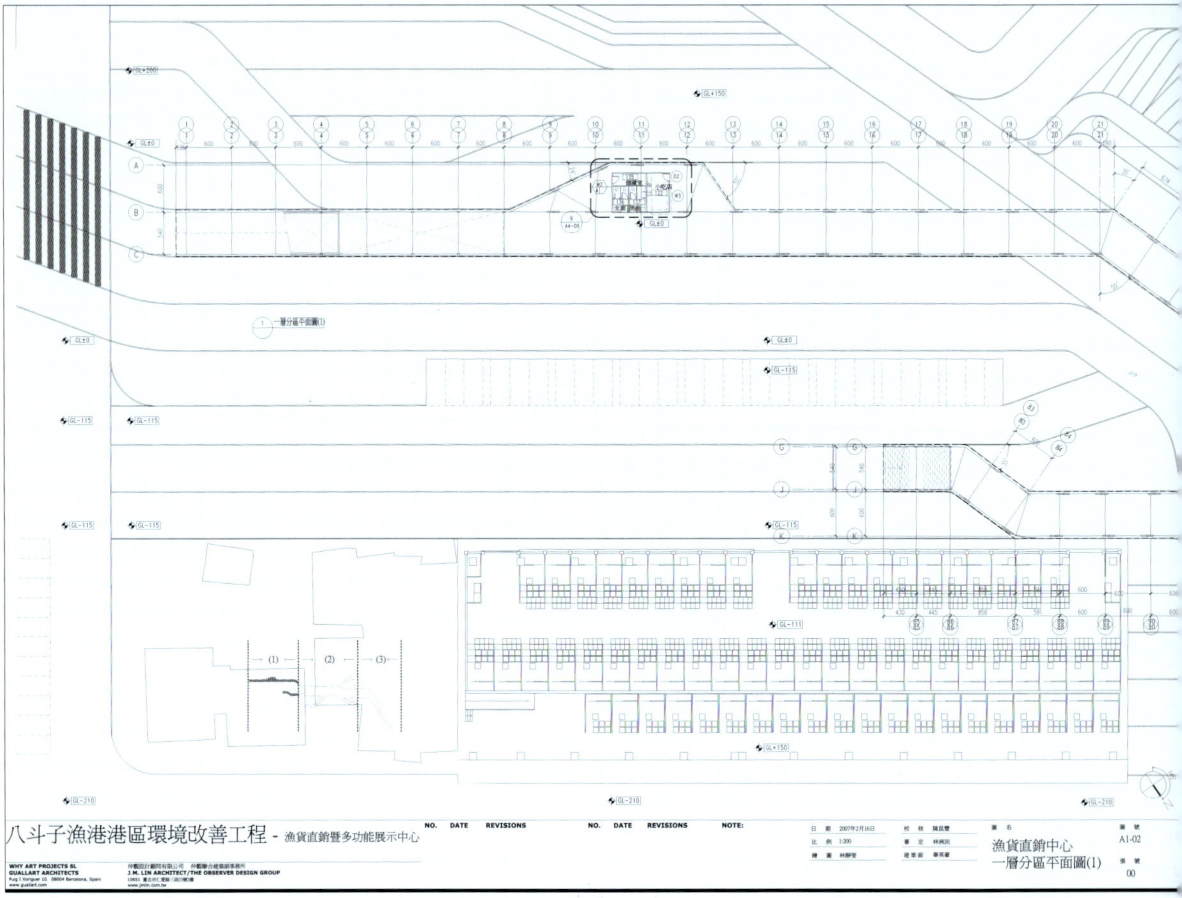

八斗子漁港港區環境改善工程 - 漁貨直銷暨多功能展示中心
NO. DATE REVISIONS
NO. DATE REVISIONS
NOTE:
WHY ART PROJECTS SL
GUALLART ARCHITECTS
J.M. LIN ARCHITECT/THE OBSERVER DESIGN GROUP
漁貨直銷中心
一層分區平面圖(1)
A1-02

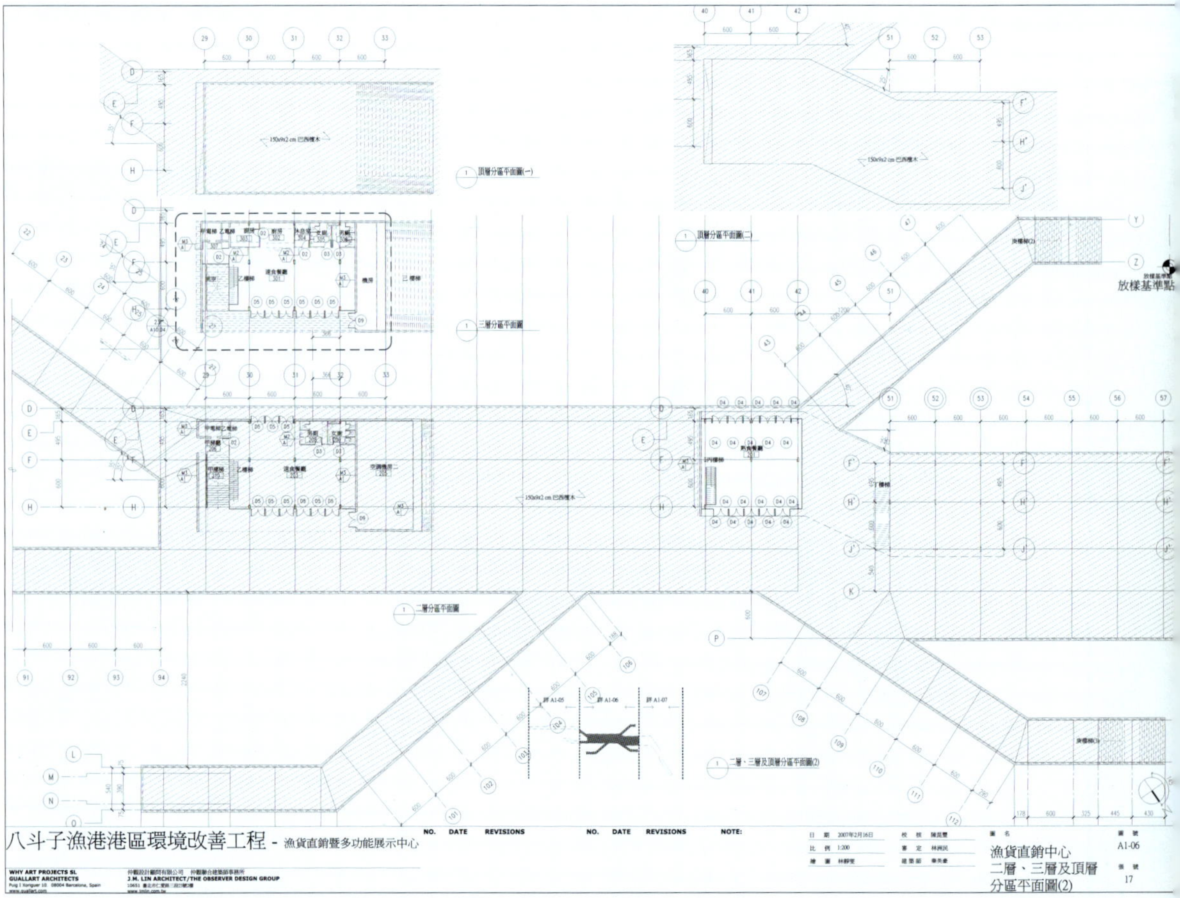

八斗子漁港港區環境改善工程 - 漁貨直銷暨多功能展示中心
NO. DATE REVISIONS
NO. DATE REVISIONS
NOTE:
WHY ART PROJECTS SL
GUALLART ARCHITECTS
J.M. LIN ARCHITECT/THE OBSERVER DESIGN GROUP
漁貨直銷中心
二層、三層及頂層
分區平面圖(2)
A1-06
放樣基準點

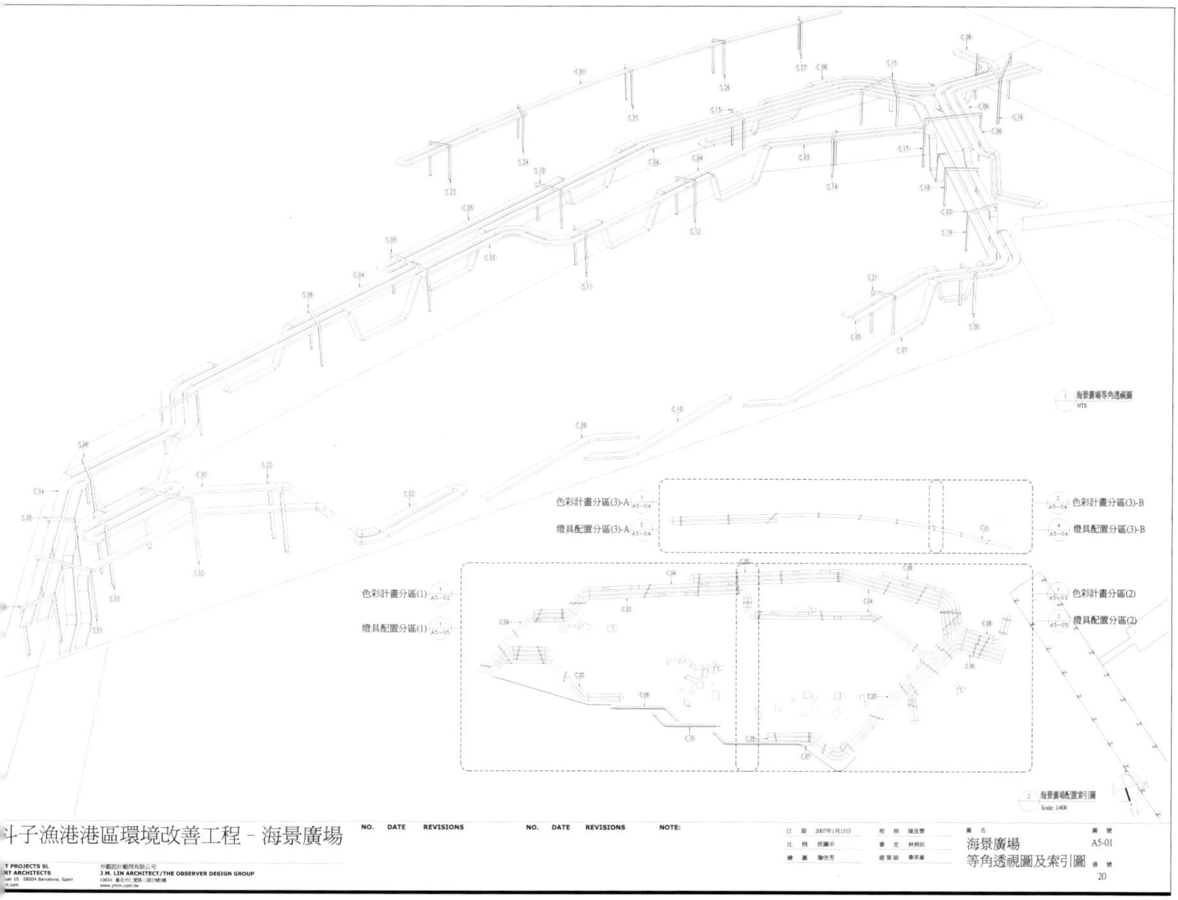
海景廣場等角透視圖
NTS
色彩計畫分區(3)-A
燈具配置分區(3)-A
色彩計畫分區(3)-B
燈具配置分區(3)-B
色彩計畫分區(1)
燈具配置分區(1)
色彩計畫分區(2)
燈具配置分區(2)
海景廣場配置索引圖
Scale: 1/400
斗子漁港港區環境改善工程 - 海景廣場
NO. DATE REVISIONS
NO. DATE REVISIONS
NOTE:
T PROJECTS SL
RT ARCHITECTS
uer 10 08004 Barcelona, Spain
仲觀設計顧問有限公司
J.M. LIN ARCHITECT/THE OBSERVER DESIGN GROUP
www.jmlin.com.tw
2007年1月15日
依圖示
海景廣場
等角透視圖及索引圖
A5-01
20

rock_01
rock_02
rock_03
rock_04
rock_05
rock_06
rock_07
rock_08
rock_09
rock_10
rock_11
rock_12
rock_13
rock_14
rock_15
rock_16
rock_17

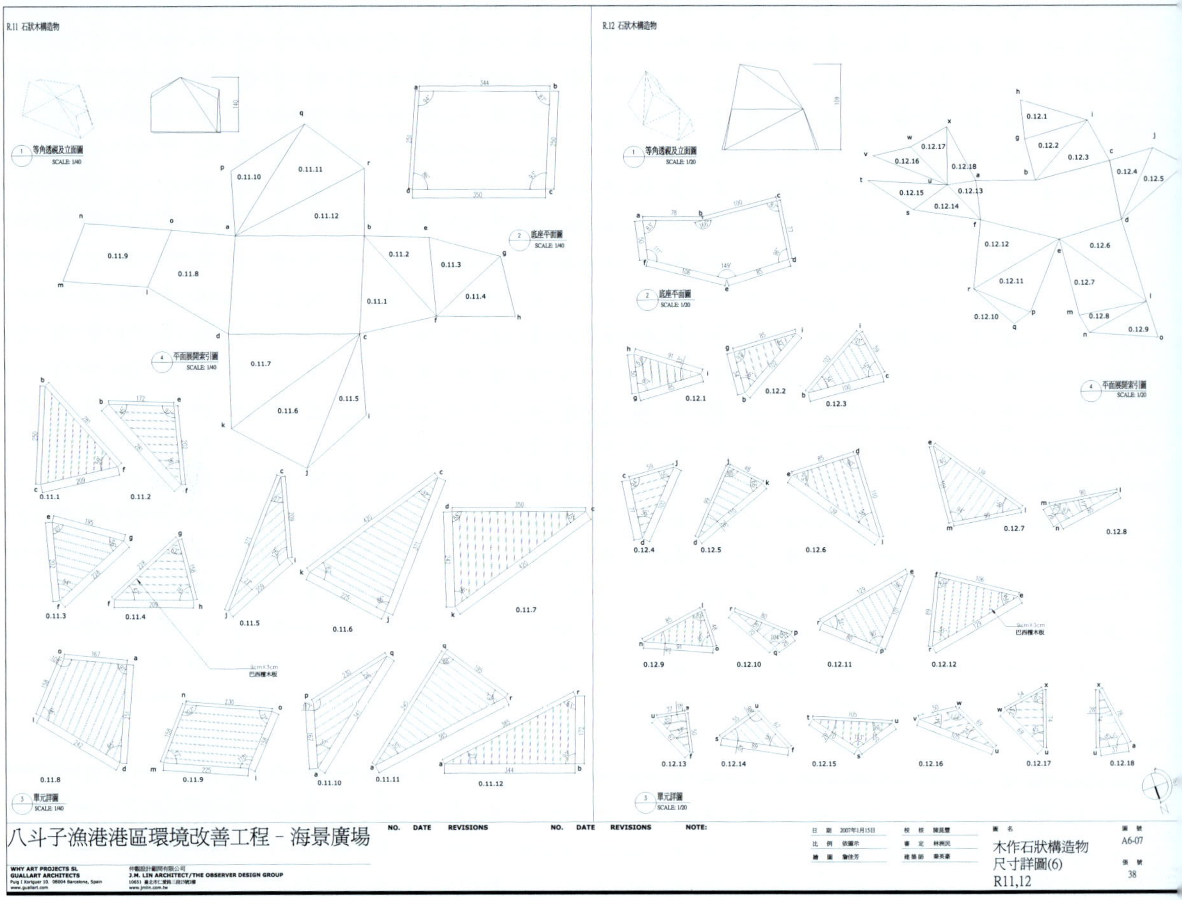
R.11 石狀木構造物
R.12 石狀木構造物
等角透視及立面圖
底座平面圖
平面展開索引圖
單元詳圖
八斗子漁港港區環境改善工程 - 海景廣場
NO. DATE REVISIONS
NO. DATE REVISIONS
NOTE:
WHY ART PROJECTS SL
GUALLART ARCHITECTS
Puig I Xoriguer 10. 08004 Barcelona, Spain
www.guallart.com
仲觀設計顧問有限公司
J.M. LIN ARCHITECT/THE OBSERVER DESIGN GROUP
www.jmlin.com.tw
日期 2007年1月15日
比例 依圖示
木作石狀構造物
尺寸詳圖(6)
R11,12
A6-07
38

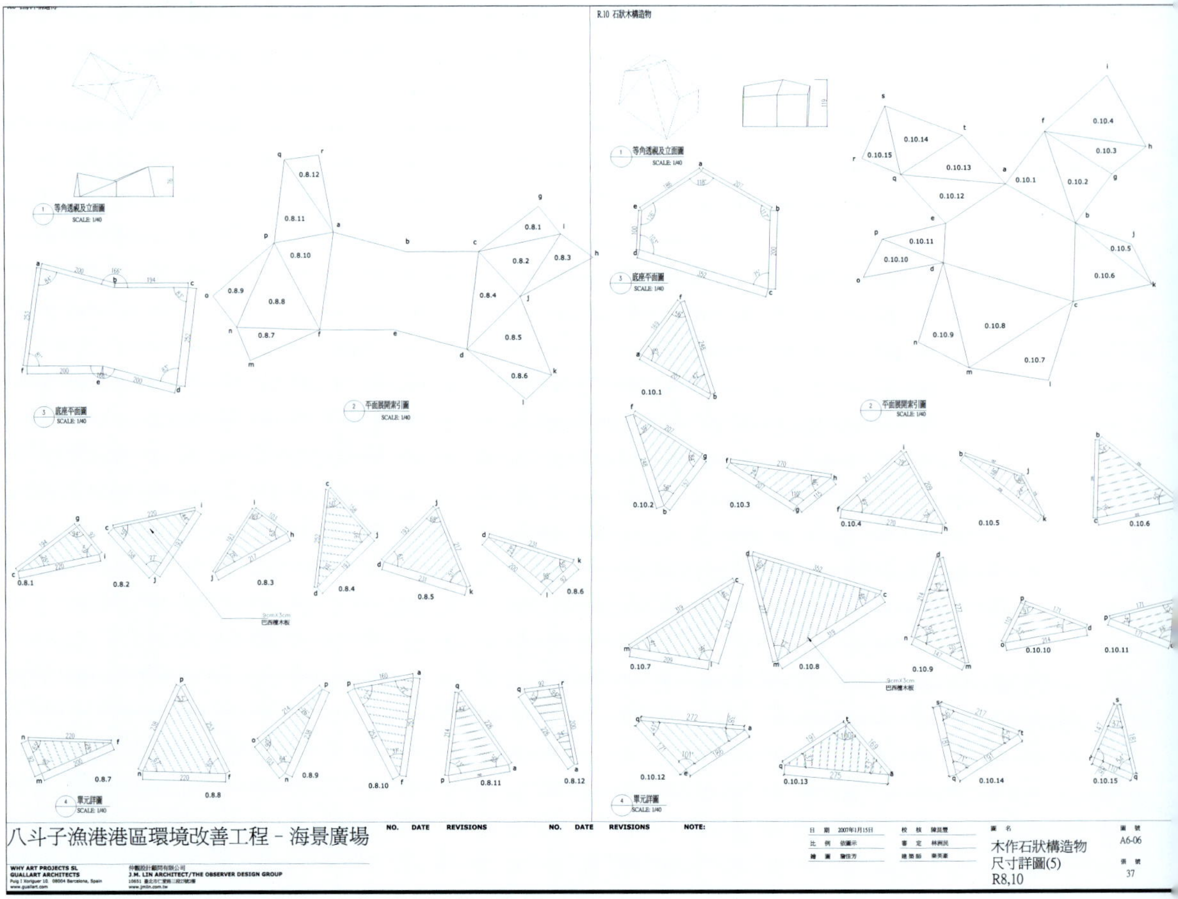
R.10 石狀木構造物
等角透視及立面圖
底座平面圖
平面展開索引圖
單元詳圖
八斗子漁港港區環境改善工程 - 海景廣場
NO. DATE REVISIONS
NO. DATE REVISIONS
NOTE:
WHY ART PROJECTS SL
GUALLART ARCHITECTS
Puig I Xoriguer 10. 08004 Barcelona, Spain
www.guallart.com
仲觀設計顧問有限公司
J.M. LIN ARCHITECT/THE OBSERVER DESIGN GROUP
www.jmlin.com.tw
日期 2007年1月15日
比例 依圖示
木作石狀構造物
尺寸詳圖(5)
R8,10
A6-06
37

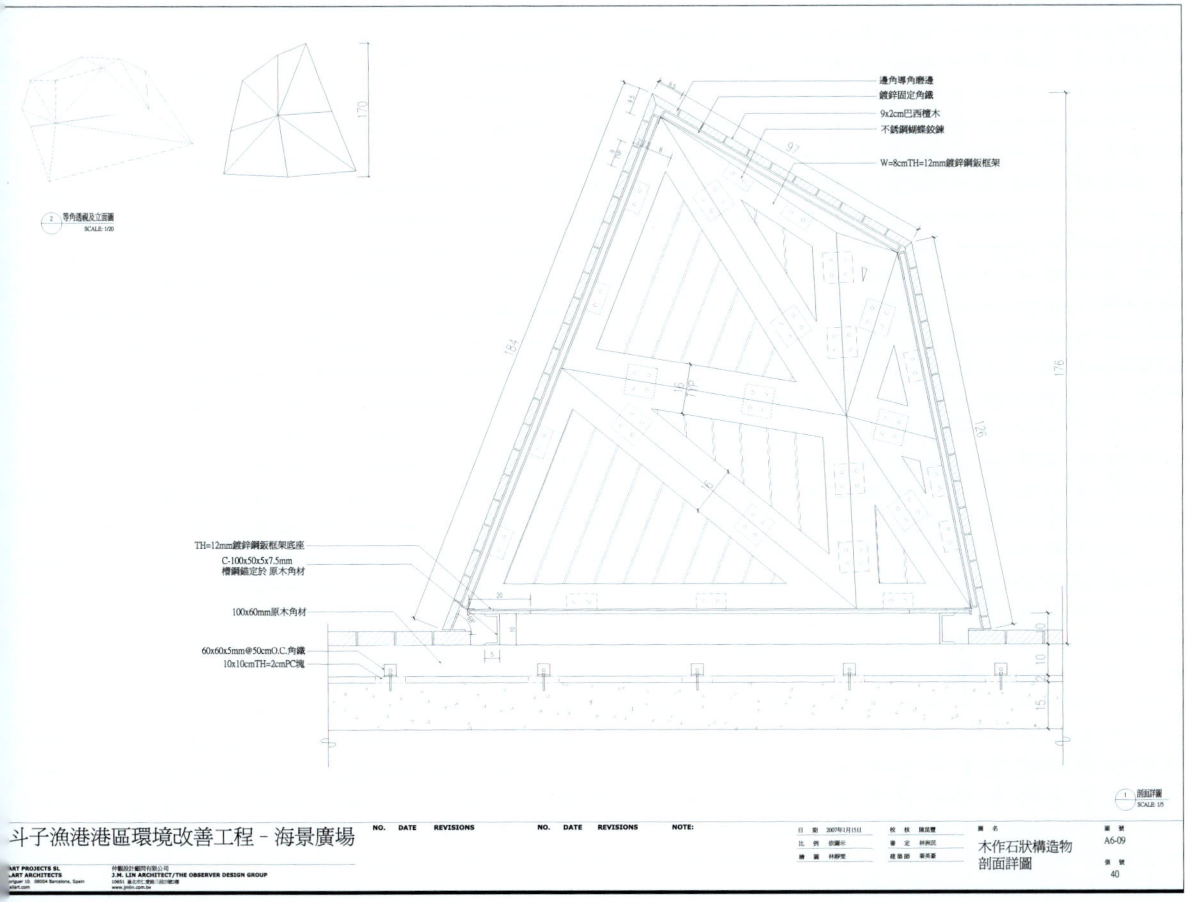

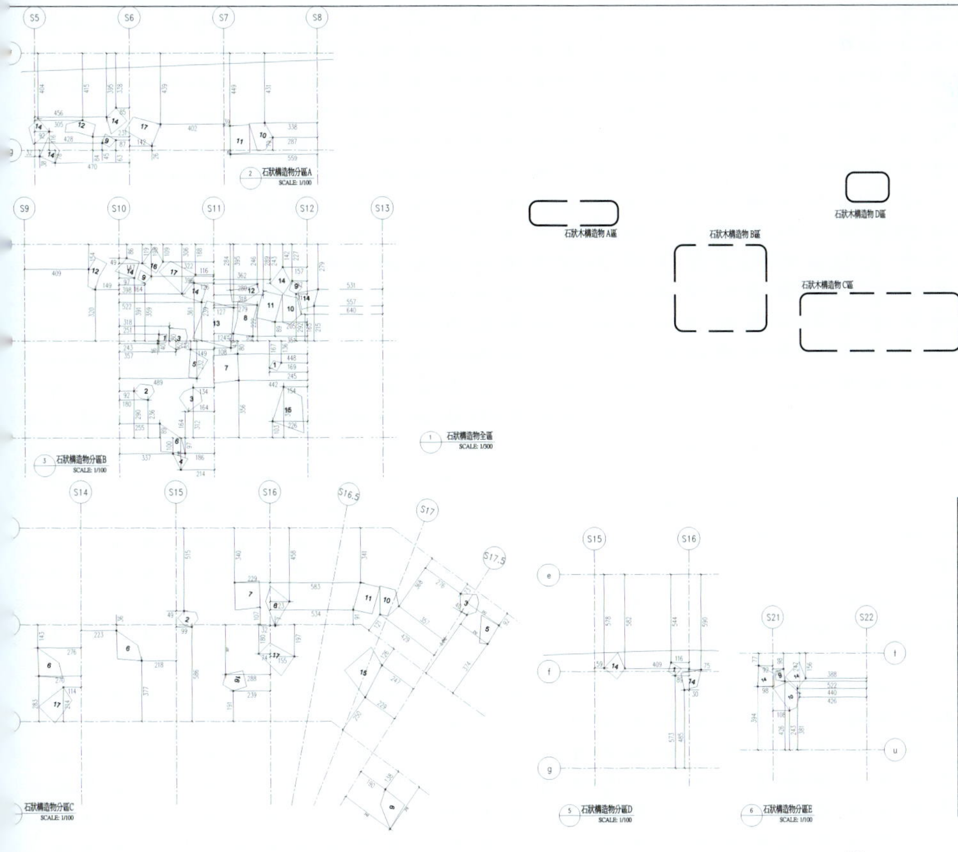

	A區	B區	C區	D區	E區	總數
R1		2		1		3
R2		1	1			2
R3		2	1			3
R4		1				1
R5		1	1			2
R6		1	4			5
R7		1	1			2
R8		1				1
R9	1	2			1	4
R10	1	1	1		1	4
R11	1	1	1			3
R12	1	2				3
R13		1		2	2	5
R14	3	4				7
R15		1	1			2
R16		1	1			2
R17	1	1	2			4

各區石狀構造物計量表 (總計53座)

木作石狀構造物
平面配置圖

A6-01

32

FUGEE PORT

In association with J.M. Lin/ The Observer Design Group

富基渔港

协同仲观设计顾问有限公司联合设计

TOURIST PROPOSAL FOR FUGEE
EXCURSION TO THE NEW TAIWANESE TRADITION

Fugee is a small fishing port with a special geographical significance:
It is on the northern cape of the Island of Taiwan.
Our proposal for Fugee is an especially respectful intervention in the environment that will offer a ne vision of Taiwanese traditions, culture and cuisine.
The proposal is structured in three parts:

1. PORT

Constructing a series of individualized roofs to create an interior microclimate that will protect the fis market, the restaurant and the shops.
The roof, with water, will give shelter from the sun and help to purify the air of the smell of fish. There will be distributed access for public, fishermen anc for delivery.

2. NORTH ITINERARY

Excursion to the north with three enclaves:
2a. Artificial seawater swimming pools (as in the Cnary Islands) to enable a richer spectrum of leisure use of the port.
2b. Panoramic belvedere viewing platform looking to the north, with a wooden deck and a scale mode of Taiwan.
2c. Snack bar by the beach.

3. FUTURE

In the future a new prestige hotel by the port, beac hotels and a series of public amenities will encourage residence in this exceptional setting.

UMBRELLA STRUCTURE

The structures of the roofs of the market and restaurants will serve to create a microclimate that uses natural air flow to reduce the ambient temperature by ventilation.
The section and plan adopt the geometrical principles of volcanic stone.

PROCESSING WASTEWATER

The contact of the water with shellfish produces contamination.
In order to purify it is proposed a natural system to accomplish in the roof of the market, by a process of decanting into four tanks with aquatic plants to oxygenate the water and reduce the level of bacteria.

•Gravel
•Schoenoplectrus
•Typha
•Phragmites
•Gravel

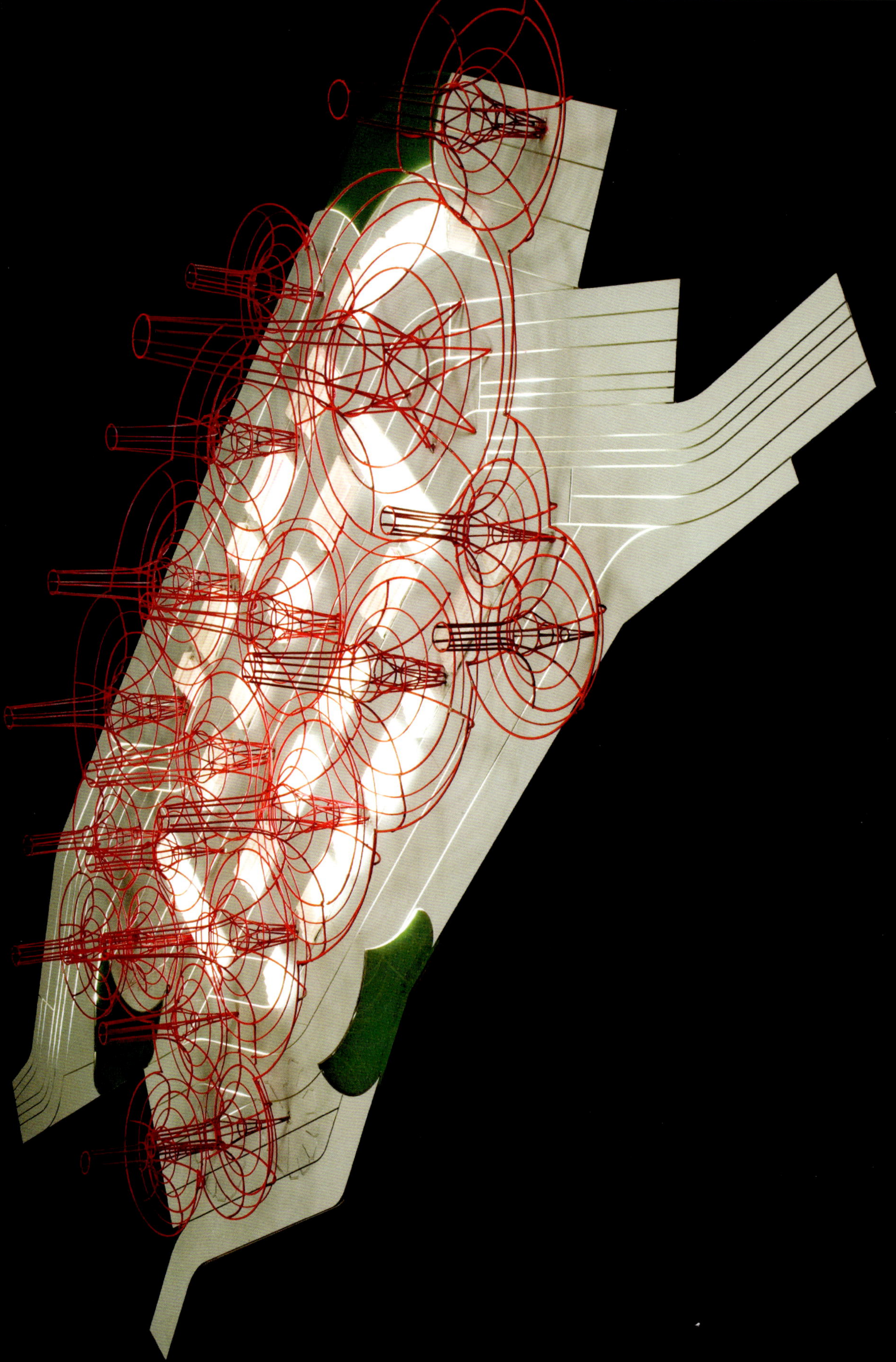

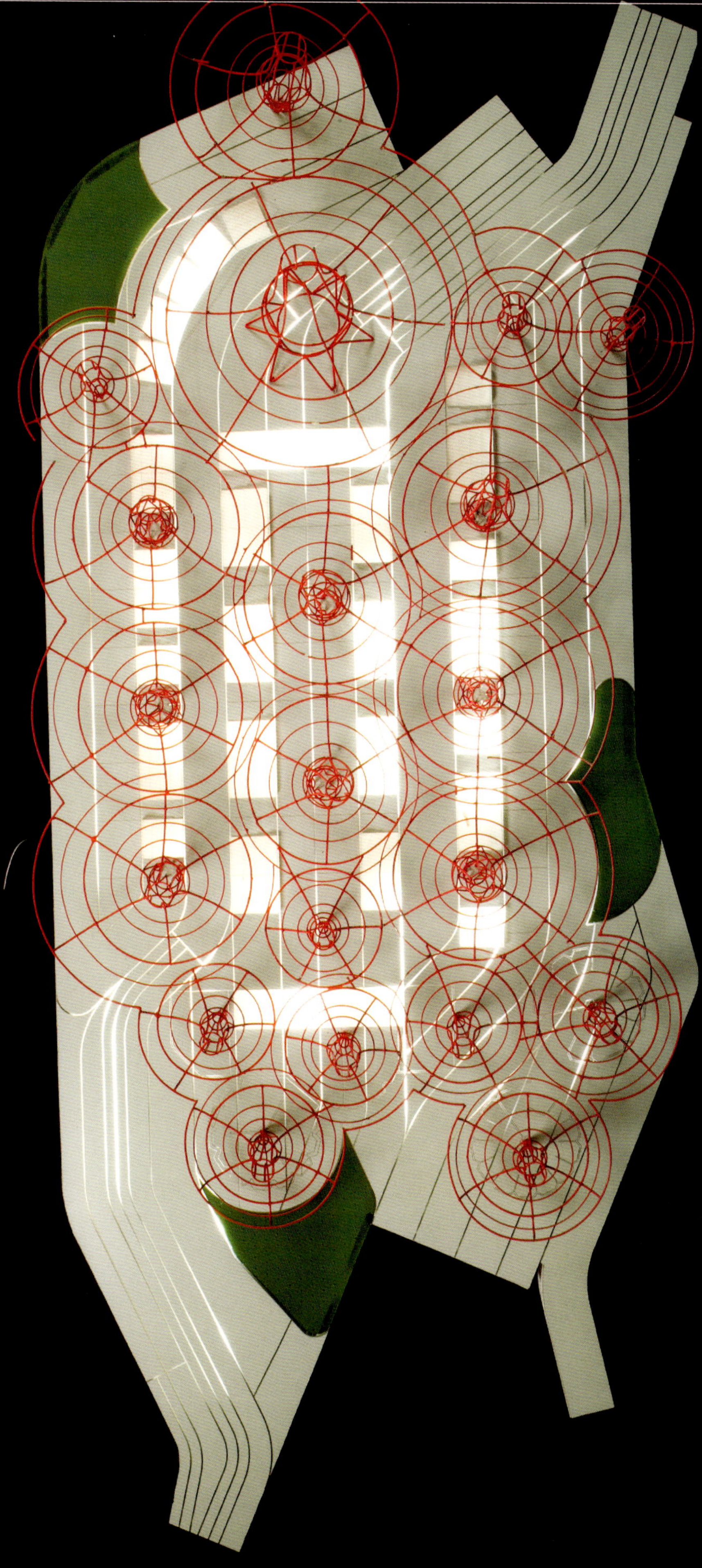

富基旅游地预案
台湾新传统观览

富基是一个带有特殊地理重要性的小渔港：
它位于整个台湾岛的北海角。
我们为富基提供的方案是经过特殊的考量后对于环境的介入，它将会提供一个审视台湾传统、文化和美食的全新视角。
方案由三部分构成：

港口
通过建造一系列相互独立的顶棚来创造一种内部的小气候，用以保护水产品市场、餐厅和商店。
带有水的顶棚在遮蔽太阳光直射的同时也能净化水产品的异味。这也就保证了公众、捕捞者和运输从业者的互通性。

北部湾
北部湾包含三个围和层次：
2a. 人工海水游泳池（就像在Canary岛的那样），保证港口休闲方式具有一个丰富的量级。
2b. 朝向北面的全景观景台，同时带有木制甲板和一个缩小的台湾模型。
2c. 海边的休闲小食店。

未来
未来，一个紧邻港口具有竞争力的酒店，海滩酒店，以及一系列的公共休闲设施将会在某种层次上提高本地居民的生活质量。

伞状结构
市场和餐厅的顶棚结构将会创造出一种微型气候，它通过通风系统的空气流动来降低原有温度。
剖面和平面结合了火山岩石的几何特性。

废水处理
壳类水产品会导致水体污染。
为了净化水体，我们需要一个自然的系统通过流入四个水槽来融入到市场的顶棚当中，通过水生植物来使水体富氧并降低细菌水平。

台湾新传统观览
将富基港的空间性限制条件考虑在内，我们的方案希望能够
过将一个旅游措施变为现实来补充台湾主要的现有设施，探
富基传统生命力特色化的可能性，重新发现它与海洋和捕捞
的关系，并发展空间的其他可选择功能使其与东北部的绿化
间相联系。

旅游产品
这些具有操作性的原则首先确认了一项旅游产品将不会造成
店住宿设施的建造，将自身限制在可提供休闲活动的同时，
免使港口及其相关设施去自然化。这些提案保证了包括传统
式及其他各种方法下的捕捞活动的可持续性，诸如休闲垂钓
及租借渔船进行乘船旅行等与主题相切合的活动。

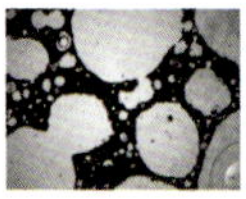

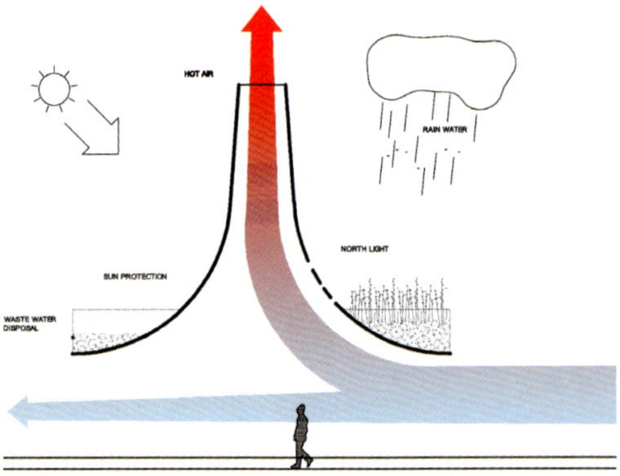

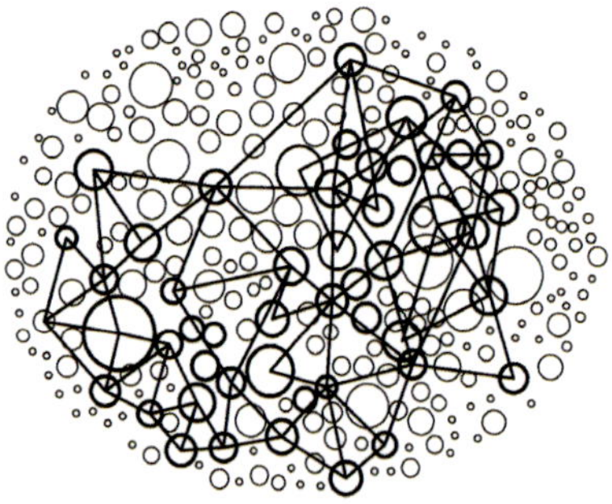

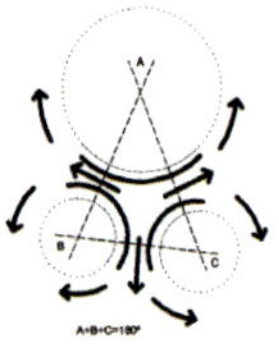

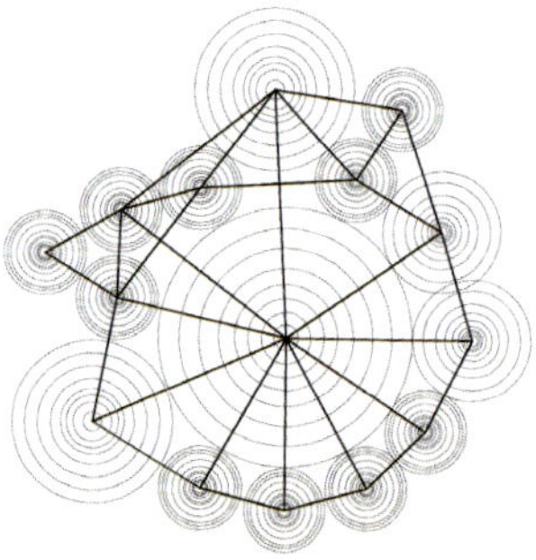

XCURSION TO THE NEW TAIWANESE TRADITION
king into account the spatial limitations of the
ort of Fugee, our proposal envisages a tourist
peration designed to complement Taiwan's prin-
pal existing facilities, exploring the possibilities
Fugee's traditional productive specialization, its
lationship to the sea and fishing, and developing
her alternative uses of the space that they are
ked to the green space to the North-east.

PRODUCT FOR TOURIST EXCURSIONS
nese operational principles configure a tour-
t product that, in its first phase, will not entail
e construction of hotel accommodation and will
onfine itself to developing leisure provisions that
oid denaturing the Port and its constituent facili-
es. This proposal envisages the continuation of
shing activity, both in its traditional form and in
her ways, such as leisure fishing, that are fully
mpatible with the principal activity: for example,
oat trips and the hiring of fishing boats.
e organization of time sequences is structured
the basis of the potential of the two principal
aces that make up the project:

绿茵大道，北方之北
于通向北方的绿茵大道，这一项目提出对包括设置一个环绕周边的道路并设置休息区和观景点的轻微介入计划。这一设具体实现了三项介入措施：
提供海水浴的滨海游泳场将被设置在岩石区，同夏威夷以Canaries的火山岛海水游泳池相似。
北边的观景点。与阳光房相邻，位于台湾的最北点，这一观点将会被配备一套木质结构，同时还包含一个等比例缩小的台湾模型，以加强该场所在地理上的特殊角色感。
临近东北幼沙沙滩的一座海滩酒吧。

港口
一方案在港口区域所实现的是与捕捞业直接相关的功能，它将场地塑造成为饮食和手工艺的中心，并以此使得方案根植于基港口的历史当中。一系列不同规模和种类的餐馆的建成将动为不同层次的需要提供服务的鱼类零售业的发展。然而，案的目标在于富基渔港应当通过其在海产品烹饪上的特色被义为国家级的美食中心。同时，方案建议旅游纪念品商店应设有专门用来展示台湾手工艺的区域，在这里游客应当能够眼见识到手工艺匠人的工作场景。不同功能的安排每天都在个小时左右，并根据他们各自的需要而调节。

向未来的新可能性：独特的居住性休闲设施
一选择挖掘了富基渔港成为停留在其他大型休闲地区的游，当地以及来自台北的游客的短途旅行目的地的潜质。此外二阶段准备呈现的临近的海滩，虽然现在还没有进入实施阶，但已经包含了一家四星级酒店综合体的项目，该酒店可容3000床位，并还带有自己的休闲场地设施，这也使得已经进实施阶段的方案加强了在这一方面的配备。当第二阶段方案入实施以后，富基渔港将会成为一处原生态的具有传统旅业中心特质的休闲娱乐胜地，并成为台湾其他休闲娱乐设施完善部分。

A. THE GREEN PROMENADE: THE NORTH OF THE NORTH
For the Green Promenade to the North, the project proposes a very light intervention consisting in the laying out of a perimeter path with rest areas and viewing points. The scheme envisages three specific interventions:
1. Coastal swimming pools for seawater bathing, situated on the rocky zone, similar to the seawater swimming pools on volcanic islands such as Hawaii and the Canaries.
2. North viewing point. Next to the lighthouse, the most northerly point of Taiwan, the viewing point will have a wooden structure and contain a scale model of all Taiwan, reinforcing the unique geographical character of the place.
3. A beach bar near the norh east virginal beach.

B. THE PORT
In the port area the project envisages uses directly related to fishing, configuring the space as a focus for food and crafts, in this way rooting the proposal in the history of the Port of Fugee. The retail sale of fish will be complemented with the construction of a series of restaurants of different sizes and categories in order to respond to the different levels of demand; however, the aim is that Fugee should be identified as a major centre of gastronomic at the national level, through its specialization in seafood. At the same time, it is suggested that the souvenir shops should have special sections devoted to Taiwanese crafts where tourists can watch the craftsmen at work. The spectrum of uses will be active for 18 hours a day, alternating according to their nature.

NEW POSSIBILITIES FOR THE FUTRE: UNIQUE RESIDENTIAL RESORT:
This option reinforces Fugee potential to become a destination for day trips, both for the tourists staying in other larger resorts and for local people and visitors from Taipei. In addition, the presence of beaches in the vicinity suggests a second phase, not envisaged in the present project, consisting in the creation of a four-star hotel complex that would have at least 3,000 beds, with its own range of leisure provision, while drawing on the present proposal to a substantial degree in order to complement and enhance this provision. In the event that this second phase were undertaken, the Port of Fugee could become a stay-over resort without detriment to its character as a centre for traditional tourism with provision for excursions and as a complement to other resorts in Taiwan.

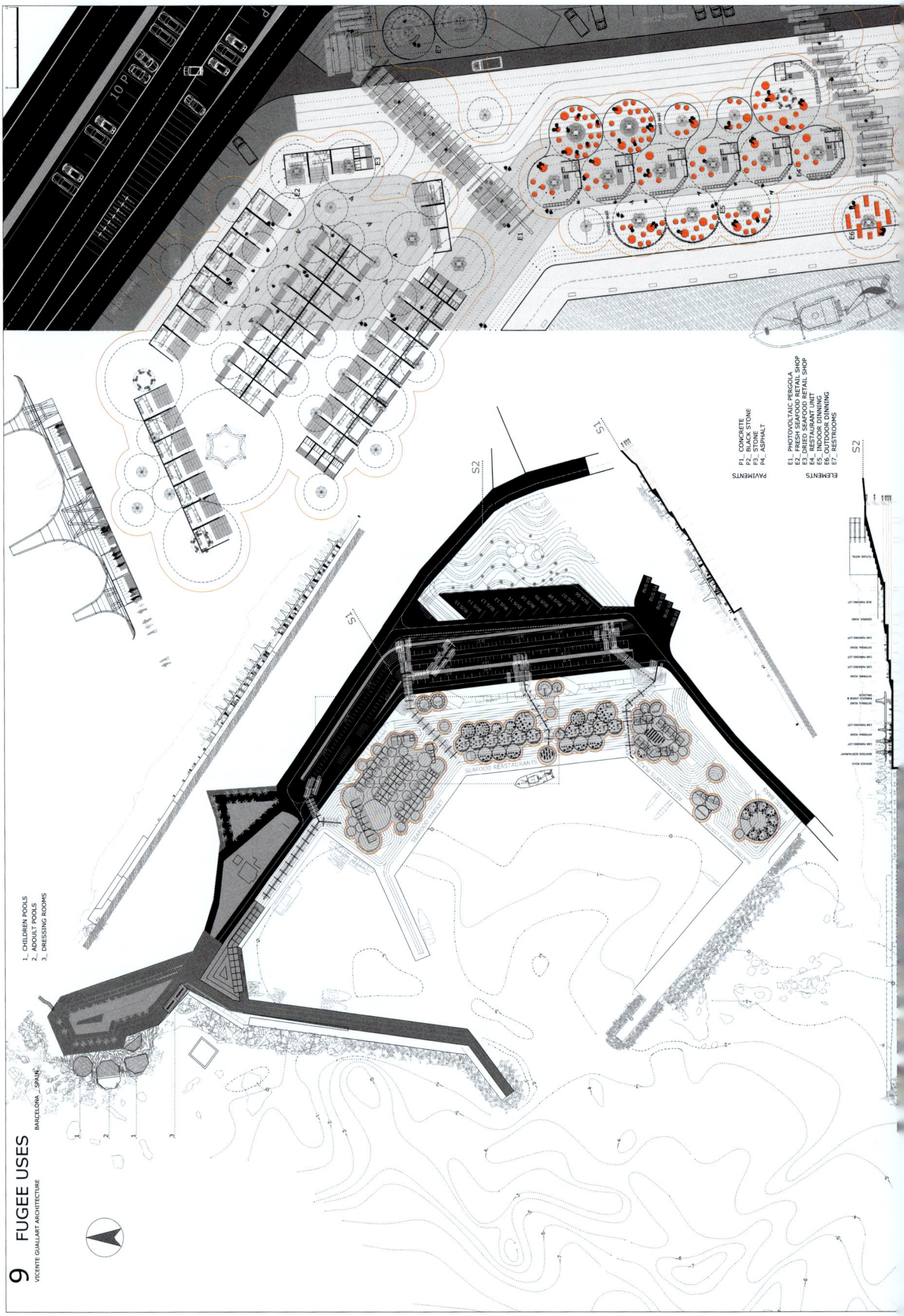
9 FUGEE USES
VICENTE GUALLART ARCHITECTURE
BARCELONA _SPAIN
1_ CHILDREN POOLS
2_ ADOULT POOLS
3_ DRESSING ROOMS
PAVIMENTS
P1_ CONCRETE
P2_ BLACK STONE
P3_ STONE
P4_ ASPHALT
ELEMENTS
E1_ PHOTOVOLTAIC PERGOLA
E2_ FRESH SEAFOOD RETAIL SHOP
E3_ DRIED SEAFOOD RETAIL SHOP
E4_ RESTAURANT UNIT
E5_ INDOOR DINNING
E6_ OUTDOOR DINNING
E7_ RESTROOMS
SEAFOOD RESTAURANTS
SEAFOOD MARKET
S1
S2

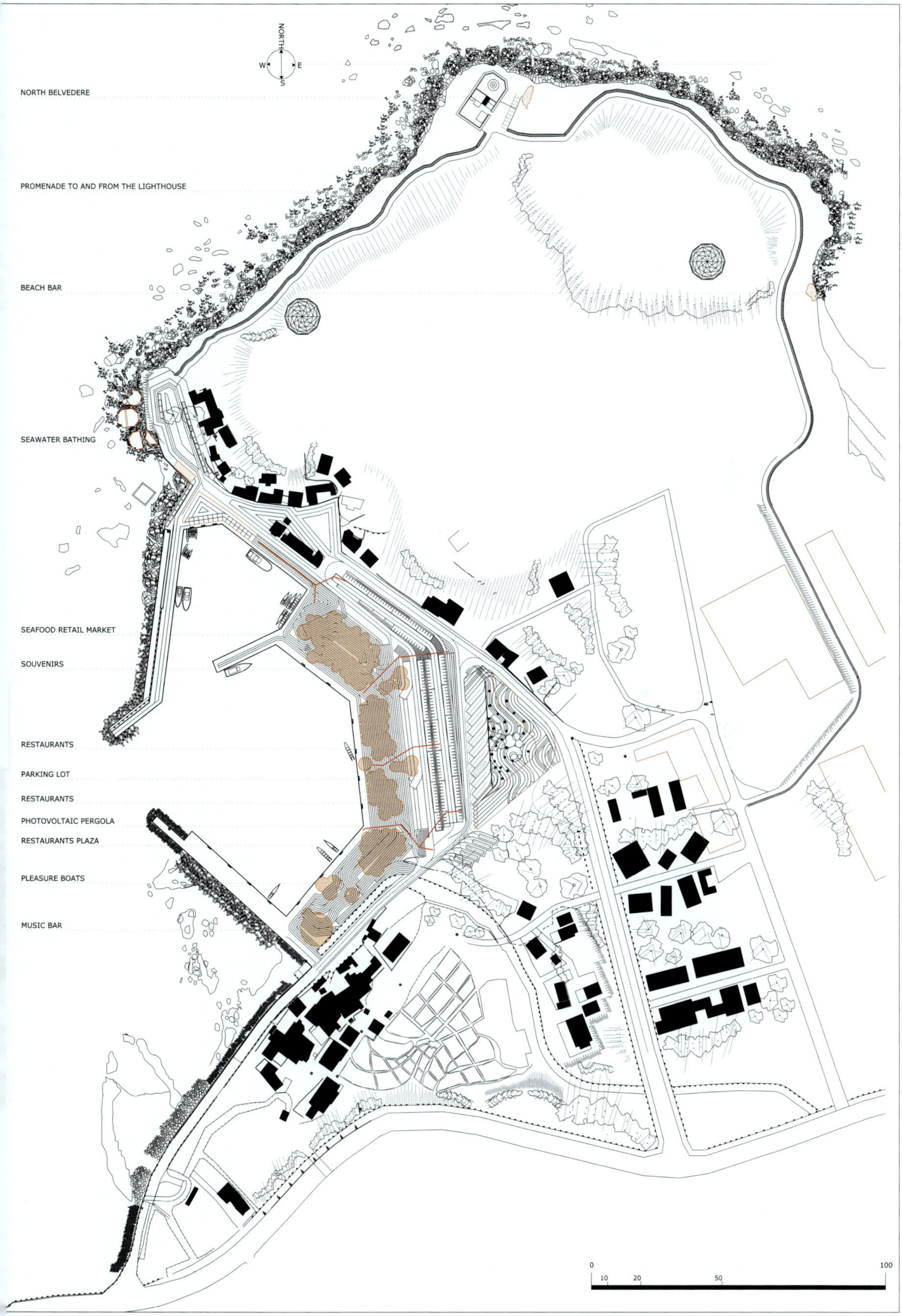
NORTH
W
E
S
NORTH BELVEDERE
PROMENADE TO AND FROM THE LIGHTHOUSE
BEACH BAR
SEAWATER BATHING
SEAFOOD RETAIL MARKET
SOUVENIRS
RESTAURANTS
PARKING LOT
RESTAURANTS
PHOTOVOLTAIC PERGOLA
RESTAURANTS PLAZA
PLEASURE BOATS
MUSIC BAR
0
10
20
50
100

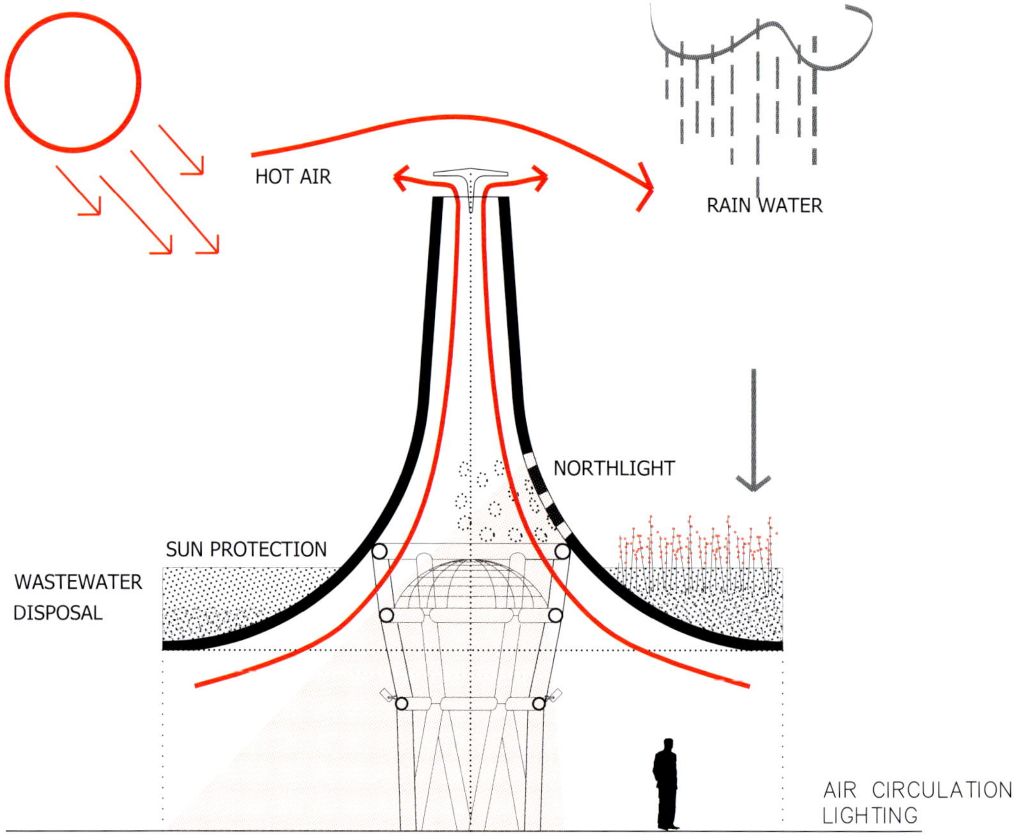
HOT AIR
RAIN WATER
NORTHLIGHT
SUN PROTECTION
WASTEWATER
DISPOSAL
AIR CIRCULATION
LIGHTING

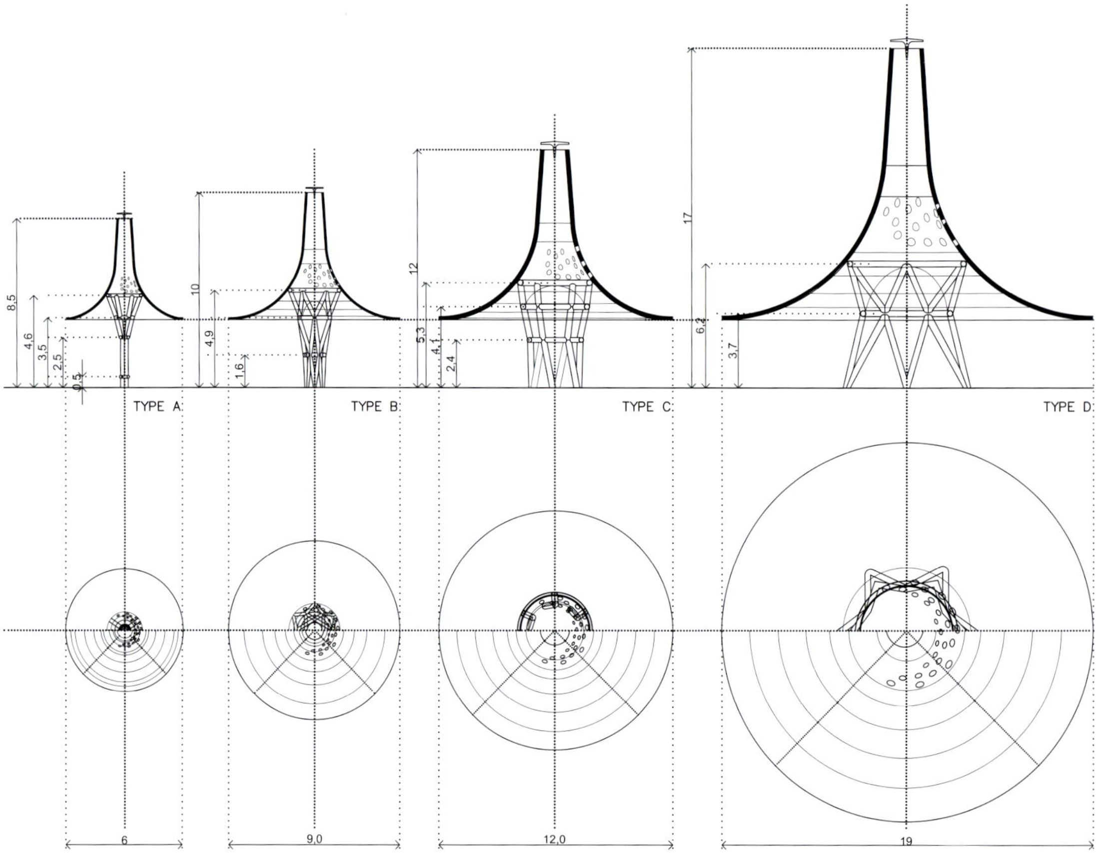
8,5
4,6
3,5
2,5
0,5
TYPE A
10
4,9
1,6
TYPE B
12
5,3
4,1
2,4
TYPE C
17
6,2
3,7
TYPE D
6
9,0
12,0
19

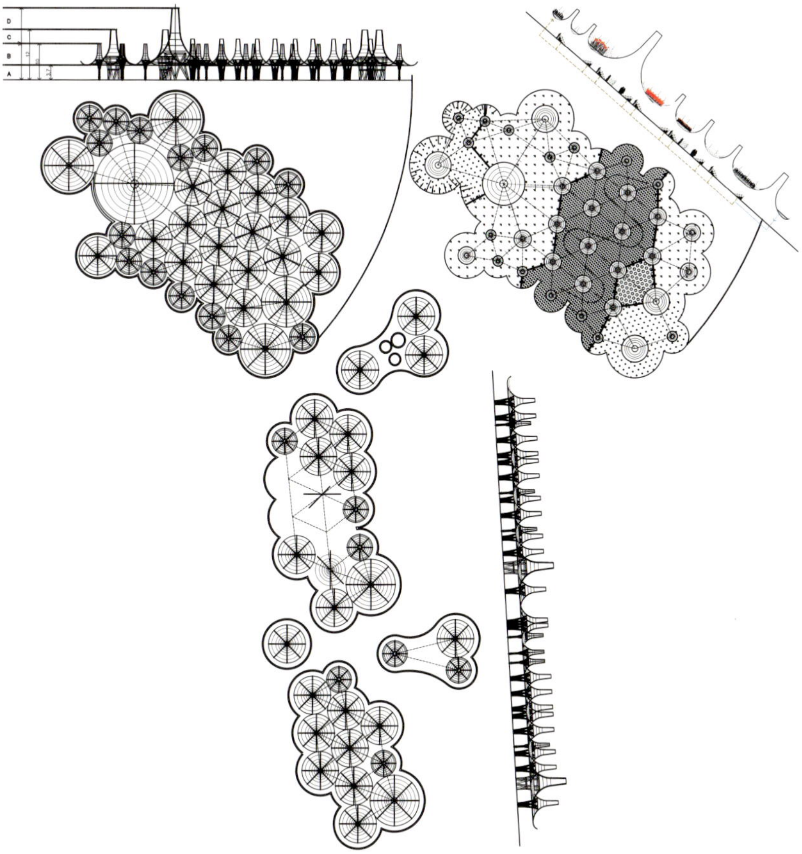

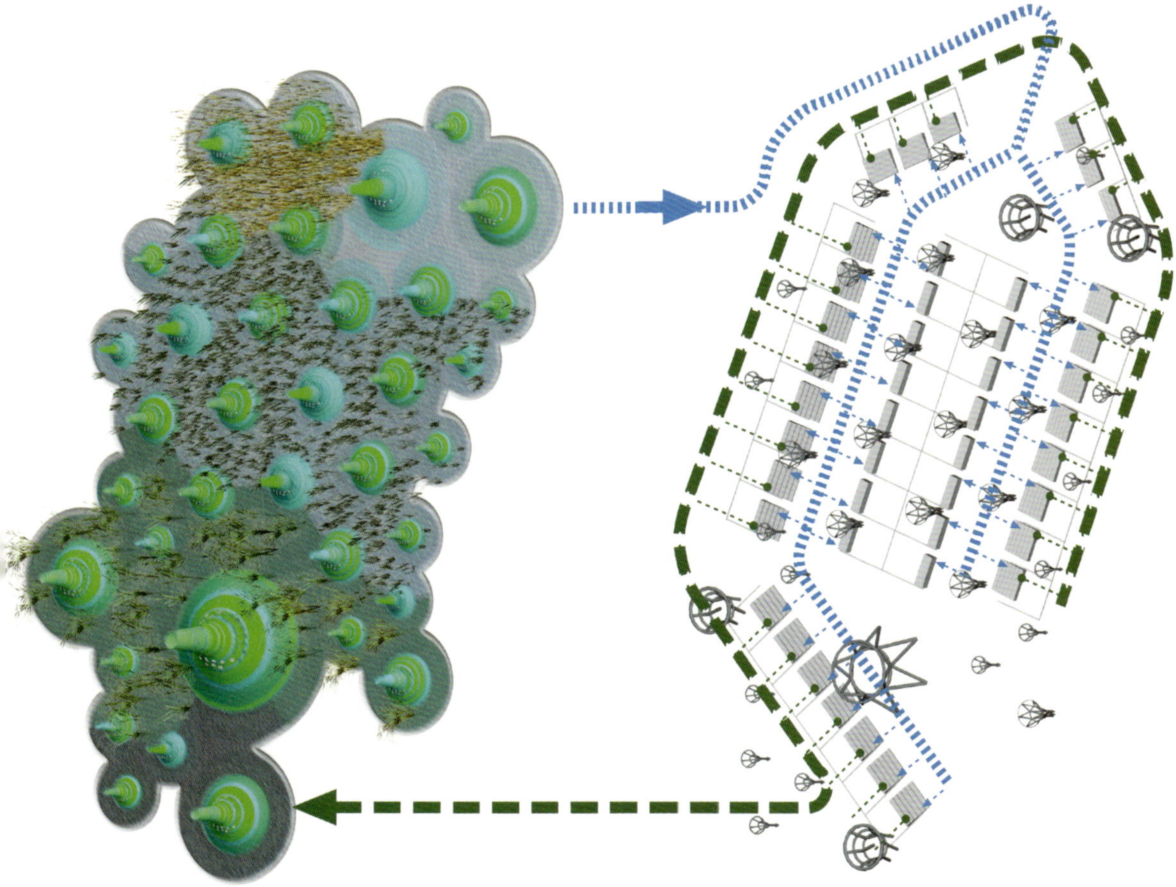

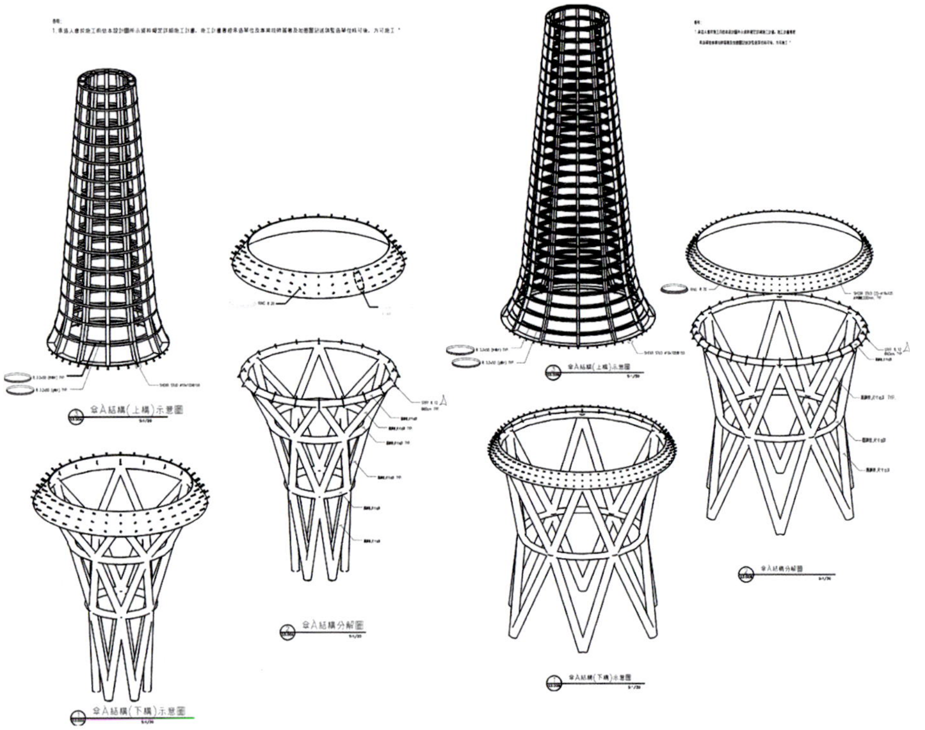
傘A結構(上構)示意圖
傘A結構分解圖
傘A結構(下構)示意圖
傘A結構(上構)示意圖
傘A結構(下構)示意圖
傘A結構分解圖

ON SALE
特價中
國民旅遊卡特約商店
海產大賣場
各種

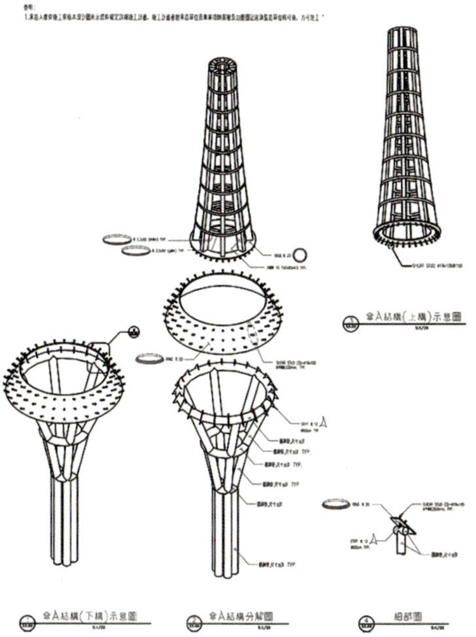
傘A結構(上構)示意圖
傘A結構(下構)示意圖
傘A結構分解圖
細部圖

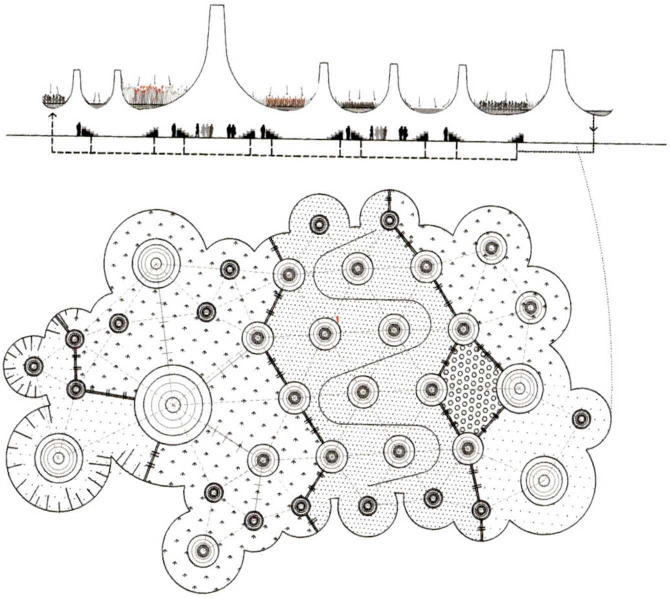

魚鬆總匯
乾料批發
39

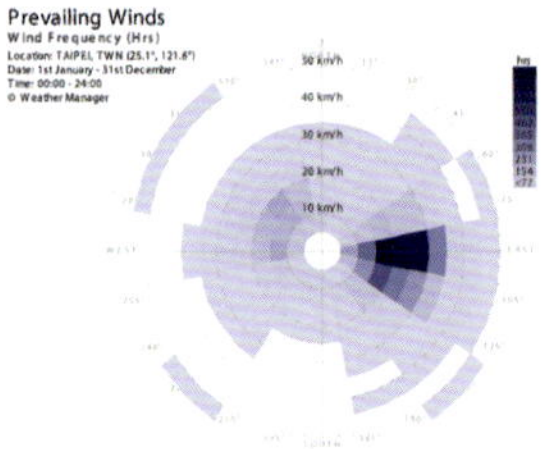
Prevailing Winds
Wind Frequency (Hrs)
Location: TAIPEI, TWN (25.1°, 121.6°)
Date: 1st January - 31st December
Time: 00:00 - 24:00
© Weather Manager
50 km/h
40 km/h
30 km/h
20 km/h
10 km/h

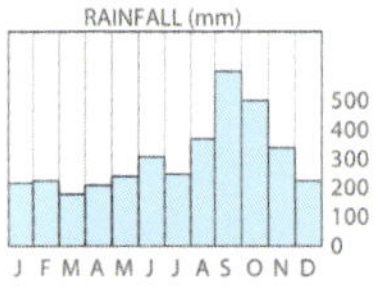
RAINFALL (mm)
500
400
300
200
100
0
J F M A M J J A S O N D

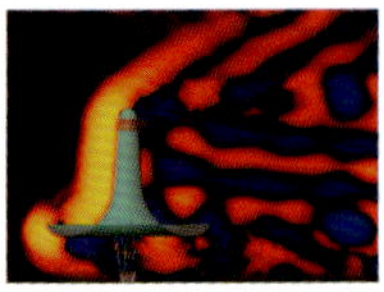

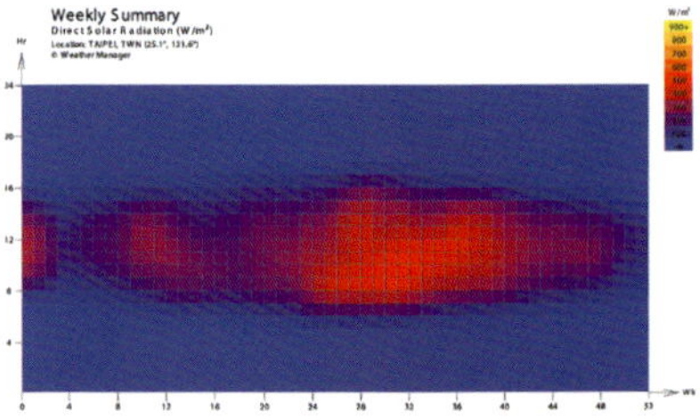
Weekly Summary
Direct Solar Radiation (W/m²)
Location: TAIPEI, TWN (25.1°, 121.6°)
© Weather Manager

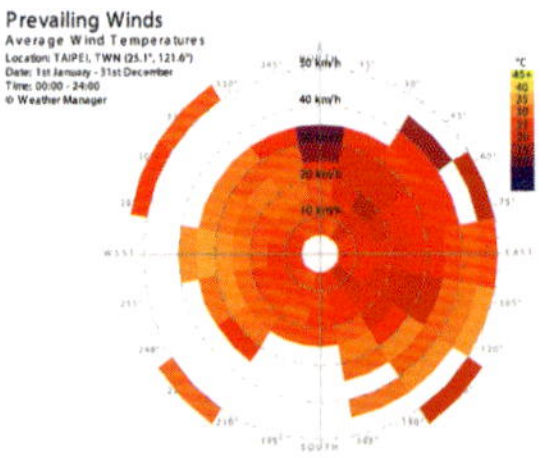
Prevailing Winds
Average Wind Temperatures
Location: TAIPEI, TWN (25.1°, 121.6°)
Date: 1st January - 31st December
Time: 00:00 - 24:00
© Weather Manager
50 km/h
40 km/h

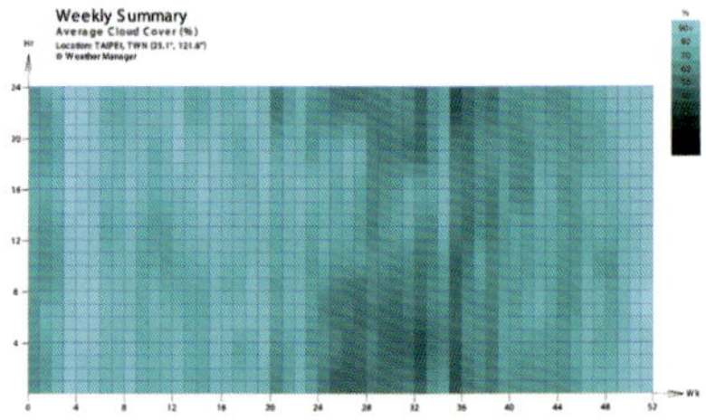
Weekly Summary
Average Cloud Cover (%)
Location: TAIPEI, TWN (25.1°, 121.6°)
© Weather Manager

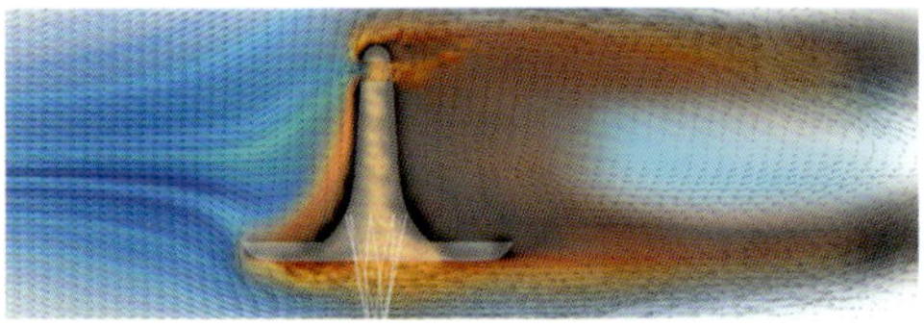

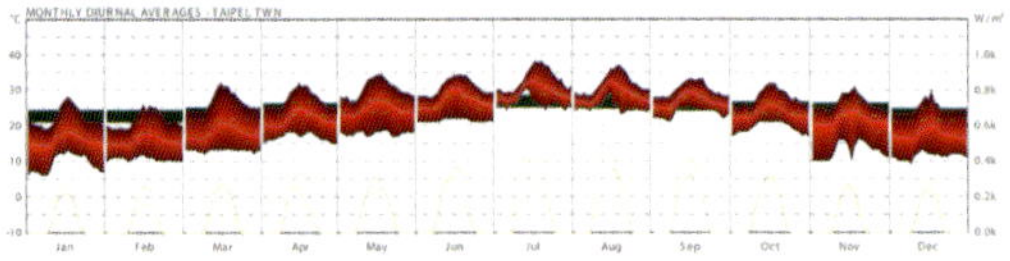
MONTHLY DIURNAL AVERAGES - TAIPEI, TWN
Jan
Feb
Mar
Apr
May
Jun
Jul
Aug
Sep
Oct
Nov
Dec

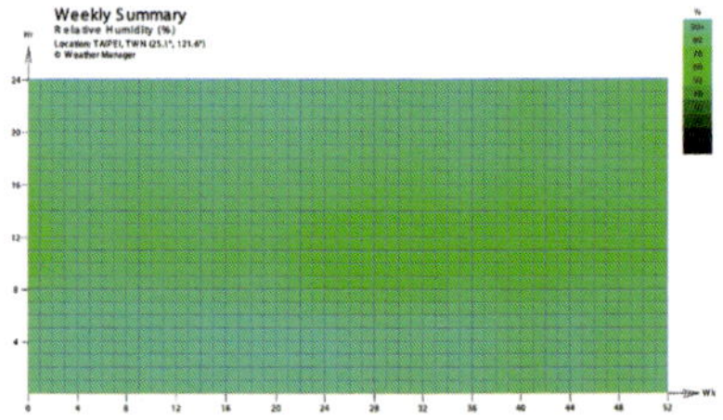
Weekly Summary
Relative Humidity (%)
Location: TAIPEI, TWN (25.1°, 121.6°)
© Weather Manager

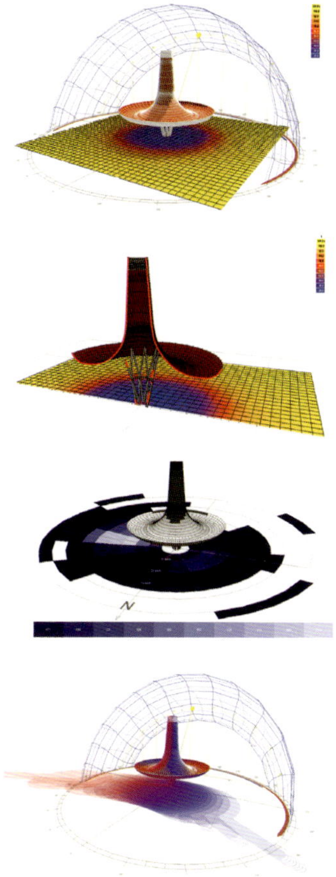
N

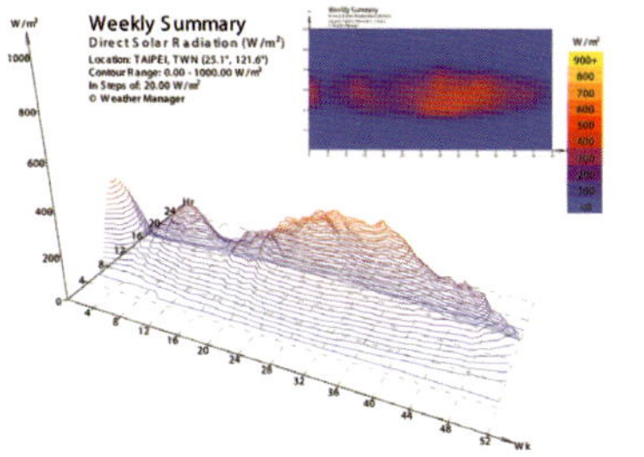
Weekly Summary
Direct Solar Radiation (W/m²)
Location: TAIPEI, TWN (25.1°, 121.6°)
Contour Range: 0.00 - 1000.00 W/m²
In Steps of: 20.00 W/m²
© Weather Manager

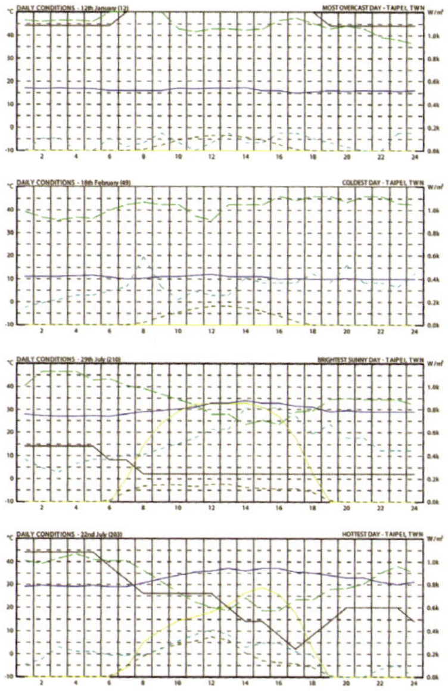
DAILY CONDITIONS - 12th January (12)
MOST OVERCAST DAY - TAIPEI, TWN
DAILY CONDITIONS - 18th February (49)
COLDEST DAY - TAIPEI, TWN
DAILY CONDITIONS - 29th July (210)
BRIGHTEST SUNNY DAY - TAIPEI, TWN
DAILY CONDITIONS - 22nd July (203)
HOTTEST DAY - TAIPEI, TWN

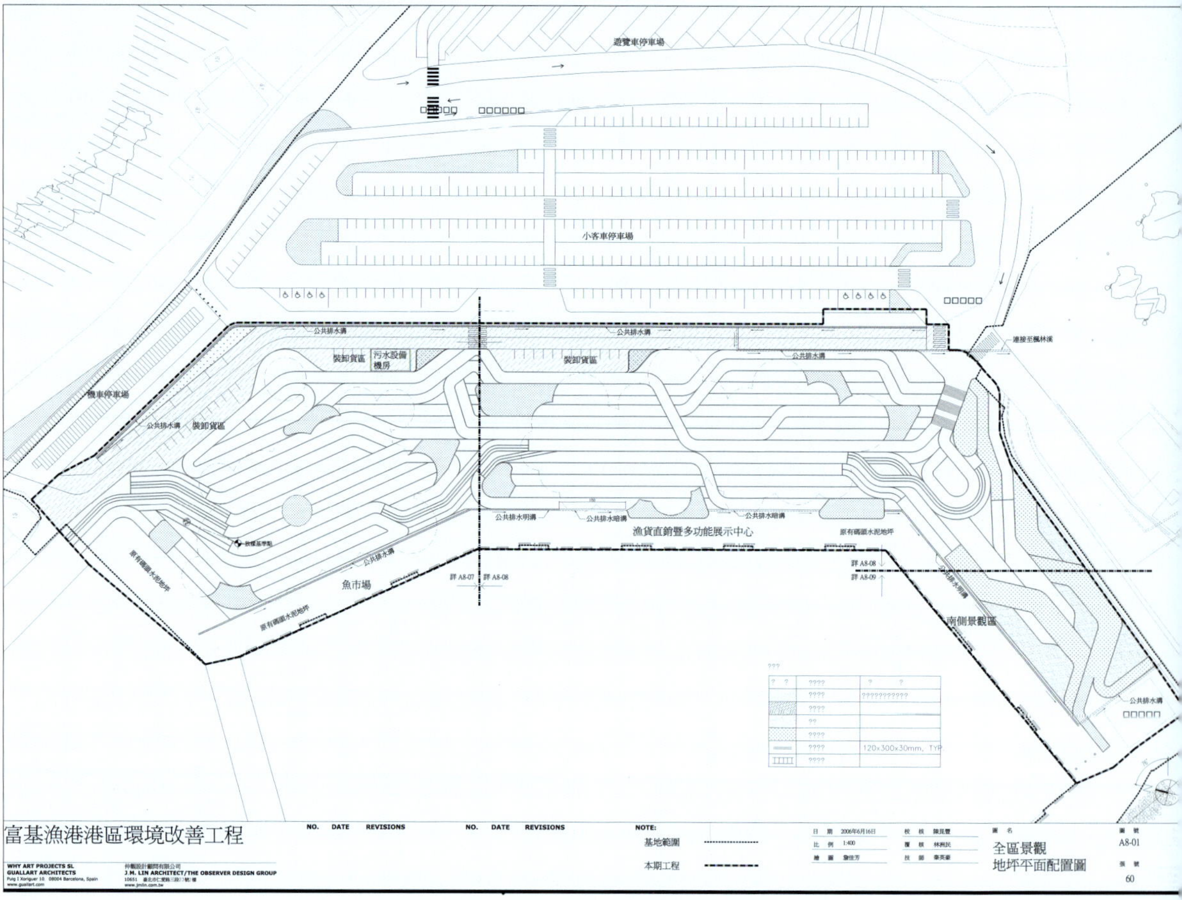
遊覽車停車場
小客車停車場
公共排水溝
裝卸貨區
污水設備機房
機車停車場
漁貨直銷暨多功能展示中心
魚市場
南側景觀區
原有碼頭水泥地坪
富基漁港港區環境改善工程
WHY ART PROJECTS SL
GUALLART ARCHITECTS
J.M. LIN ARCHITECT/THE OBSERVER DESIGN GROUP
NO. DATE REVISIONS
NOTE:
基地範圍
本期工程
全區景觀
地坪平面配置圖
A8-01
60

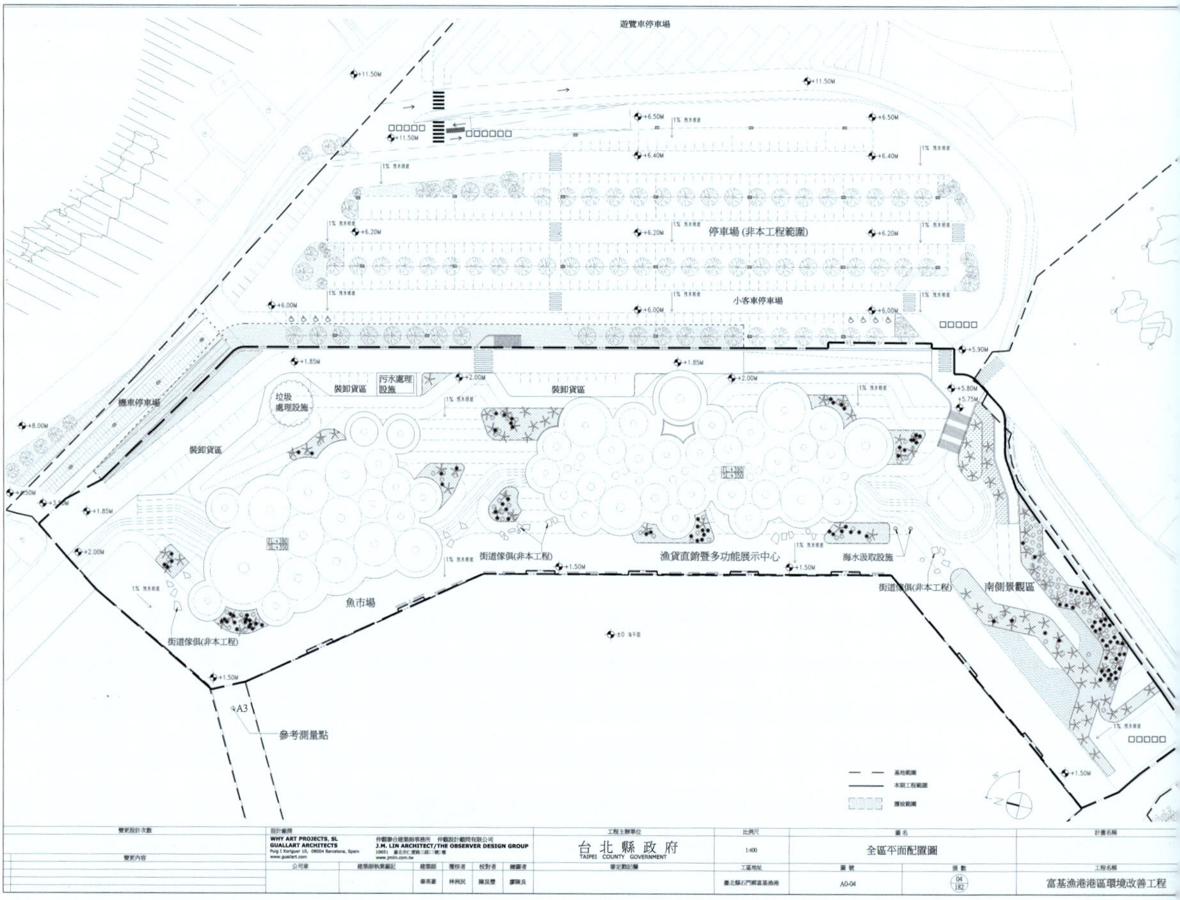
遊覽車停車場
停車場 (非本工程範圍)
小客車停車場
裝卸貨區
污水處理設施
垃圾處理設施
機車停車場
街道傢俱(非本工程)
漁貨直銷暨多功能展示中心
海水汲取設施
南側景觀區
魚市場
參考測量點
台北縣政府
TAIPEI COUNTY GOVERNMENT
全區平面配置圖
WHY ART PROJECTS, SL
GUALLART ARCHITECTS
J.M. LIN ARCHITECT/THE OBSERVER DESIGN GROUP
1:400
A0-04
富基漁港港區環境改善工程

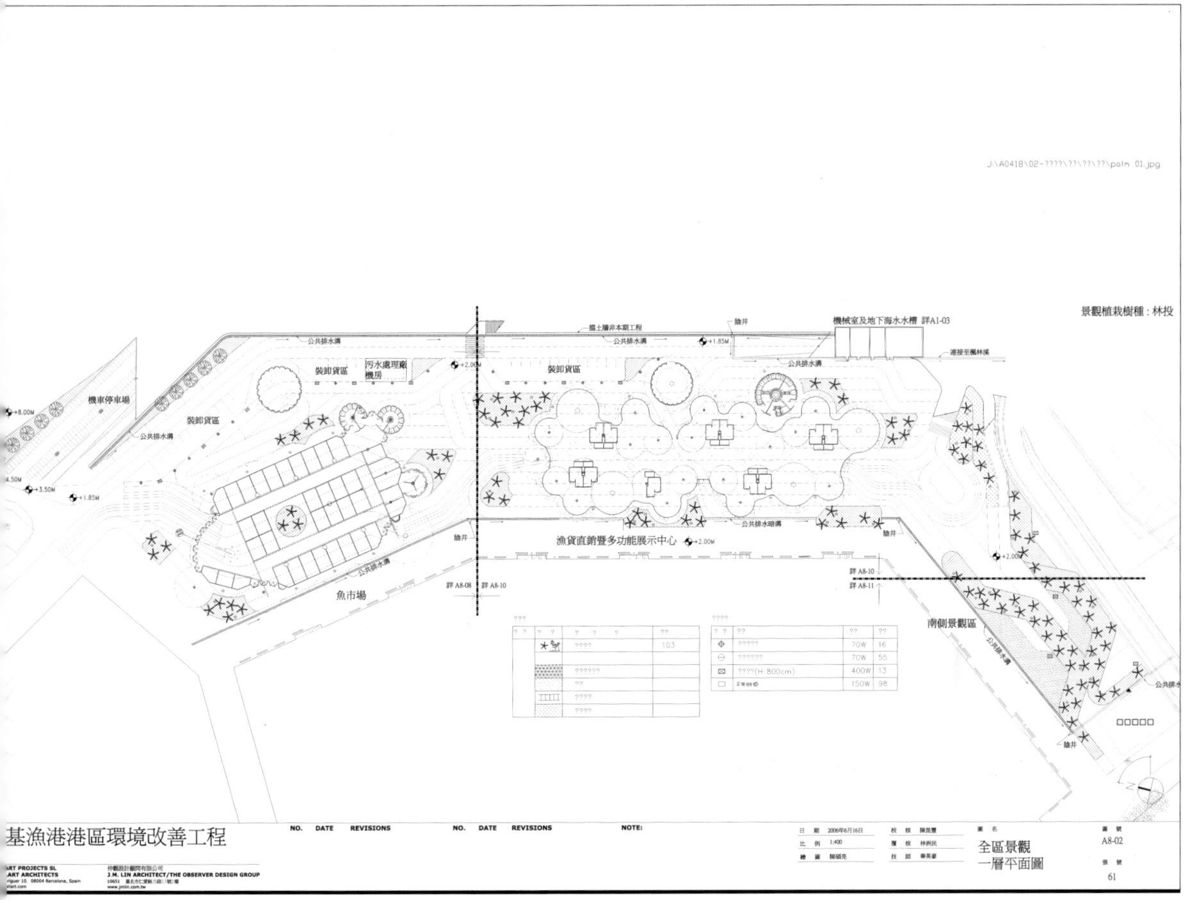
景觀植栽樹種：林投
機械室及地下海水水槽 詳A1-03
公共排水溝
裝卸貨區
污水處理廠機房
機車停車場
漁貨直銷暨多功能展示中心
魚市場
南側景觀區
詳 A8-08
詳 A8-10
詳 A8-11
基漁港港區環境改善工程
NO. DATE REVISIONS
NO. DATE REVISIONS
NOTE:
J.M. LIN ARCHITECT/THE OBSERVER DESIGN GROUP
全區景觀
一層平面圖
A8-02
61

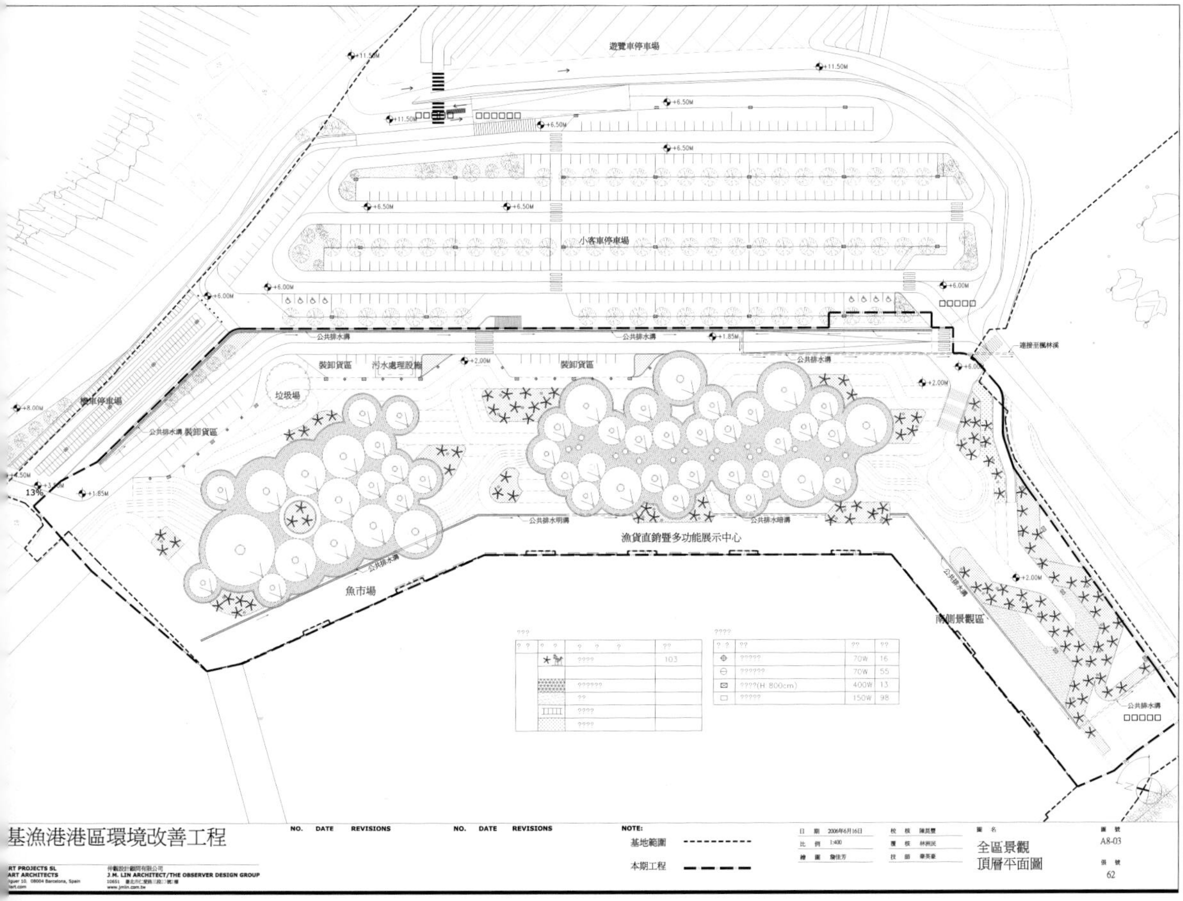
遊覽車停車場
小客車停車場
公共排水溝
裝卸貨區
污水處理設備
垃圾場
機車停車場
漁貨直銷暨多功能展示中心
魚市場
南側景觀區
基漁港港區環境改善工程
NO. DATE REVISIONS
NO. DATE REVISIONS
NOTE:
基地範圍
本期工程
J.M. LIN ARCHITECT/THE OBSERVER DESIGN GROUP
全區景觀
頂層平面圖
A8-03
62

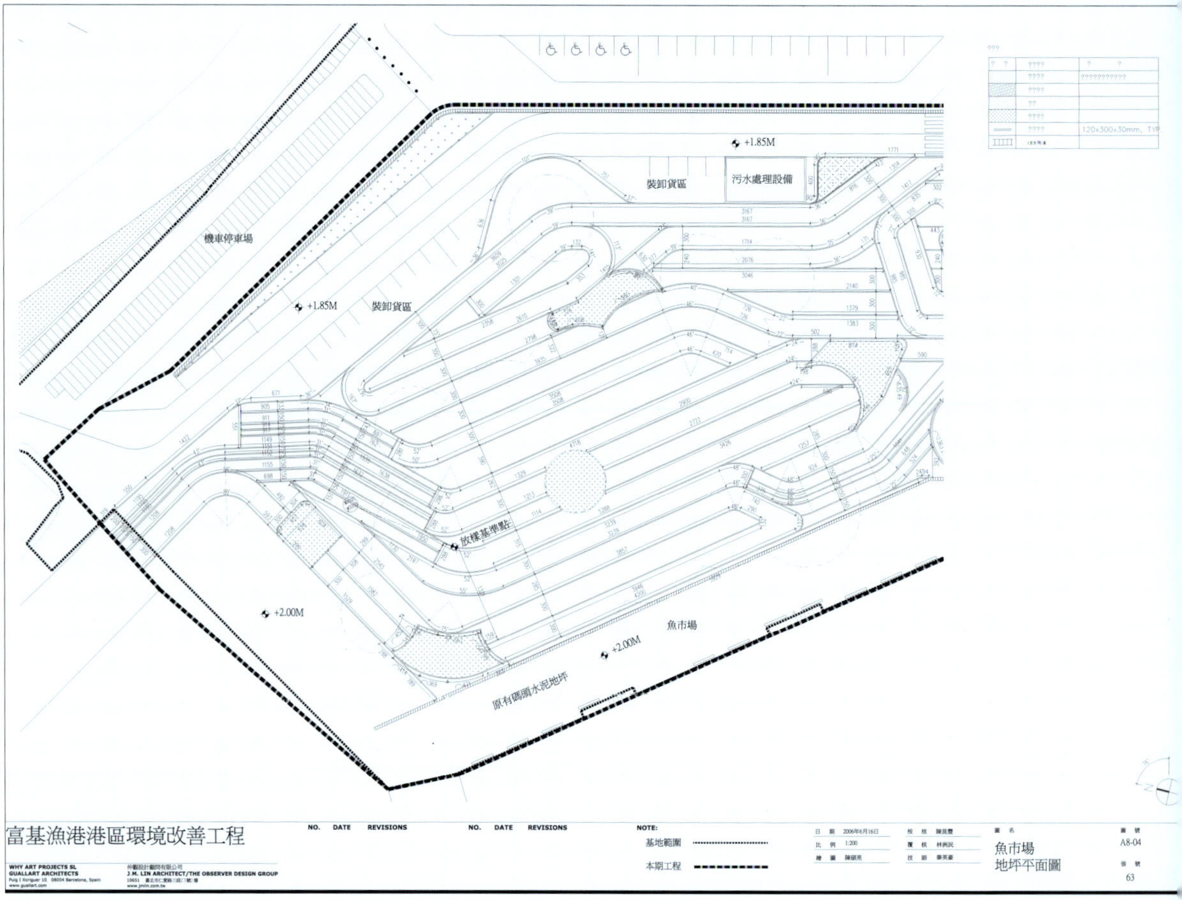

機車停車場
+1.85M
裝卸貨區
污水處理設備
放樣基準點
+2.00M
魚市場
原有碼頭水泥地坪
富基漁港港區環境改善工程
WHY ART PROJECTS SL
GUALLART ARCHITECTS
J.M. LIN ARCHITECT/THE OBSERVER DESIGN GROUP
NO. DATE REVISIONS
NOTE:
基地範圍
本期工程
魚市場
地坪平面圖
A8-04
63

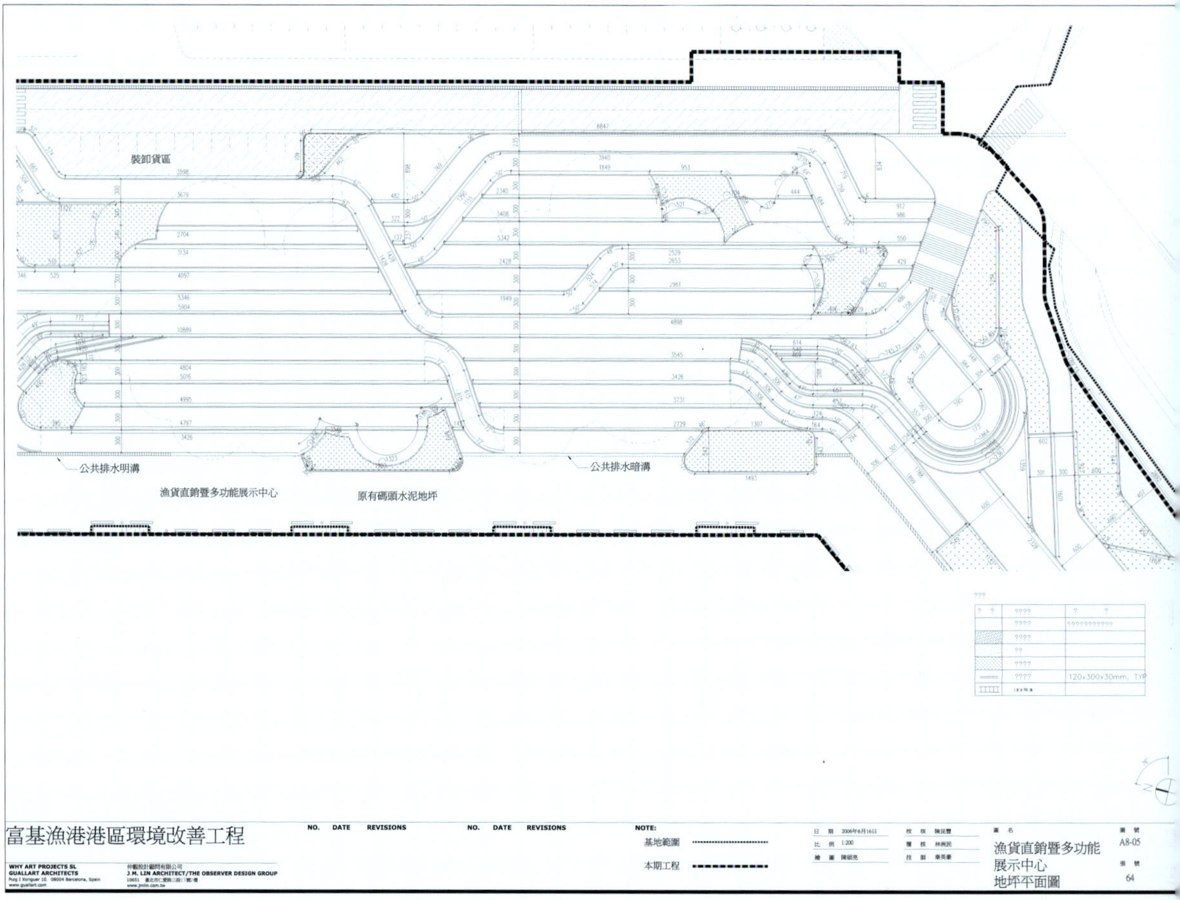

裝卸貨區
公共排水明溝
公共排水暗溝
漁貨直銷暨多功能展示中心
原有碼頭水泥地坪
富基漁港港區環境改善工程
WHY ART PROJECTS SL
GUALLART ARCHITECTS
J.M. LIN ARCHITECT/THE OBSERVER DESIGN GROUP
NO. DATE REVISIONS
NOTE:
基地範圍
本期工程
漁貨直銷暨多功能
展示中心
地坪平面圖
A8-05
64

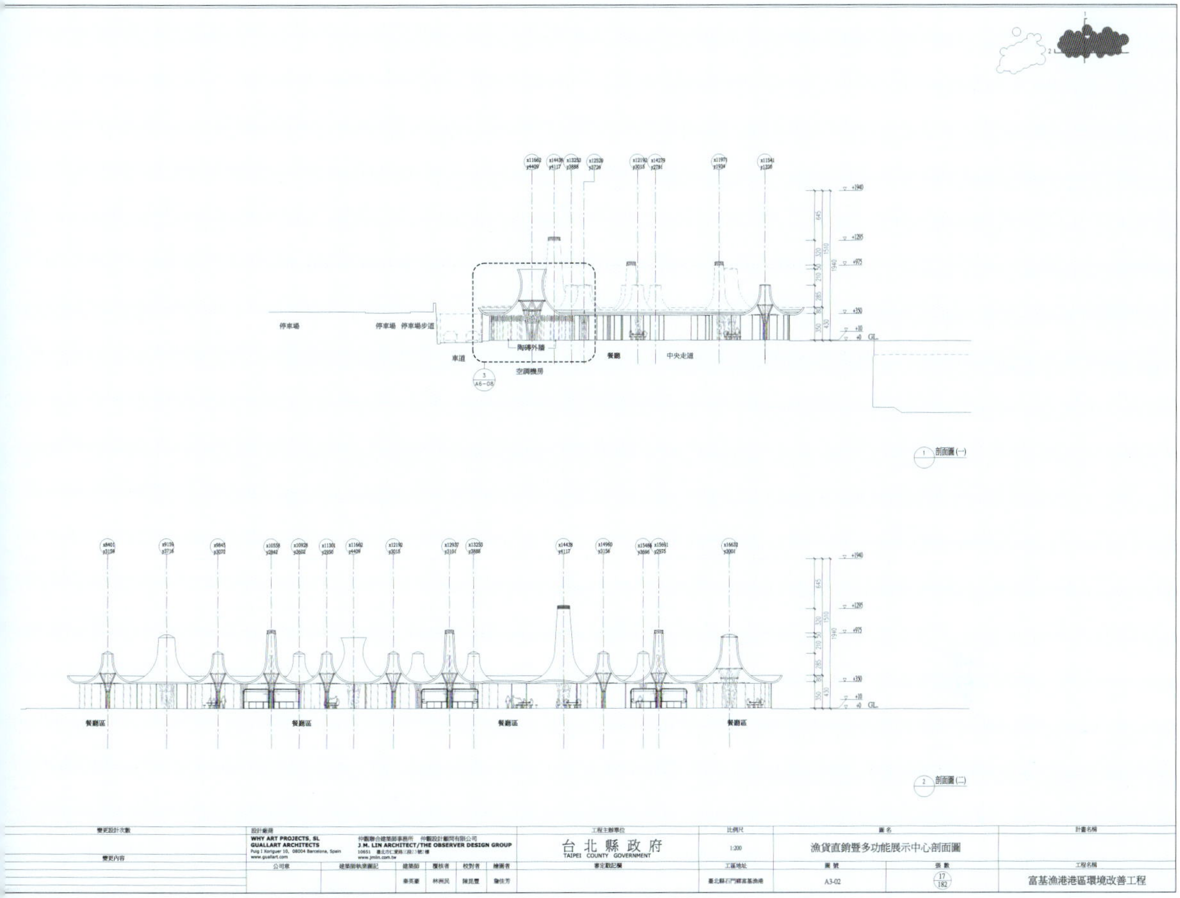
停車場
車道
空調機房
餐廳
中央走道
剖面圖(一)
餐廳區
剖面圖(二)
WHY ART PROJECTS, SL
GUALLART ARCHITECTS
J.M. LIN ARCHITECT/THE OBSERVER DESIGN GROUP
台北縣政府
TAIPEI COUNTY GOVERNMENT
1:200
漁貨直銷暨多功能展示中心剖面圖
A3-02
富基漁港港區環境改善工程

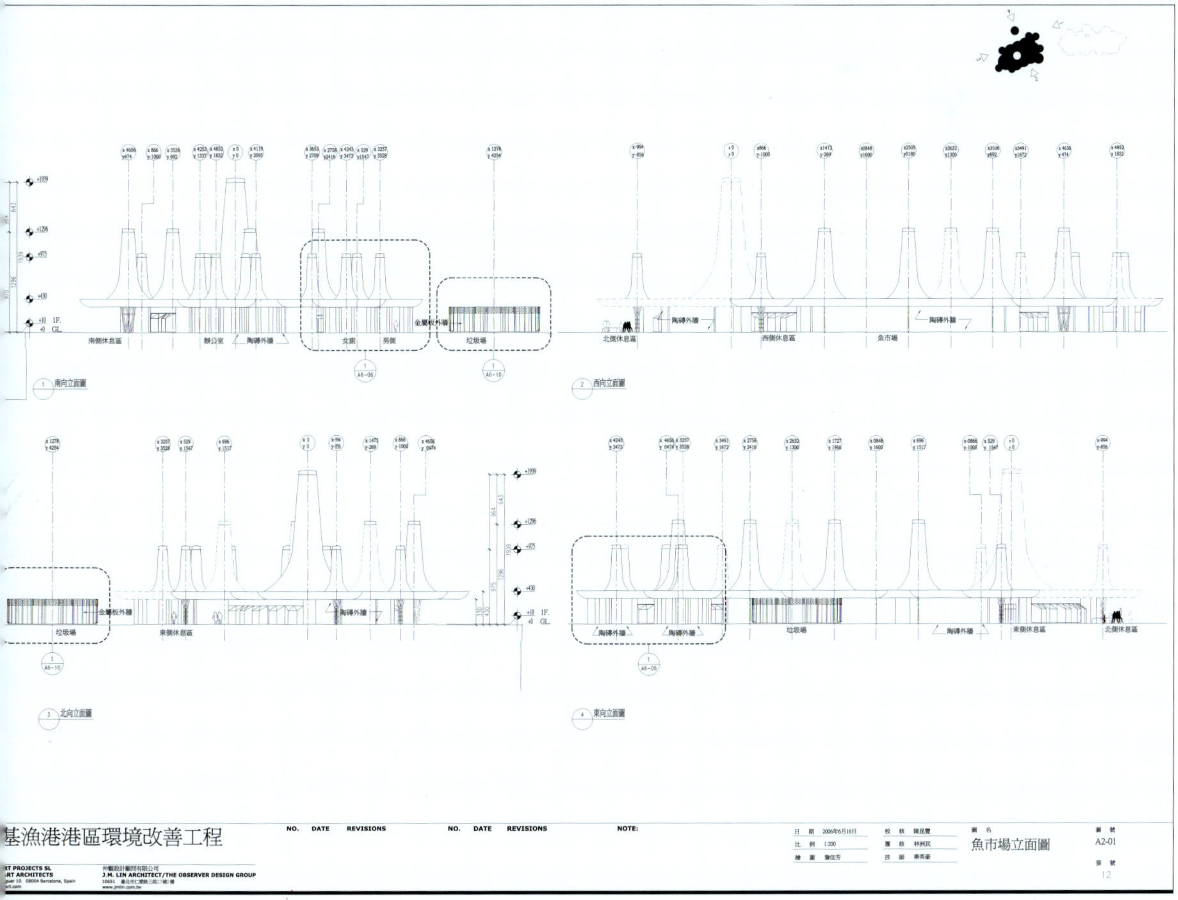
南側休息區
辦公室
陶磚外牆
女廁
男廁
垃圾場
南向立面圖
北側休息區
西側休息區
魚市場
西向立面圖
東側休息區
北向立面圖
東向立面圖
基漁港港區環境改善工程
NO. DATE REVISIONS
NOTE:
J.M. LIN ARCHITECT/THE OBSERVER DESIGN GROUP
1:200
魚市場立面圖
A2-01

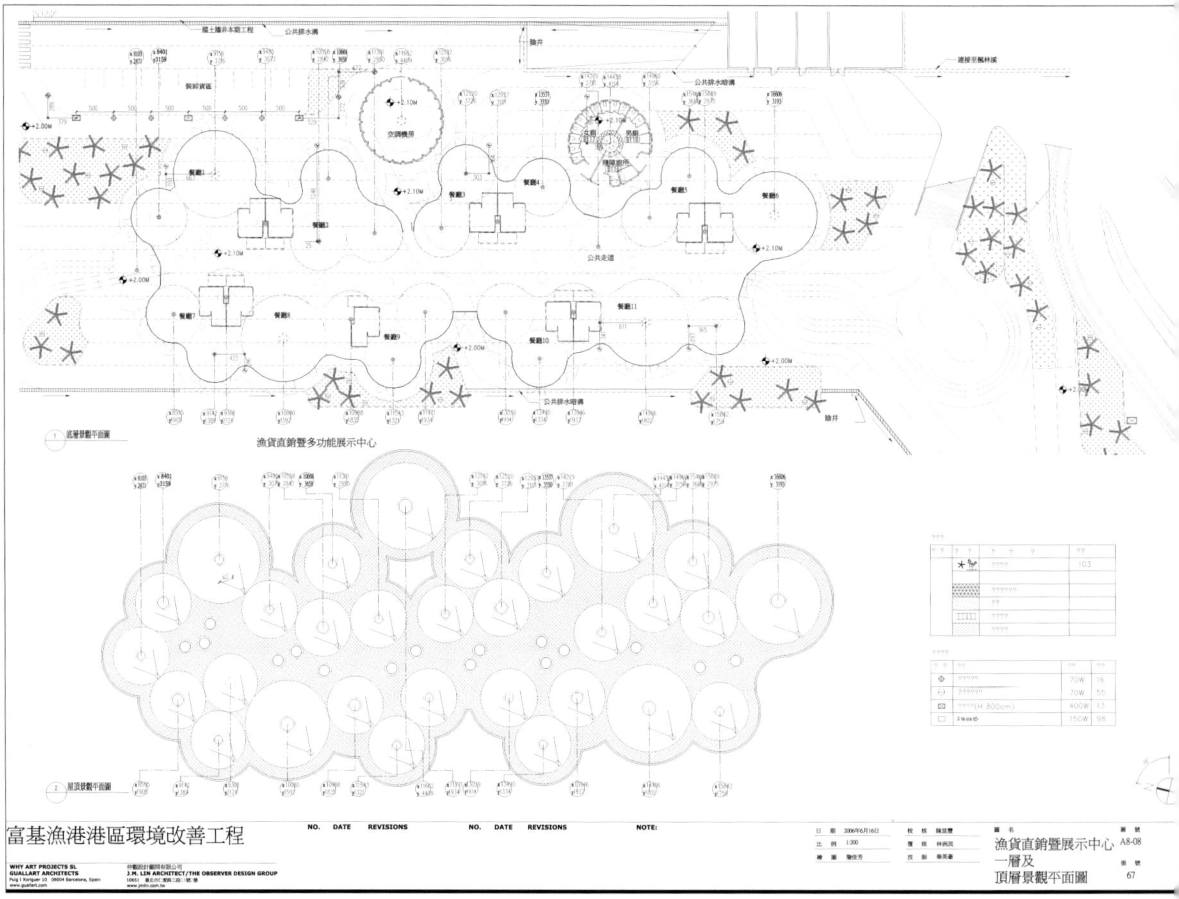
底層景觀平面圖
漁貨直銷暨多功能展示中心
屋頂景觀平面圖
富基漁港港區環境改善工程
WHY ART PROJECTS SL
GUALLART ARCHITECTS
J.M. LIN ARCHITECT/THE OBSERVER DESIGN GROUP
NO. DATE REVISIONS
NOTE:
漁貨直銷暨展示中心
一層及
頂層景觀平面圖
A8-08
67

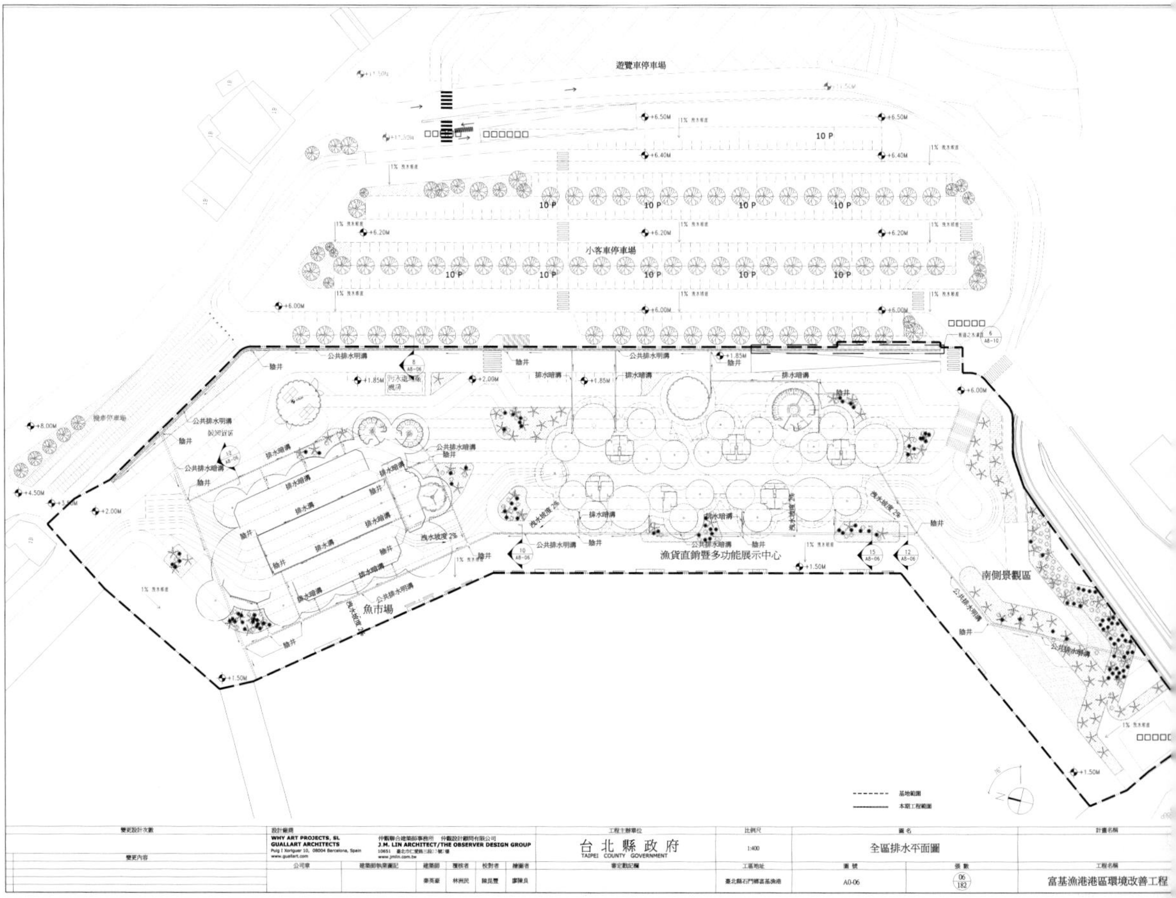
遊覽車停車場
小客車停車場
漁貨直銷暨多功能展示中心
魚市場
南側景觀區
WHY ART PROJECTS, SL
GUALLART ARCHITECTS
J.M. LIN ARCHITECT/THE OBSERVER DESIGN GROUP
台北縣政府
TAIPEI COUNTY GOVERNMENT
1:400
全區排水平面圖
A0-06
富基漁港港區環境改善工程

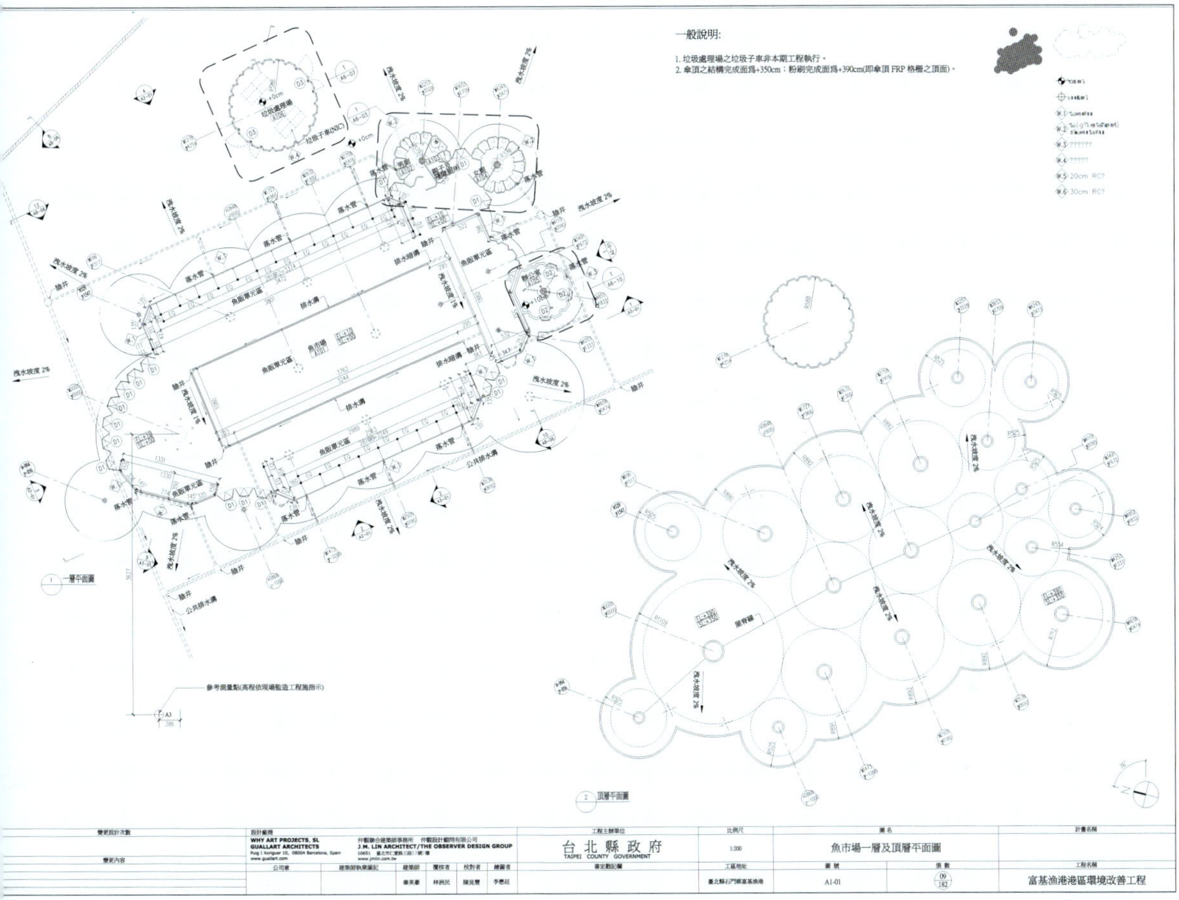
一般說明:
1. 垃圾處理場之垃圾子車非本期工程執行。
2. 傘頂之結構完成面爲+350cm；粉刷完成面爲+390cm(即傘頂 FRP 格柵之頂面)。
1 一層平面圖
2 頂層平面圖
WHY ART PROJECTS, SL
GUALLART ARCHITECTS
J.M. LIN ARCHITECT/THE OBSERVER DESIGN GROUP
台北縣政府
TAIPEI COUNTY GOVERNMENT
1:200
魚市場一層及頂層平面圖
A1-01
富基漁港港區環境改善工程

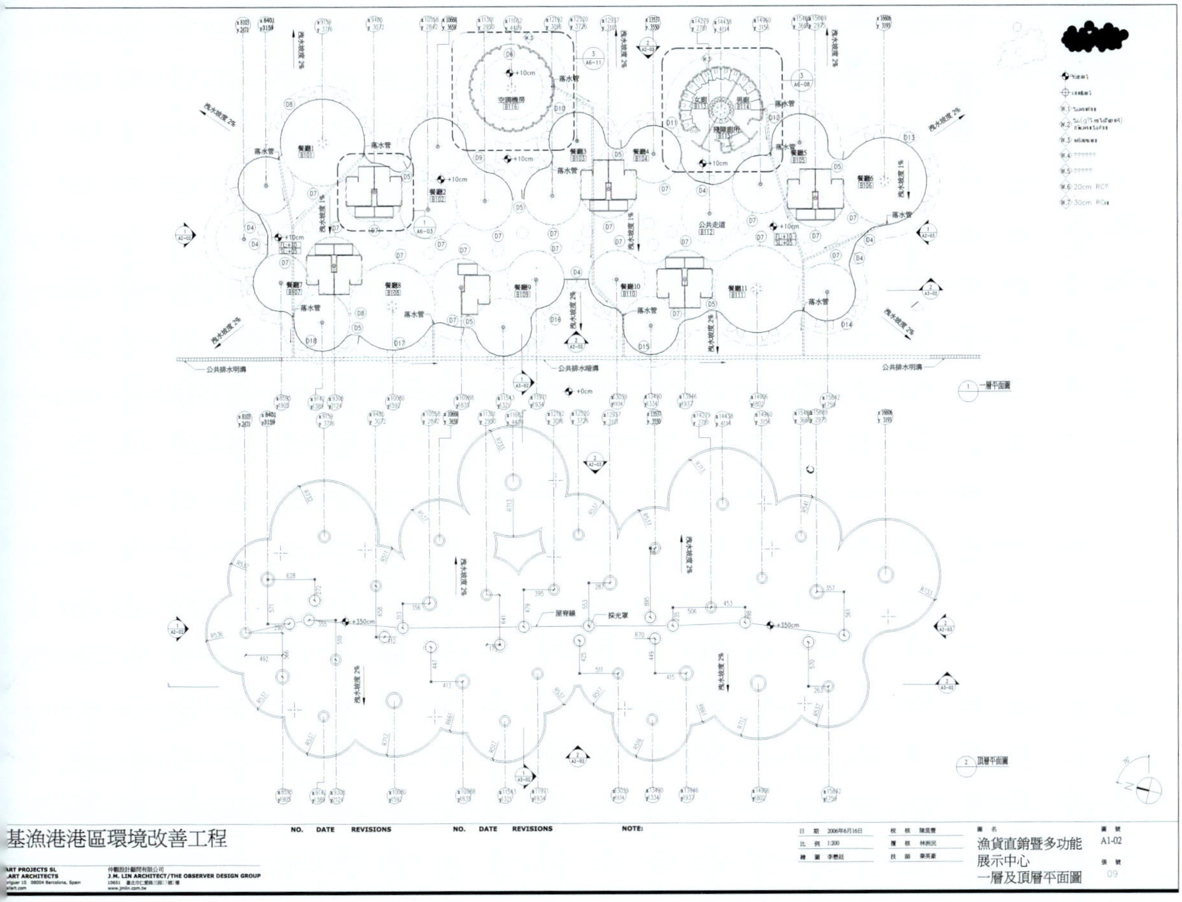
1 一層平面圖
2 頂層平面圖
基漁港港區環境改善工程
NO. DATE REVISIONS
NO. DATE REVISIONS
NOTE:
J.M. LIN ARCHITECT/THE OBSERVER DESIGN GROUP
漁貨直銷暨多功能
展示中心
一層及頂層平面圖
A1-02

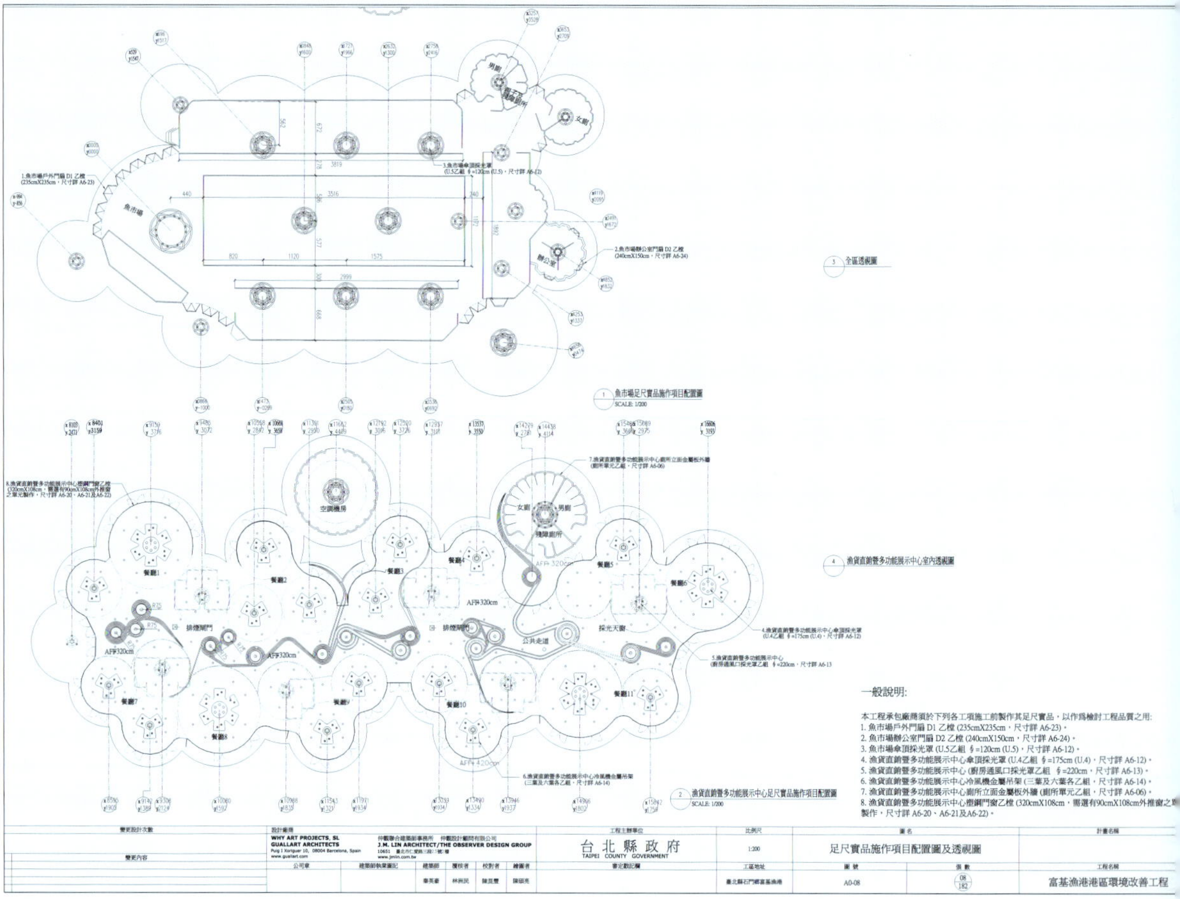
1 魚市場足尺實品施作項目配置圖
SCALE: 1/200
2 漁貨直銷暨多功能展示中心足尺實品施作項目配置圖
SCALE: 1/200
3 全區透視圖
4 漁貨直銷暨多功能展示中心室內透視圖
一般說明:
本工程承包廠商須於下列各工項施工前製作其足尺實品，以作爲檢討工程品質之用:
1. 魚市場戶外門扇 D1 乙樘 (235cmX235cm，尺寸詳 A6-23)。
2. 魚市場辦公室門扇 D2 乙樘 (240cmX150cm，尺寸詳 A6-24)。
3. 魚市場傘頂採光罩 (U.5乙組 ∮=120cm (U.5)，尺寸詳 A6-12)。
4. 漁貨直銷暨多功能展示中心傘頂採光罩 (U.4乙組 ∮=175cm (U.4)，尺寸詳 A6-12)。
5. 漁貨直銷暨多功能展示中心 (廚房通風口採光罩乙組 ∮=220cm，尺寸詳 A6-13)。
6. 漁貨直銷暨多功能展示中心冷風機金屬吊架 (三葉及六葉各乙組，尺寸詳 A6-14)。
7. 漁貨直銷暨多功能展示中心廁所立面金屬板外牆 (廁所單元乙組，尺寸詳 A6-06)。
8. 漁貨直銷暨多功能展示中心塑鋼門窗乙樘 (320cmX108cm，需選有90cmX108cm外推窗之製作，尺寸詳 A6-20、A6-21及A6-22)。
WHY ART PROJECTS, SL
GUALLART ARCHITECTS
J.M. LIN ARCHITECT/THE OBSERVER DESIGN GROUP
台北縣政府
TAIPEI COUNTY GOVERNMENT
1:200
足尺實品施作項目配置圖及透視圖
A0-08
08/182
富基漁港港區環境改善工程

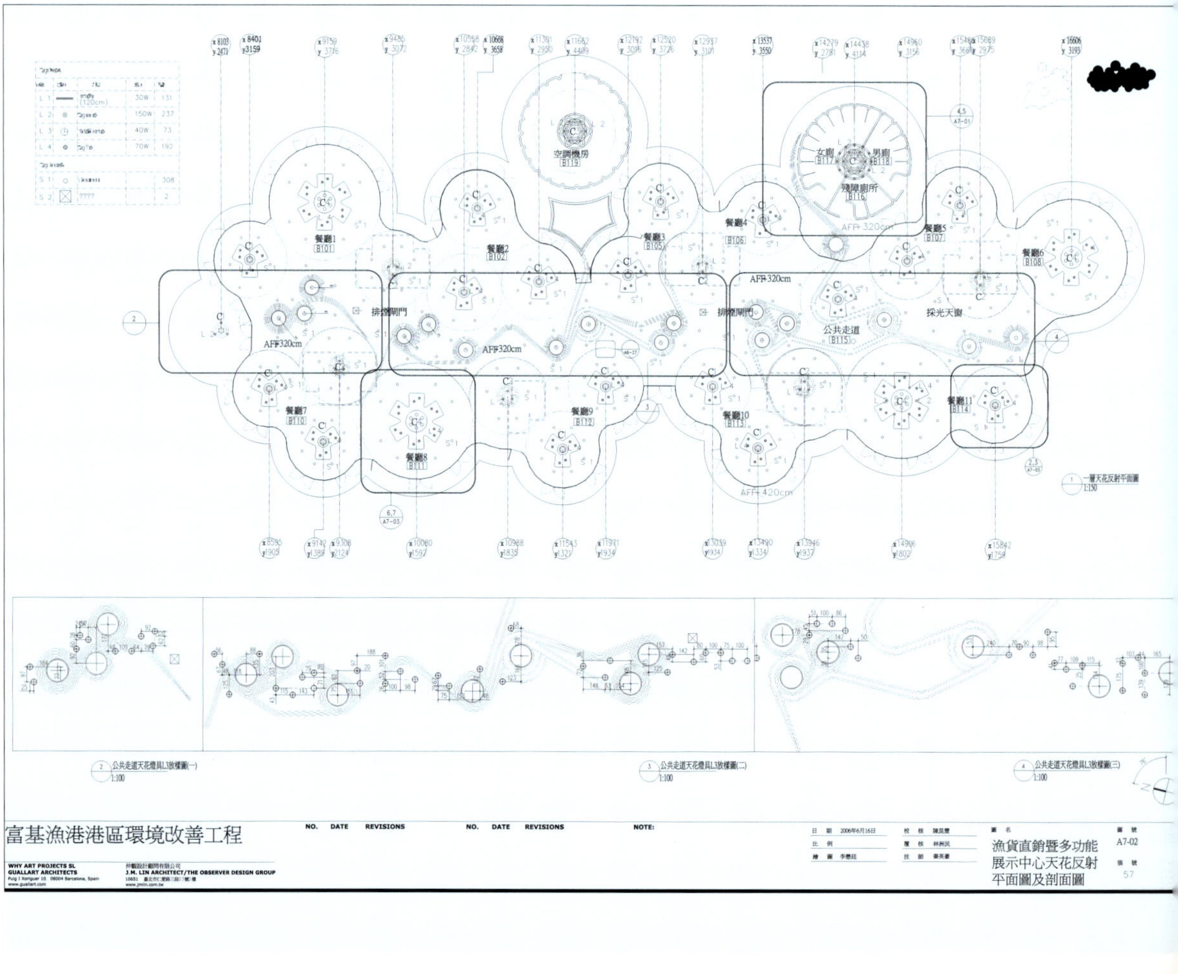
空調機房
女廁
男廁
殘障廁所
餐廳1
餐廳2
餐廳3
餐廳4
餐廳5
餐廳6
餐廳7
餐廳8
餐廳9
餐廳10
餐廳11
排煙閘門
公共走道
採光天窗
AFF320cm
AFF420cm
1 一層天花反射平面圖
1:150
2 公共走道天花燈具L3放樣圖(一)
1:100
3 公共走道天花燈具L3放樣圖(二)
1:100
4 公共走道天花燈具L3放樣圖(三)
1:100
富基漁港港區環境改善工程
WHY ART PROJECTS SL
GUALLART ARCHITECTS
J.M. LIN ARCHITECT/THE OBSERVER DESIGN GROUP
NO. DATE REVISIONS
NOTE:
2006年6月16日
漁貨直銷暨多功能展示中心天花反射平面圖及剖面圖
A7-02
57

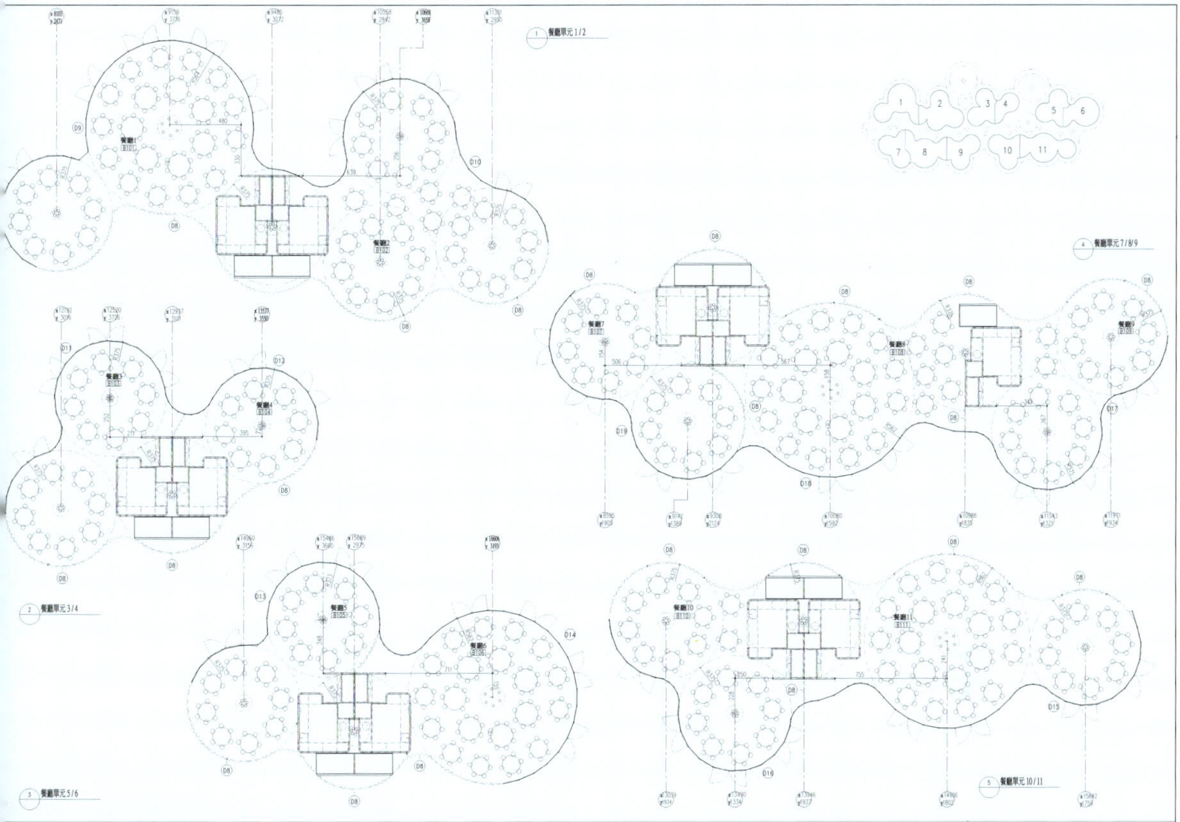

基漁港港區環境改善工程

NO. DATE REVISIONS NO. DATE REVISIONS NOTE:

日 期 2006年6月16日 校 核 陳昆豐
比 例 1:100 覆 核 林洲民
繪 圖 詹佳芳 技 師 秦英豪

圖 名 漁貨直銷 暨多功能展示中心 餐廳單元平面詳圖

圖 號 A6-01 張 號 27

RT PROJECTS SL
ART ARCHITECTS
guer 10. 08004 Barcelona, Spain
lart.com

仲觀設計顧問有限公司
J.M. LIN ARCHITECT/THE OBSERVER DESIGN GROUP
10651 臺北市仁愛路三段23號7樓
www.jmlin.com.tw

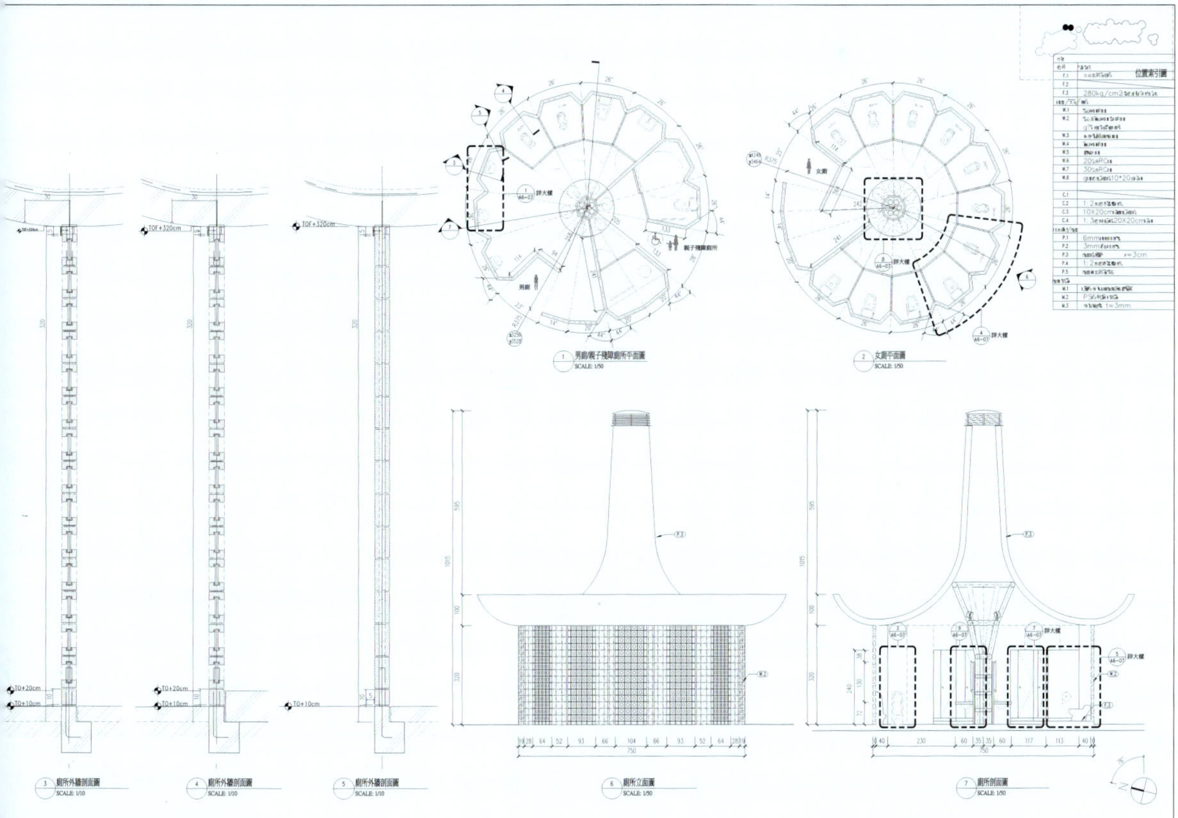

基漁港港區環境改善工程

NO. DATE REVISIONS NO. DATE REVISIONS NOTE:

日 期 2006年6月16日 校 核 陳昆豐
比 例 依圖示 覆 核 林洲民
繪 圖 陳頌亮 技 師 秦英豪

圖 名 魚市場廁所詳圖(一)

圖 號 A6-03 張 號 29

RT PROJECTS SL
ART ARCHITECTS
guer 10. 08004 Barcelona, Spain
art.com

仲觀設計顧問有限公司
J.M. LIN ARCHITECT/THE OBSERVER DESIGN GROUP
10651 臺北市仁愛路三段23號7樓
www.jmlin.com.tw

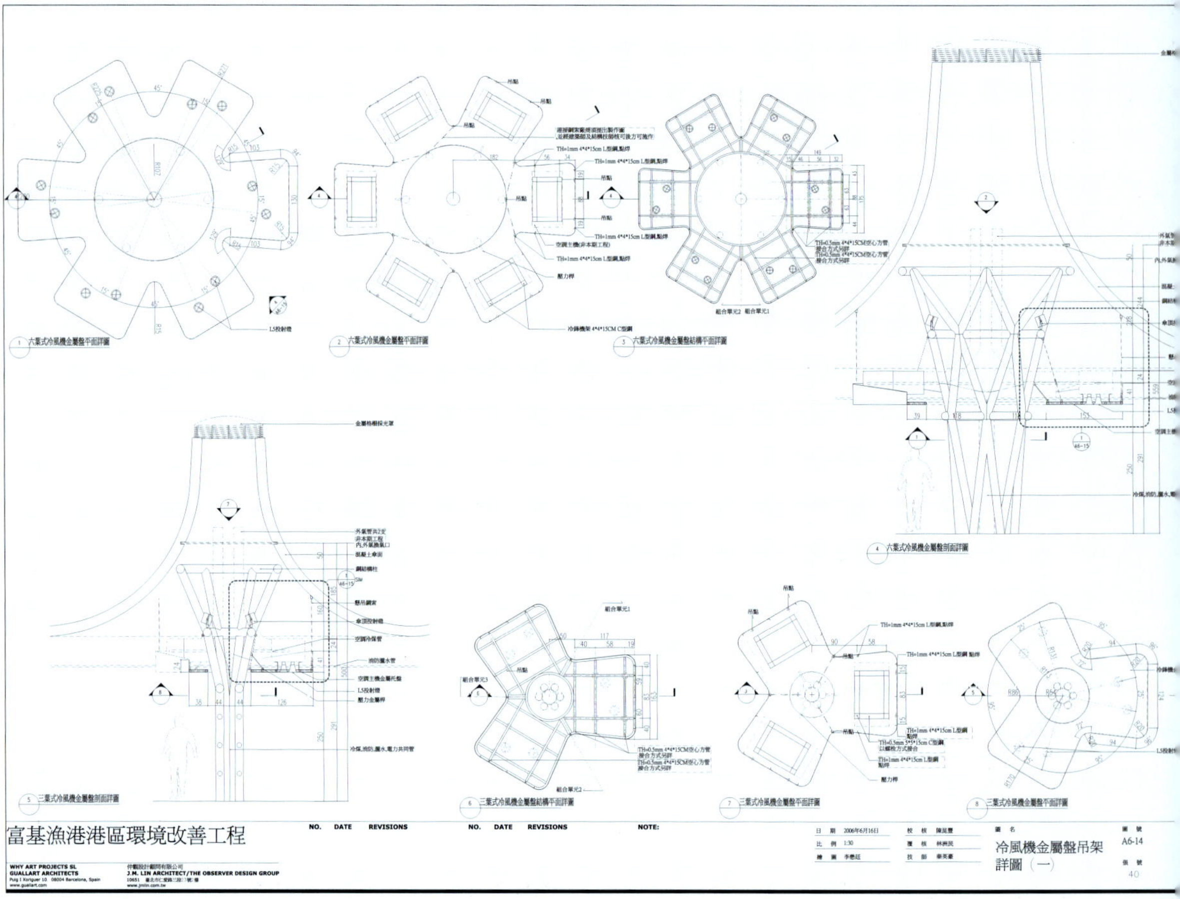
六葉式冷風機金屬盤平面詳圖
六葉式冷風機金屬盤平面詳圖
六葉式冷風機金屬盤結構平面詳圖
六葉式冷風機金屬盤剖面詳圖
三葉式冷風機金屬盤剖面詳圖
三葉式冷風機金屬盤結構平面詳圖
三葉式冷風機金屬盤平面詳圖
三葉式冷風機金屬盤平面詳圖
富基漁港港區環境改善工程
NO. DATE REVISIONS
NO. DATE REVISIONS
NOTE:
WHY ART PROJECTS SL
GUALLART ARCHITECTS
J.M. LIN ARCHITECT/THE OBSERVER DESIGN GROUP
冷風機金屬盤吊架詳圖（一）
A6-14
40

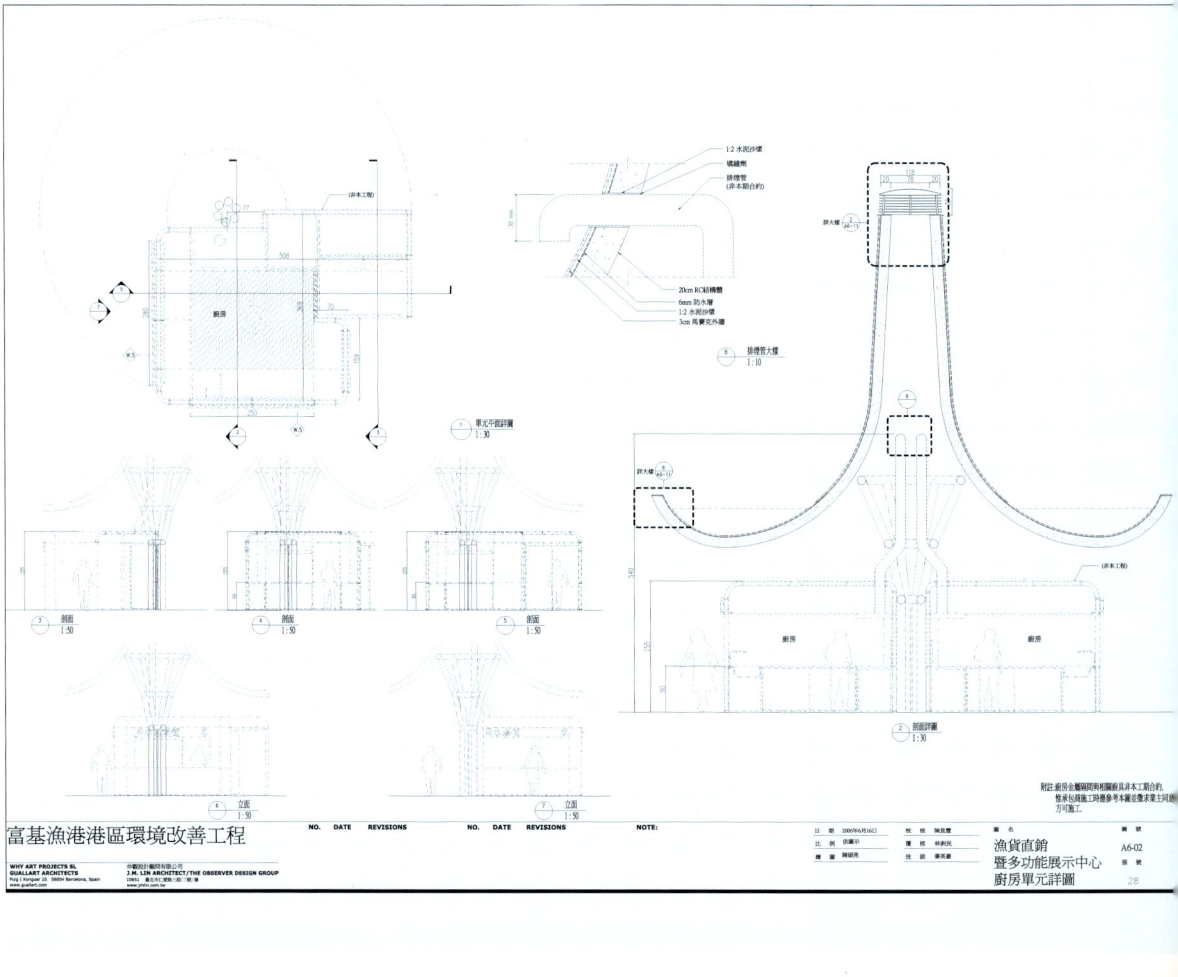
單元平面詳圖 1:30
排煙管大樣 1:10
剖面詳圖 1:30
剖面 1:50
剖面 1:50
剖面 1:50
立面 1:50
立面 1:50
富基漁港港區環境改善工程
NO. DATE REVISIONS
NO. DATE REVISIONS
NOTE:
WHY ART PROJECTS SL
GUALLART ARCHITECTS
J.M. LIN ARCHITECT/THE OBSERVER DESIGN GROUP
漁貨直銷
暨多功能展示中心
廚房單元詳圖
A6-02
28

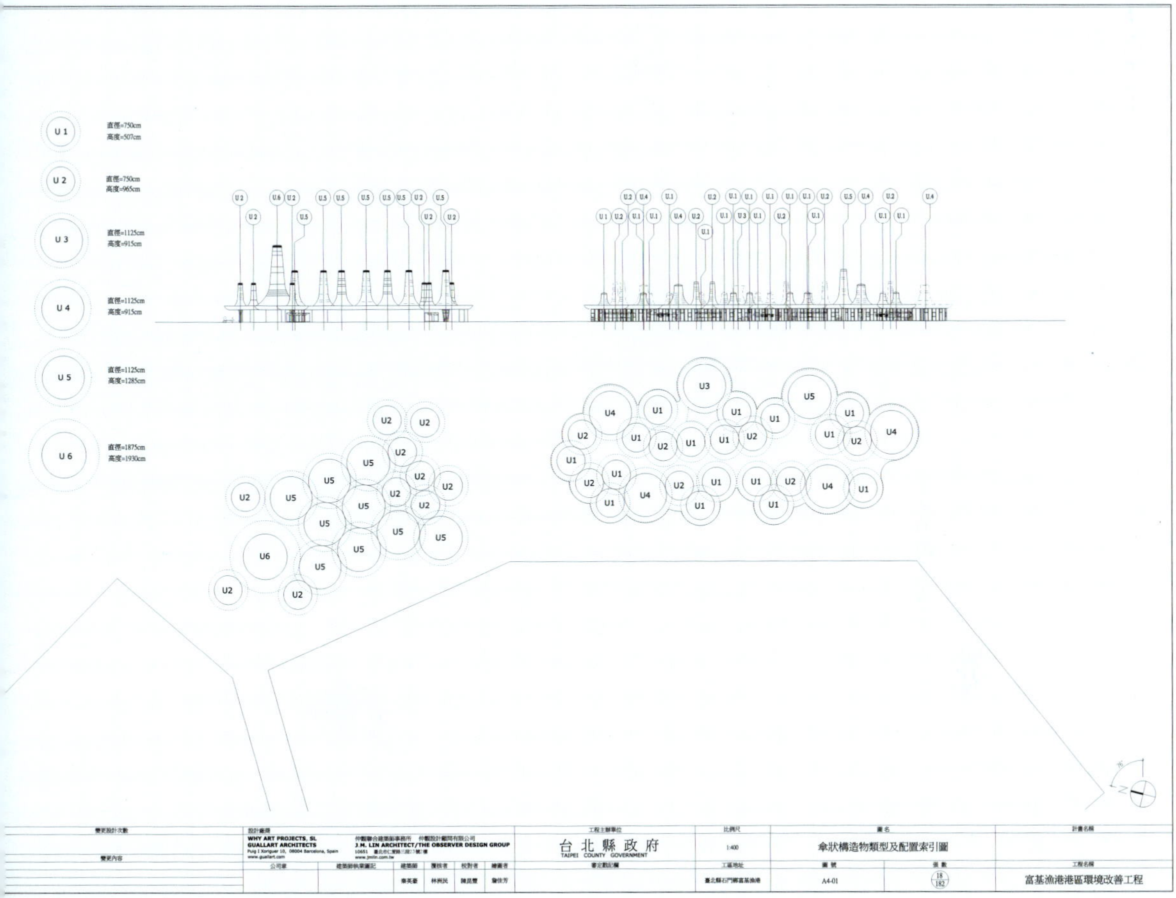
U 1 直徑=750cm 高度=507cm
U 2 直徑=750cm 高度=965cm
U 3 直徑=1125cm 高度=915cm
U 4 直徑=1125cm 高度=915cm
U 5 直徑=1125cm 高度=1285cm
U 6 直徑=1875cm 高度=1930cm
台 北 縣 政 府
TAIPEI COUNTY GOVERNMENT
1:400
傘狀構造物類型及配置索引圖
A4-01
富基漁港港區環境改善工程

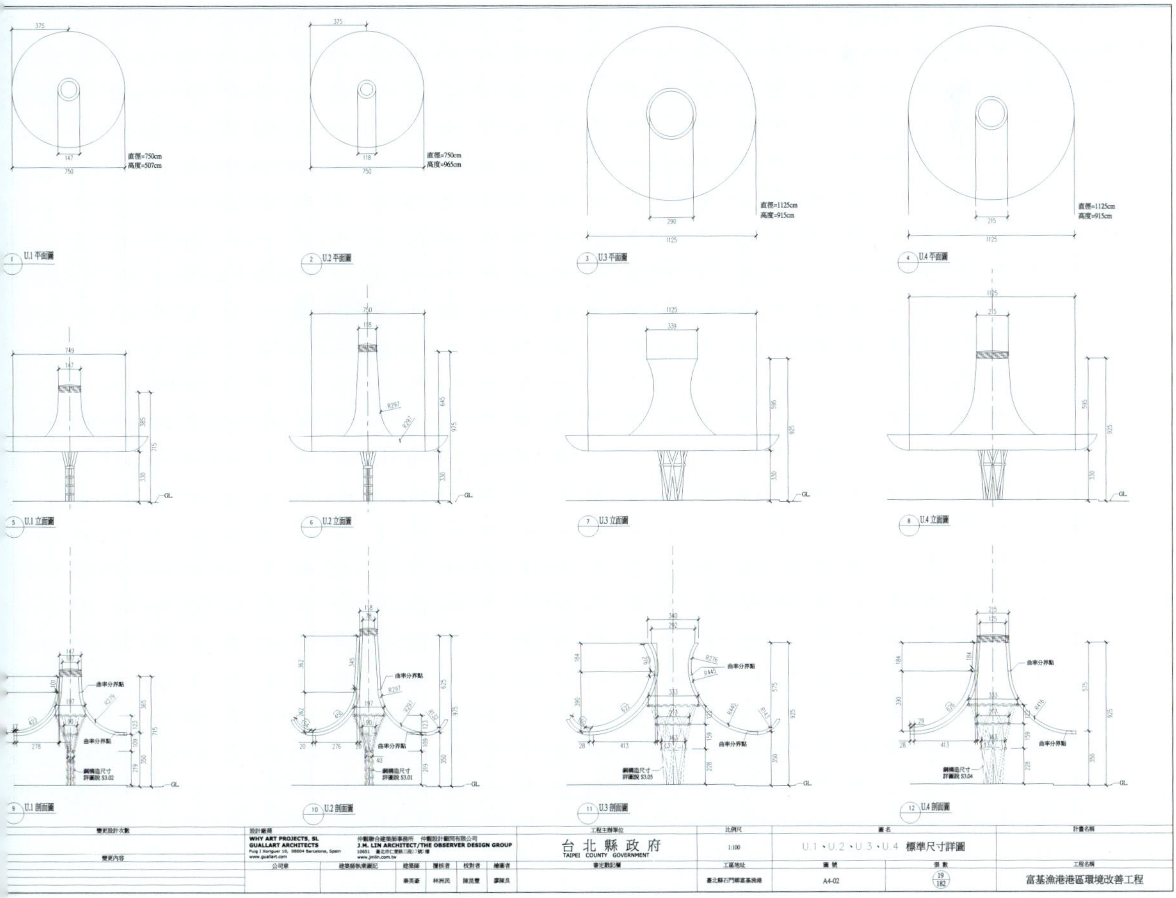
U.1 平面圖
U.2 平面圖
U.3 平面圖
U.4 平面圖
U.1 立面圖
U.2 立面圖
U.3 立面圖
U.4 立面圖
U.1 剖面圖
U.2 剖面圖
U.3 剖面圖
U.4 剖面圖
台 北 縣 政 府
TAIPEI COUNTY GOVERNMENT
1:100
U.1、U.2、U.3、U.4 標準尺寸詳圖
A4-02
富基漁港港區環境改善工程

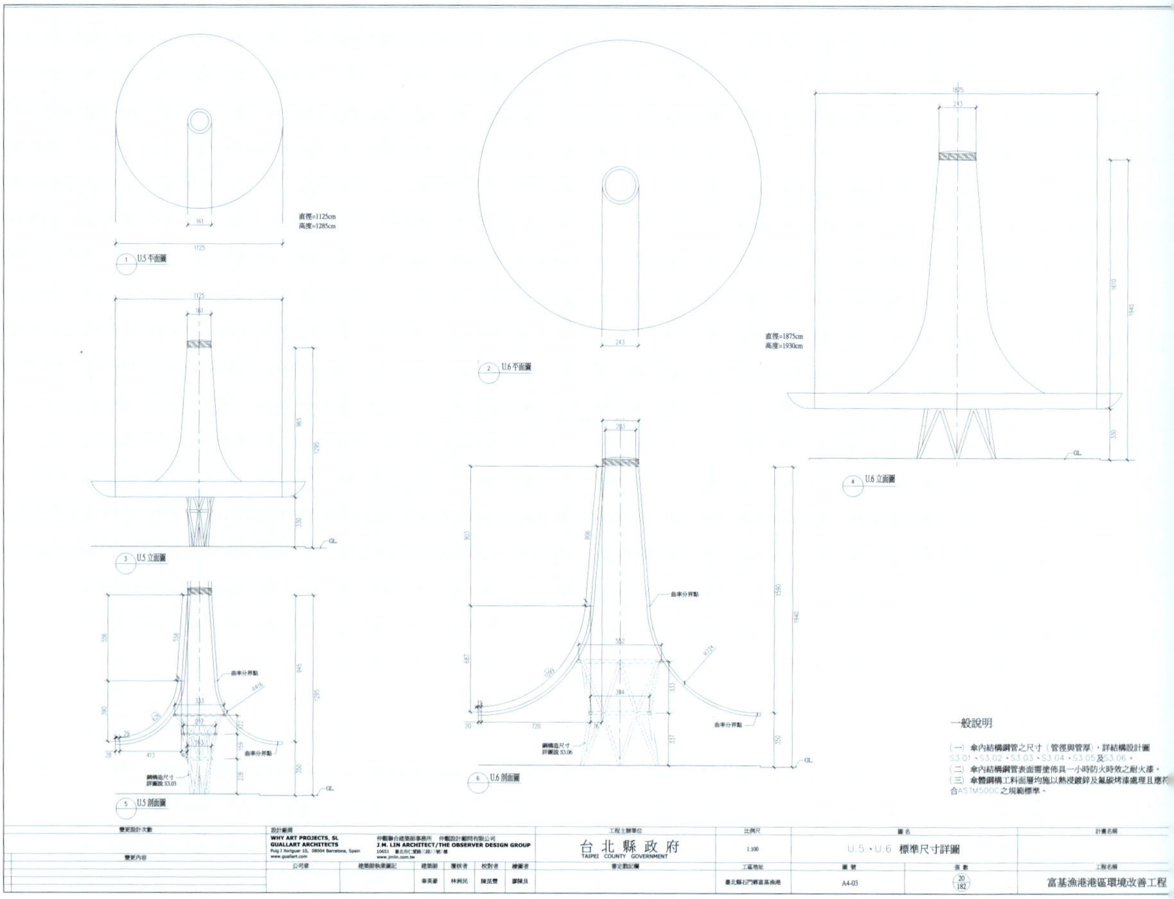
直徑=1125cm
高度=1285cm
1 U.5 平面圖
2 U.6 平面圖
直徑=1875cm
高度=1930cm
3 U.5 立面圖
4 U.6 立面圖
曲率分界點
鋼構造尺寸
詳圖說 S3.03
5 U.5 剖面圖
鋼構造尺寸
詳圖說 S3.06
6 U.6 剖面圖
GL
一般說明
(一) 傘內結構鋼管之尺寸(管徑與管厚),詳結構設計圖S3.01、S3.02、S3.03、S3.04、S3.05及S3.06。
(二) 傘內結構鋼管表面需塗佈具一小時防火時效之耐火漆。
(三) 傘體鋼構工料面層均施以熱浸鍍鋅及氟碳烤漆處理且應符合ASTM500C之規範標準。
WHY ART PROJECTS, SL
GUALLART ARCHITECTS
Puig i Xoriguer 10, 08004 Barcelona, Spain
www.guallart.com
仲觀聯合建築師事務所 仲觀設計顧問有限公司
J.M. LIN ARCHITECT/THE OBSERVER DESIGN GROUP
www.jmlin.com.tw
台北縣政府
TAIPEI COUNTY GOVERNMENT
1:100
U.5、U.6 標準尺寸詳圖
臺北縣石門鄉富基漁港
A4-03
20/182
富基漁港港區環境改善工程

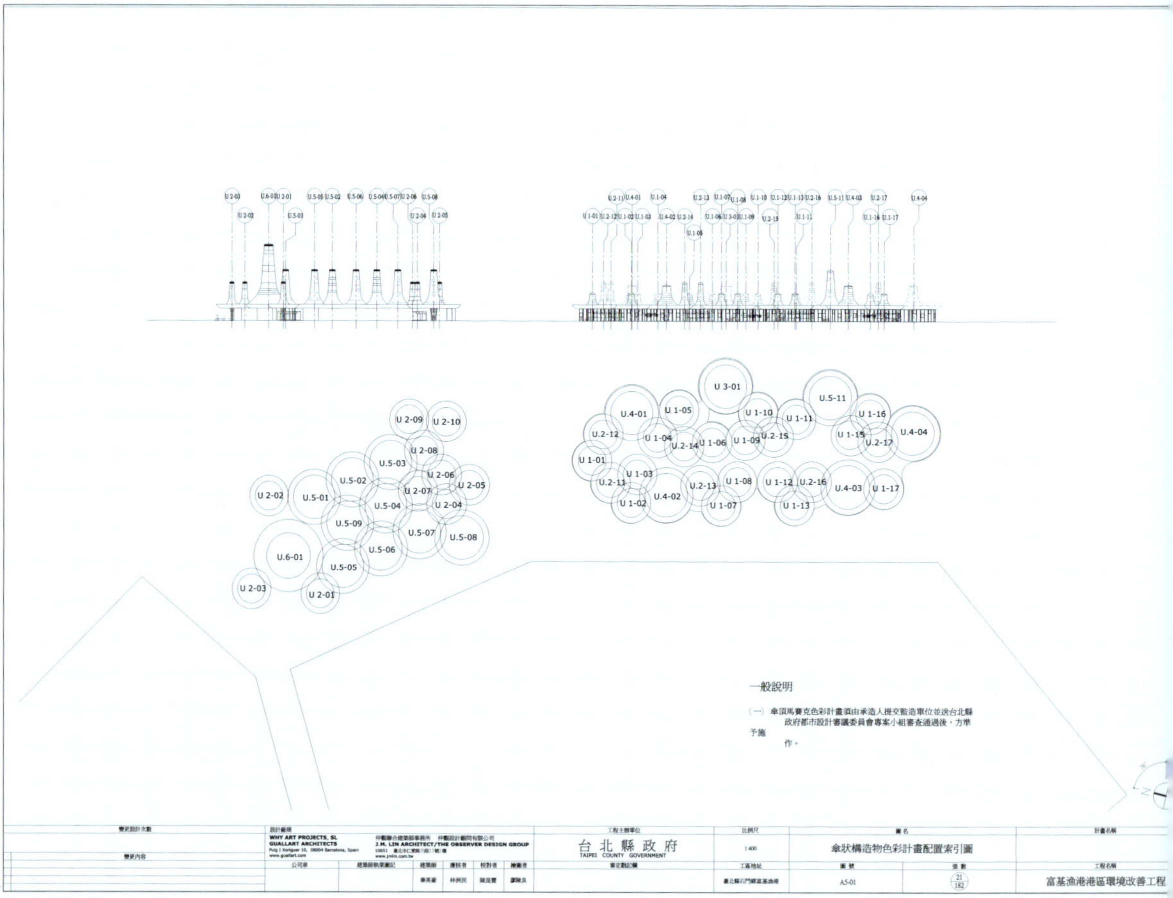
U.6-01
U.5-01
U.5-02
U.5-03
U.5-04
U.5-05
U.5-06
U.5-07
U.5-08
U.5-09
U 2-01
U 2-02
U 2-03
U 2-04
U 2-05
U 2-06
U 2-07
U 2-08
U 2-09
U 2-10
U 3-01
U.5-11
U.4-01
U 1-05
U 1-10
U 1-11
U 1-16
U.4-04
U.2-12
U 1-04
U.2-14
U 1-06
U 1-09
U.2-15
U 1-15
U.2-17
U 1-01
U 1-03
U.2-11
U 1-02
U.4-02
U.2-13
U 1-08
U 1-07
U 1-12
U.2-16
U 1-13
U.4-03
U 1-17
一般說明
(一) 傘頂馬賽克色彩計畫須由承造人提交監造單位並送台北縣政府都市設計審議委員會專案小組審查通過後,方準予施作。
WHY ART PROJECTS, SL
GUALLART ARCHITECTS
Puig i Xoriguer 10, 08004 Barcelona, Spain
www.guallart.com
仲觀聯合建築師事務所 仲觀設計顧問有限公司
J.M. LIN ARCHITECT/THE OBSERVER DESIGN GROUP
www.jmlin.com.tw
台北縣政府
TAIPEI COUNTY GOVERNMENT
1:400
傘狀構造物色彩計畫配置索引圖
臺北縣石門鄉富基漁港
A5-01
21/182
富基漁港港區環境改善工程

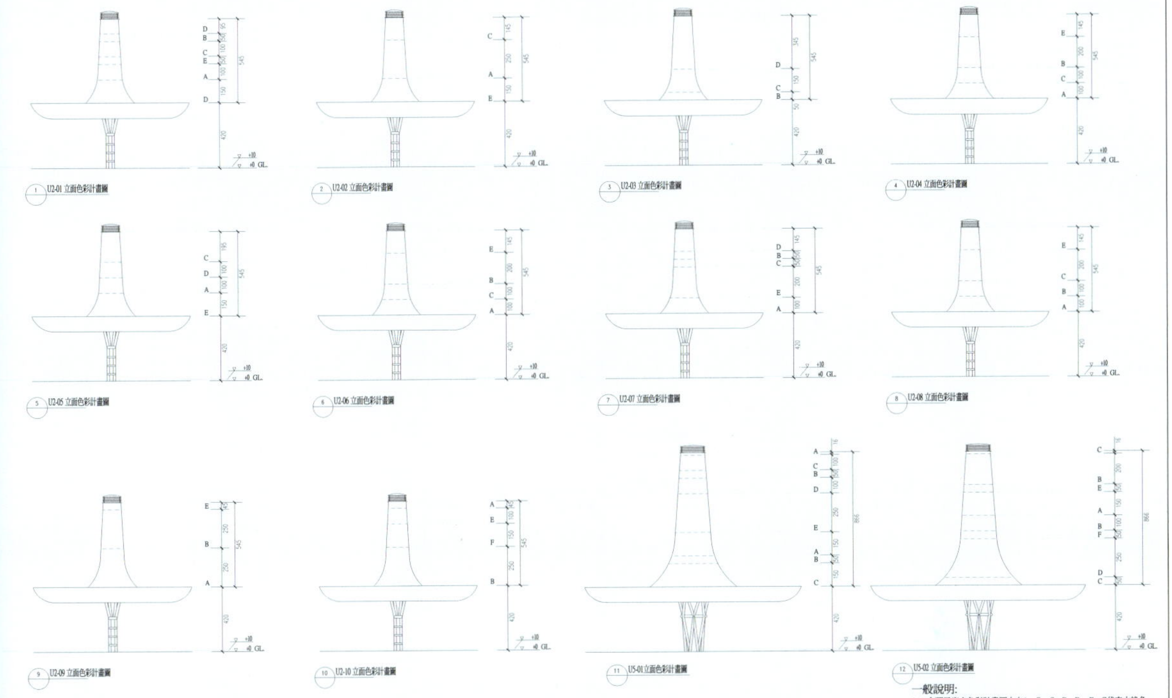

一般說明:

(一) 傘頂馬賽克色彩計畫圖中之A、B、C、D、E、F、G代表由綠色到黃色漸層之釉色，共計七色。承造人於施工前應送監造單位審核釉色及拼排方式，並提交台北縣政府都市設計審議委員會專案小組審查通過，方准予施作。

(二) 傘頂馬賽克圖磚之拼排，承造人需繪製製作圖並送監造單位審核。

(三)馬賽克備品顏色及儲放位置由業主指示。

變更設計次數	設計廠商	工程主辦單位	比例尺	圖名	計畫名稱
變更內容	WHY ART PROJECTS, SL GUALLART ARCHITECTS Puig I Xoriguer 10, 08004 Barcelona, Spain www.guallart.com / 仲觀聯合建築師事務所 仲觀設計顧問有限公司 J.M. LIN ARCHITECT/THE OBSERVER DESIGN GROUP 10651 www.jmlin.com.tw	台北縣政府 TAIPEI COUNTY GOVERNMENT	1:100	傘面馬賽克色彩計畫圖 (1)	
	公司章 / 建築師執業圖記 / 建築師 / 覆核者 / 校對者 / 繪圖者	審定戳記欄	工區地址	圖號 A5-02 / 張數 22/182	工程名稱 富基漁港港區環境改善工程

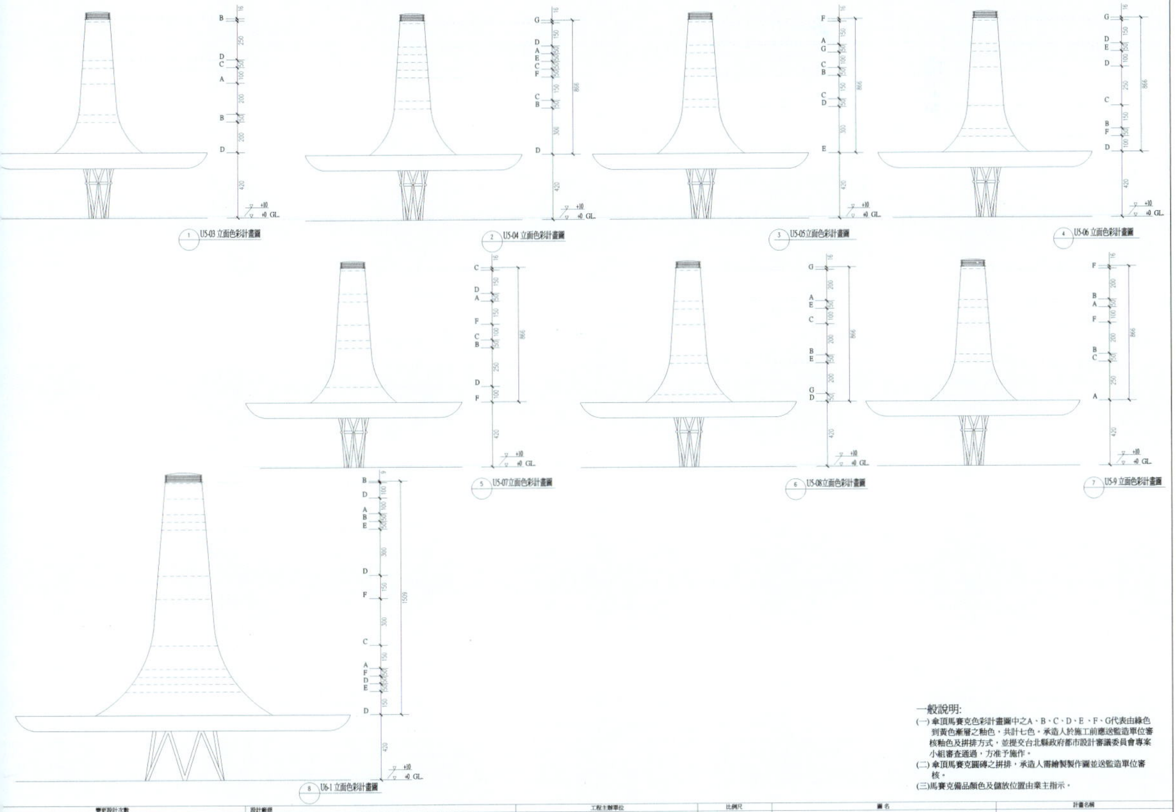

一般說明:

(一) 傘頂馬賽克色彩計畫圖中之A、B、C、D、E、F、G代表由綠色到黃色漸層之釉色，共計七色。承造人於施工前應送監造單位審核釉色及拼排方式，並提交台北縣政府都市設計審議委員會專案小組審查通過，方准予施作。

(二) 傘頂馬賽克圖磚之拼排，承造人需繪製製作圖並送監造單位審核。

(三)馬賽克備品顏色及儲放位置由業主指示。

變更設計次數	設計廠商	工程主辦單位	比例尺	圖名	計畫名稱
變更內容	WHY ART PROJECTS, SL GUALLART ARCHITECTS Puig I Xoriguer 10, 08004 Barcelona, Spain www.guallart.com / 仲觀聯合建築師事務所 仲觀設計顧問有限公司 J.M. LIN ARCHITECT/THE OBSERVER DESIGN GROUP 10651 www.jmlin.com.tw	台北縣政府 TAIPEI COUNTY GOVERNMENT	1:100	傘面馬賽克色彩計畫圖 (2)	
	公司章 / 建築師執業圖記 / 建築師 / 覆核者 / 校對者 / 繪圖者	審定戳記欄	工區地址	圖號 A5-03 / 張數 23/182	工程名稱 富基漁港港區環境改善工程

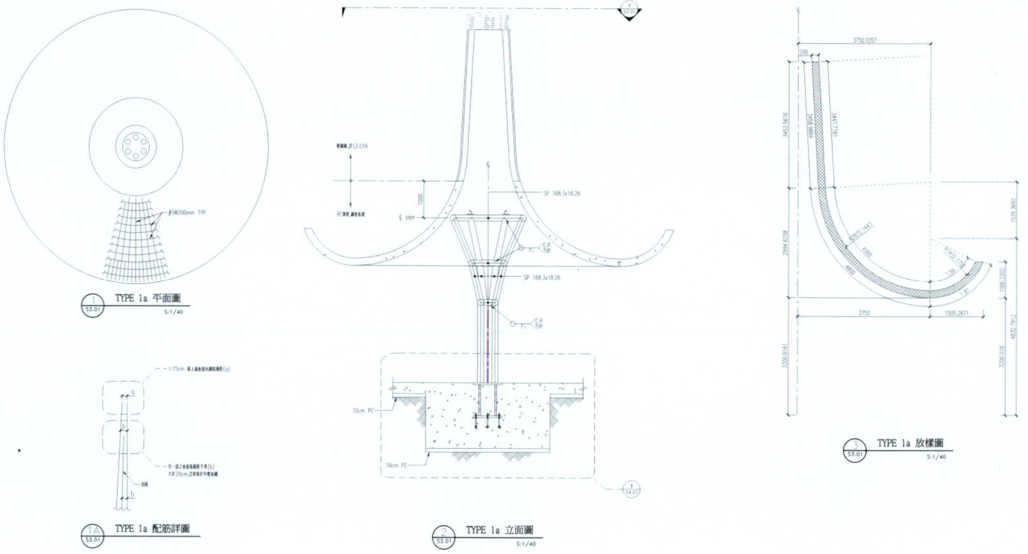

1:40 傘狀結構單元(TYPE 1a) 詳圖(一)

S3.01 20

Xref J:\9309\ENSTRU\DwFm.dwg

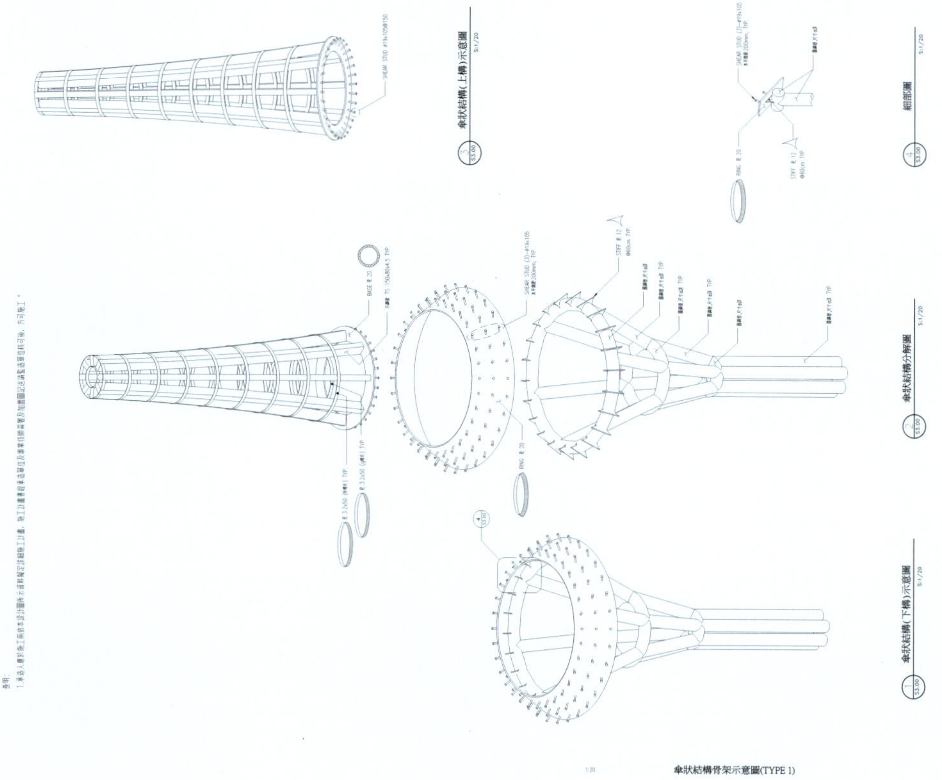

1:20 傘狀結構骨架示意圖(TYPE 1)

S3.00 17

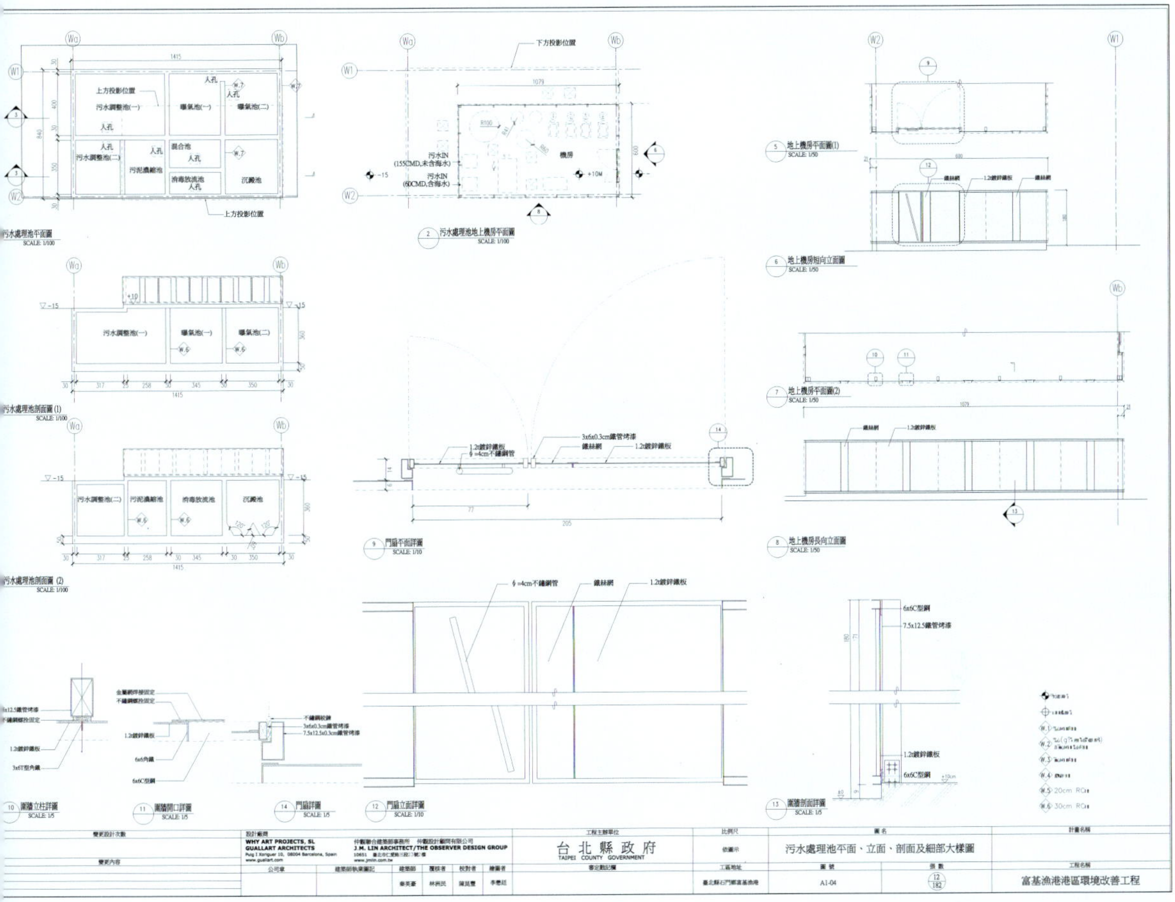
污水處理池平面圖
SCALE: 1/100
污水處理池地上機房平面圖
SCALE: 1/100
地上機房平面圖(1)
SCALE: 1/50
地上機房短向立面圖
SCALE: 1/50
污水處理池剖面圖 (1)
SCALE: 1/100
地上機房平面圖(2)
SCALE: 1/50
污水處理池剖面圖 (2)
SCALE: 1/100
門扇平面詳圖
SCALE: 1/10
地上機房長向立面圖
SCALE: 1/50
圍牆立柱詳圖
SCALE: 1/5
圍牆開口詳圖
SCALE: 1/5
門鎖詳圖
SCALE: 1/5
門扇立面詳圖
SCALE: 1/10
圍牆剖面詳圖
SCALE: 1/5
WHY ART PROJECTS, SL
GUALLART ARCHITECTS
J.M. LIN ARCHITECT/THE OBSERVER DESIGN GROUP
台北縣政府
TAIPEI COUNTY GOVERNMENT
污水處理池平面、立面、剖面及細部大樣圖
A1-04
富基漁港港區環境改善工程

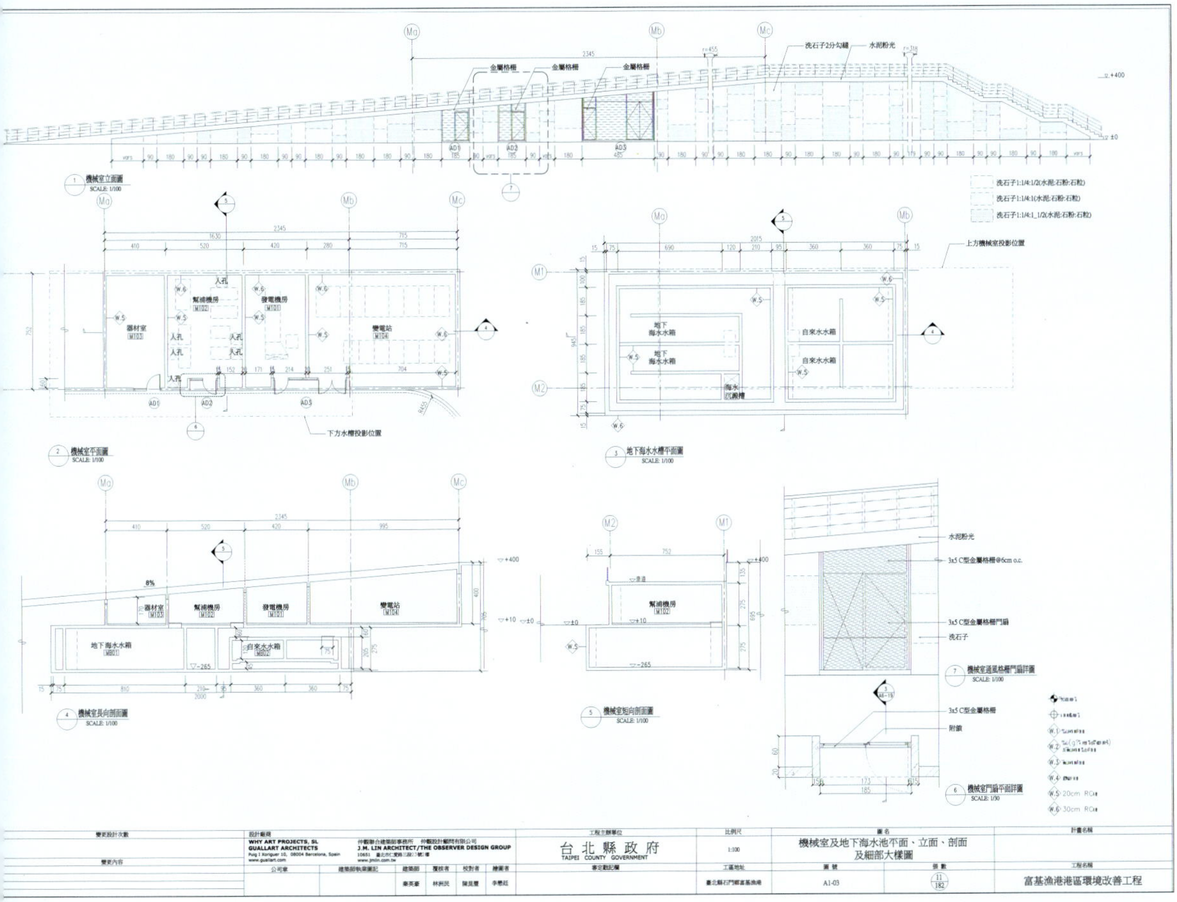
機械室立面圖
SCALE: 1/100
洗石子1:1/4:1/2(水泥:石粉:石粒)
洗石子1:1/4:1(水泥:石粉:石粒)
洗石子1:1/4:1_1/2(水泥:石粉:石粒)
機械室平面圖
SCALE: 1/100
地下海水水槽平面圖
SCALE: 1/100
機械室長向剖面圖
SCALE: 1/100
機械室短向剖面圖
SCALE: 1/100
機械室通風格柵門扇立面詳圖
SCALE: 1/100
機械室門扇平面詳圖
SCALE: 1/30
WHY ART PROJECTS, SL
GUALLART ARCHITECTS
J.M. LIN ARCHITECT/THE OBSERVER DESIGN GROUP
台北縣政府
TAIPEI COUNTY GOVERNMENT
機械室及地下海水池平面、立面、剖面
及細部大樣圖
A1-03
富基漁港港區環境改善工程

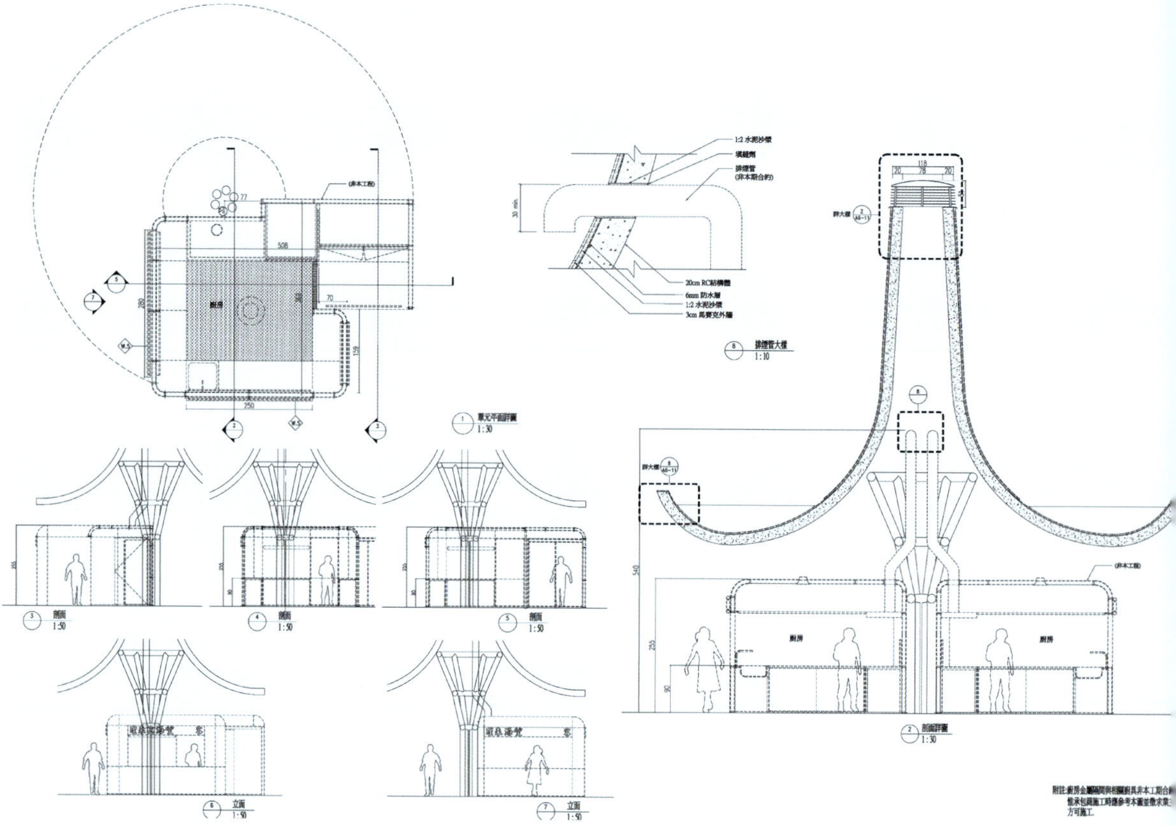
1:2 水泥沙漿
20cm RC結構體
6mm 防水層
排煙管大樣
1:10
單元平面詳圖
1:30
剖面
1:50
剖面
1:50
剖面
1:50
立面
1:50
立面
1:50
廚房
廚房
剖面詳圖
1:30

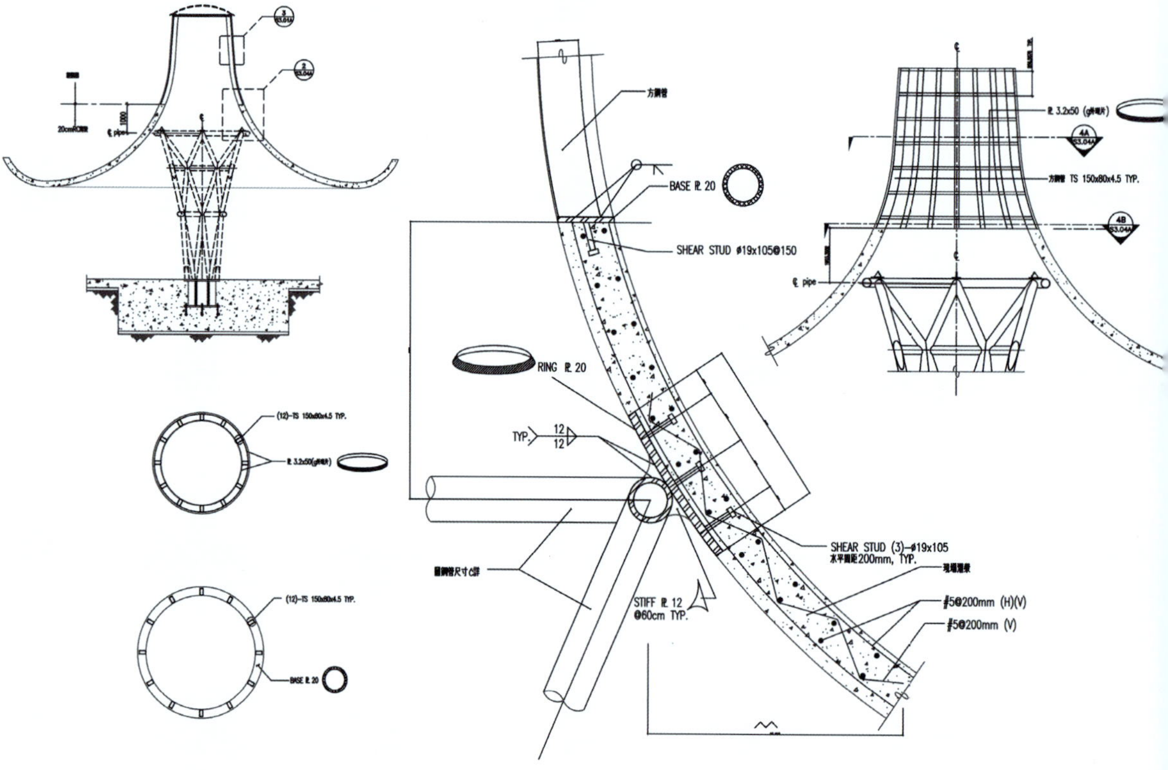
方鋼管
BASE PL 20
SHEAR STUD ø19x105@150
RING PL 20
TYP.
12
12
SHEAR STUD (3)-ø19x105
水平間距200mm, TYP.
現場灌漿
#5@200mm (H)(V)
#5@200mm (V)
STIFF PL 12
@60cm TYP.
PL 3.2x50
4A
S3.04A
TS 150x80x4.5 TYP.
4B
S3.04A
℄ pipe
(12)-TS 150x80x4.5 TYP.
(12)-TS 150x80x4.5 TYP.
BASE PL 20

J4

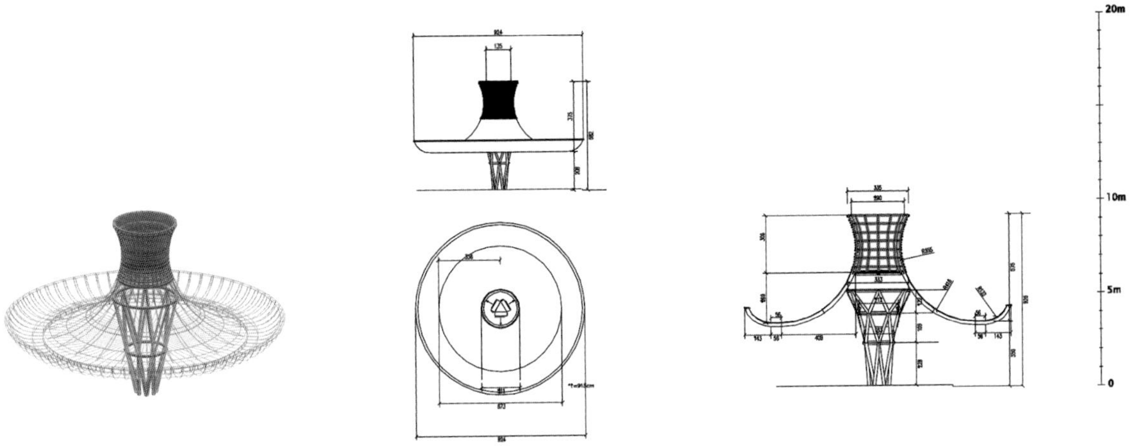

U5

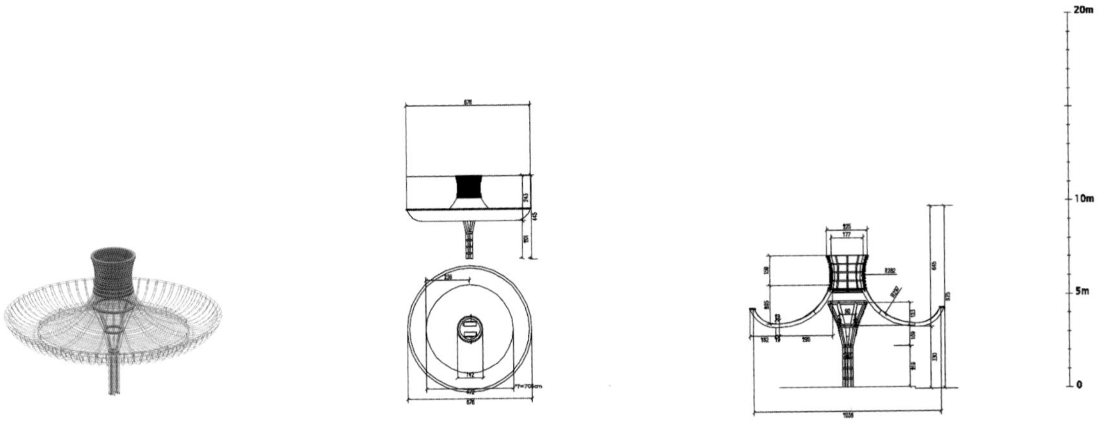

U6

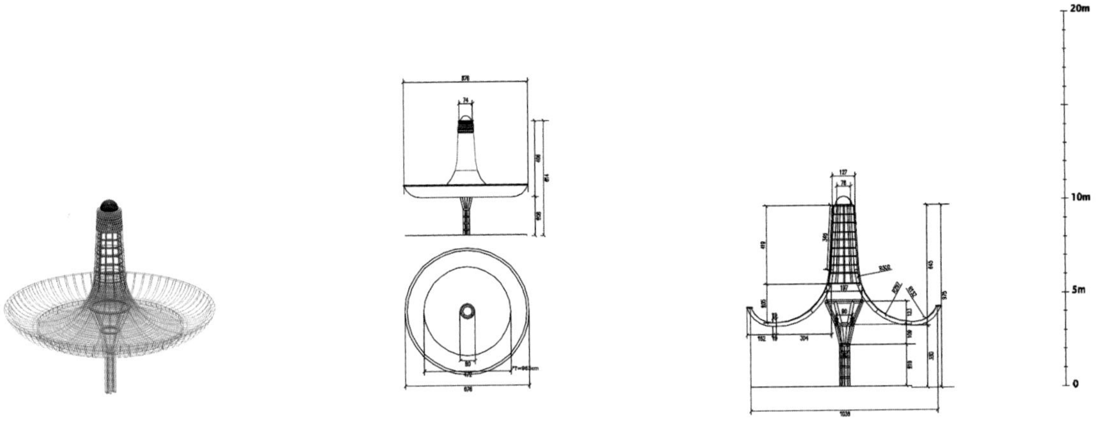

U1

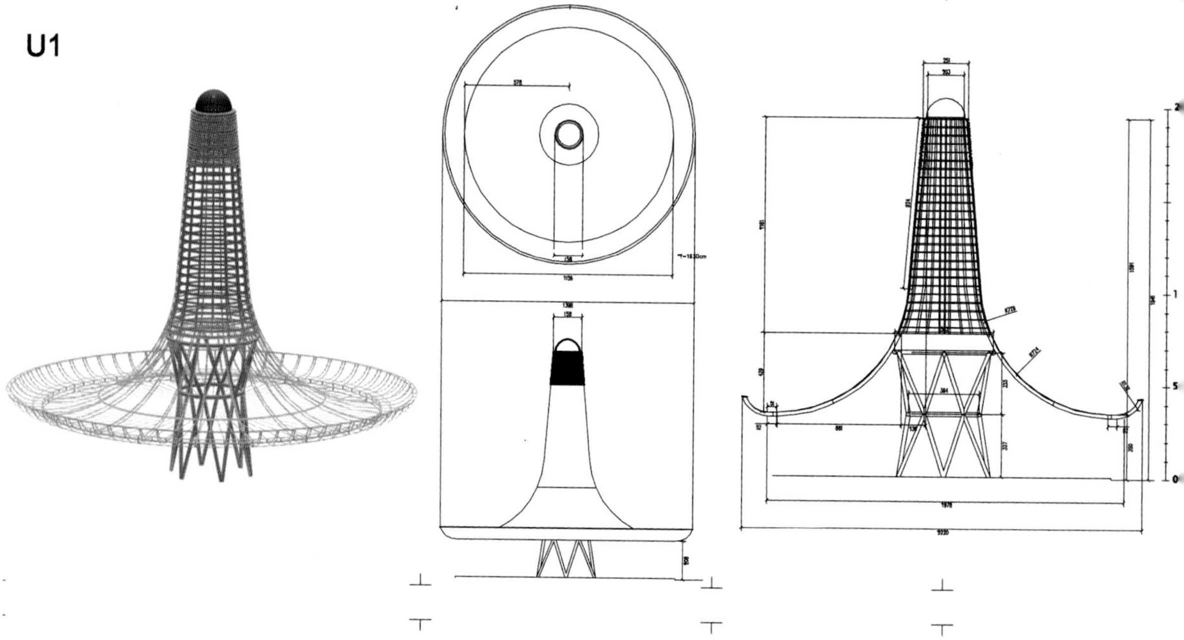

U2

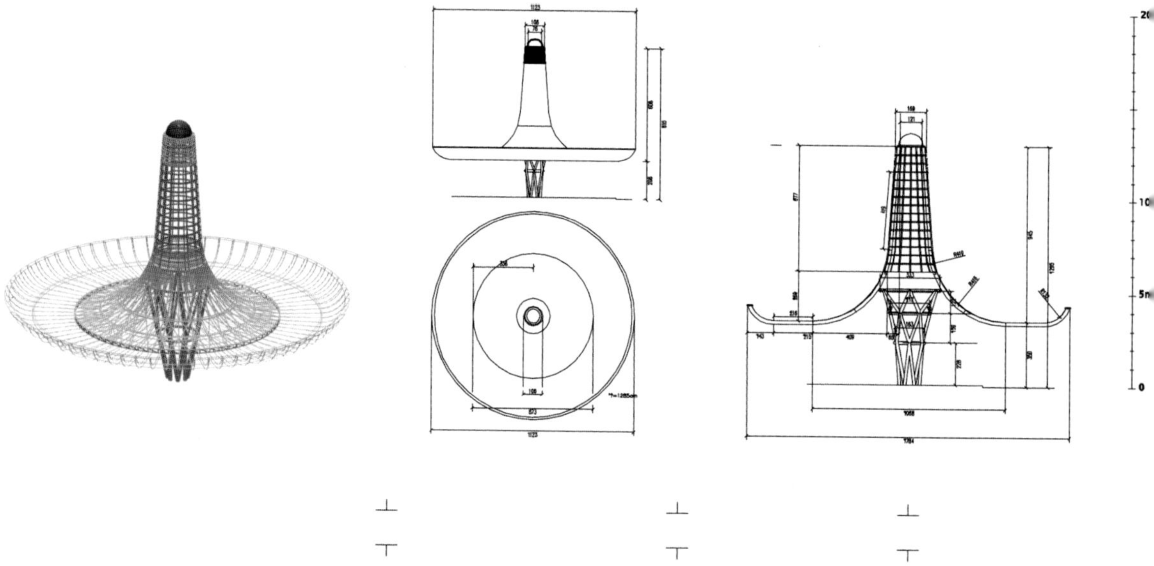

U3

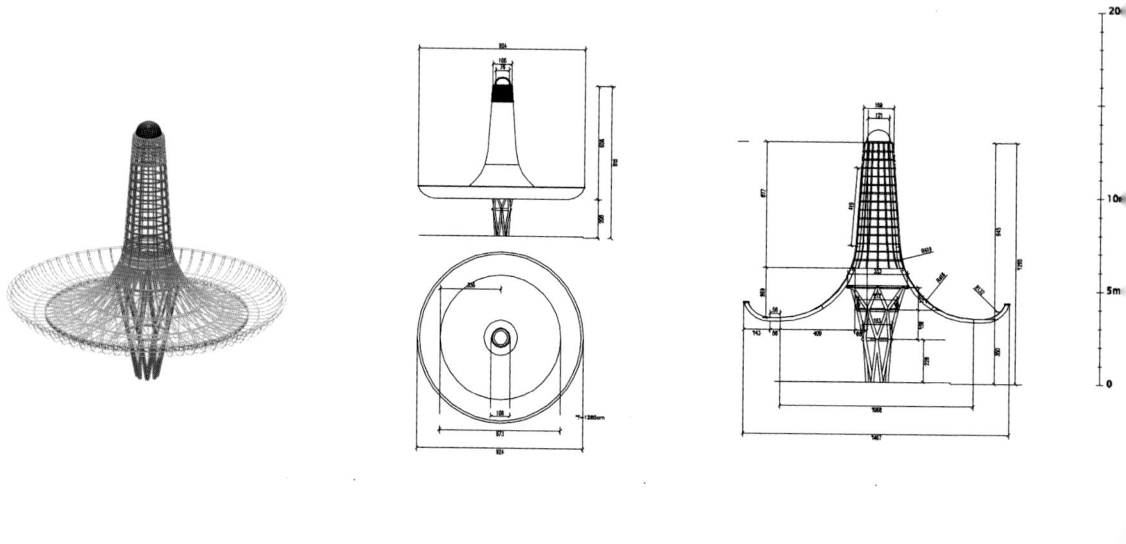

U1

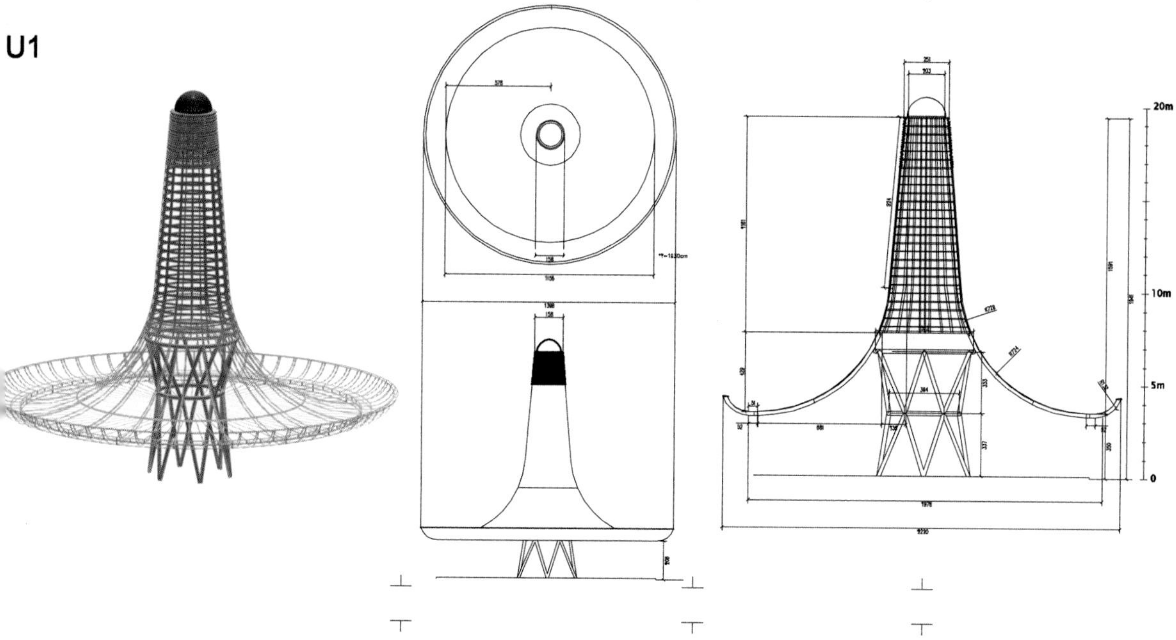

U2

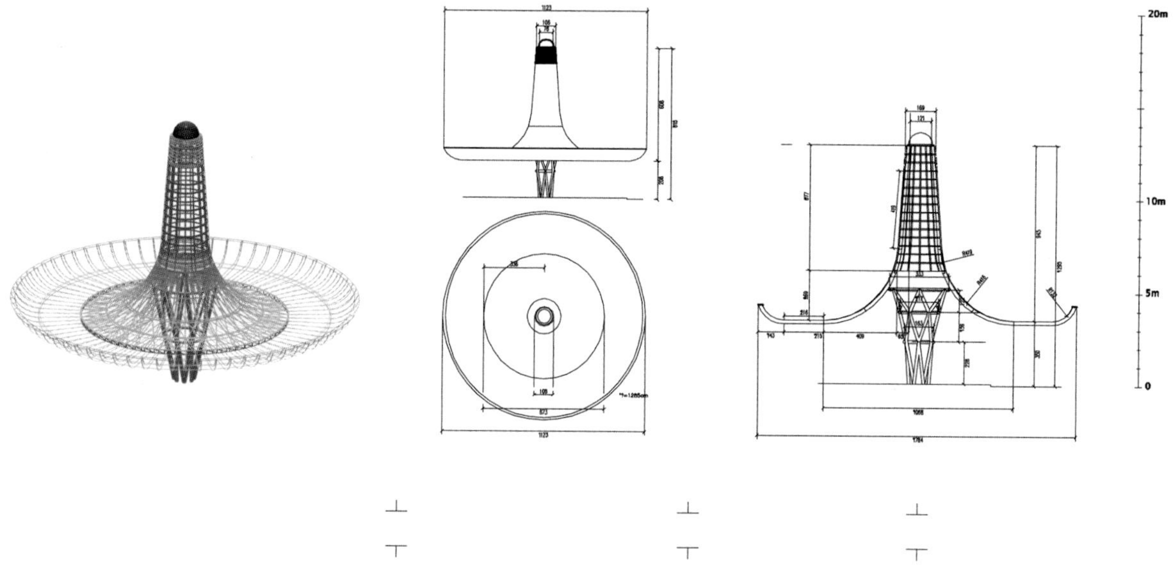

U3

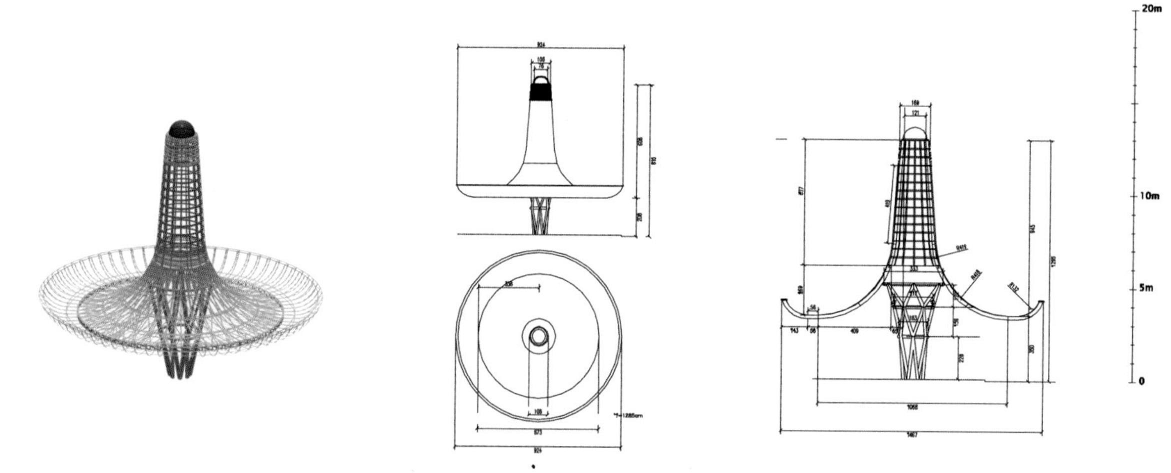

VINAROS SEA PROMENADE
VINAROZ滨海大道

The marine promenade is the urban center of Vinaroz. Is the place where the citizens gather and meet where the recognize clearly their own urban identity and where they relate to the ocean more often. It's the place where the dense and unconnected city, organizes and where the sea recuperates its unquestionable protagonist. And however it might be the encounter between the city with the beach and the sea, it's the most important and frequented space in the municipality.
The proposed arrangement solves the contradictions produced between the urban functions happening in the boarder line (specially the ones related with traffic and parking) as well as it's access condition to the beach and its condition of stay and relax.
The combination of this contradictory objectives can be solved, simple by giving the priority to the pedestrians, in the use of this space while submerging the automobiles to certain restrictions, in those zones where it's possible without bothering the general functioning of the site.
Majorly, the description and justification of the adopted decisions can be explained just by looking at the images of the project, which is why we are reducing this text to the explanation of the philosophical urban bases that stand the same.
The operation in reference is planed by a series of elements and pieces that are important to define:
-1. The type of floor convenient for the promenade -2. The compatibility between the different traffics -3. The need of generating spaces "relax islands". -4. The arrangement and preparation of the existing terraces, because they are considered indispensable for the Project. -5. The accesses to the beach. -6. The lightning systems. -7. The event's program.

滨海大道是Vinaroz的城市中心。它是市民们集会的地方，他们能够找到自己独特城市特质的地方，是他们与大海亲密接触的地方。在这里，密集而断裂的城市重新被整合起来，而海则重新成为不可置疑的主角。尽管它只是城市、海滩与海者交汇的产生物，却也无法磨灭其作为最为重要且被反复使用的市政设施。
我们提出的建议解决了城市功能相互交汇所产生的问题（尤其是那些与交通和停车相关的问题），同时为休憩和亲近海洋提供了良好的条件。
这些矛盾体的组合能通过将优先权赋予行人，同时在不影响这些区域功能的情况下将机动车控制在这些区域内来解决。已被采用的方案及其修正过程都可以通过方案图来了解，这就是为什么我们尽量通过精减后的文字来解释同样重要的城市学性基础。
项目的实施过程是通过一系列重要的元素和片断构成的：
1. 滨海大道所适用的地面类型。2. 不同种类交通工具之间兼容性。3. 生成"休憩"岛的需要。4. 由于现有台地对于该项来说是独立的，因此对于现有台地的处理和准备都是相当重的。5. 连接海滩。6. 照明系统。7. 重大事件项目。

A CONTINUOUS STONE SUPPORT

Wanting to give unity to the whole project, we proposed a continuous surface, of dark volcanic stone, that allows its use and abuse by pedestrians and cars. The chose of this material was not only taken because of it's physical characteristics and its easy maintenance but because it's an element from which we can build an identity program of Vinaroz against the rest of the Valencia's coast.
Above this surface we locate different Wood platforms that allows the stay areas to interact with the surrounding "chiringuitos" and the different access to the beach. These platforms are placed over the stone so that they can be easily changed; of place, organization and size.

CIRCULATION

Vinaroz is a population that was developed between the N 340 highway and the sea. These two territorial parallel elements had determined (and in certain cases limited) its growth and development, and, over time had created certain internal functional problems that reach their maximum expression in the ocean borderline because of the many and different activities that take place in one same spot.
The marine promenade it's a key element for the traffic mobility in the urban system of Vinaroz. To the marine promenade, of 780m long face almost 100 buildings, that require specific traffic needs, and at least 35 restaurants that need daily supply and broad access for their clients.
This is why even if the priority of the project is given to the pedestrians, we needed to keep the traffic circulation along the whole project, to give access to the houses and restaurants along the promenade. The existing parkings were kept and their access was improved and certain areas were kept for future parkings to satisfy the future need. Given the different urban configuration of the promenade in the south part (where we find parallel streets) and north part (where we find that all of the streets are perpendicular) we propose an underground parking in the northern part of the promenade (in case the budget will permit it) that can work as a traffic tunnel, that assures the traffic circulation at any time along the project.

TIME MANAGEMENT

The city is an organism that develops over time in a cyclic way.
The Mediterranean climate provides tour stations, that in a very specific way conditions the functioning of the territory along the year, specially in the locations where tourism is one of it's main incomes. The density reached during the summer drops dramatically over autumn, winter and spring, the good news is that this impact diminishes as the different Valencia's municipals improve their behavior durin the low seasons.
In any case, the differences still exist, and in the other hand the publics spaces have a very differen typology, the intensity and dynamisms acquired over the summer changes into a complete calmnes and tranquility during the rest of the year. This is why it's mandatory to project a marine promenade capable to adapt to the different use regimes and the different circumstances happening throughout the year. A marine promenade that, apart its basic functions, goes beyond a merely scenario conditio and becomes a wonderful flexible platform capabl of answering to a more wide demand of events an sporadic uses.

THE SECTION ARRANGEMENT

We propose a new section arrangement of the promenade , using the palm trees as the main stru tural element. This way, the section obtained will have the traffic circulation in the central part whil the northern and southern parts can be added in case the pedestrian flow diminishes and the need real.

URBAN GRID

GEOMETRY

The promenade is 792 meter long and a total surface of 30.500 squared meters. To organize this surface, we created a linear 3 meters grid that run equally along the surface.

MOBILITY

We drew a new mobility for the promenade, wher both, pedestrians and vehicles can run in a very compatible way. The urban attractions (school an market), the parkings, the restaurants, the square and the access to the beach are attractions at the same time natural to their surroundings and their centrality.

TRAM

Overlapping the circulation system with the linea grid, the trams orientate the space, and valorize their dimension in function to the principal movements that occur in the site. We have now a small grid, prepared for the pedestrian near to the beac and a bigger one as it approaches the city.

ELEMENTS

Over the promenade we find a series of elements a rhythmic way (44 units of 18 meters) in function of their need along the project. In this way they a organized from the main buildings, the spaces of stay for children, young people and families, the urban elements and the beach accesses.

续的石材支撑

了赋予这个项目整体性，我们提出采用一种由深色火山岩构成的连续表面来供行人和机动车随时使用。我们采用这的材料并不仅仅因为其所具有的物理特性和易于保养的特，它所带来的是一种能够将Vinaroz与巴伦西亚其他任何地区分开来的特性。在这个表面之上布置了木制平台来与周围chiringuitos进行互动，并提供接触海滩的不同方式。这平台被布置在岩石表面上，用以方便改变其位置、布局以及度。

环

naroz的人口主要聚居在N340高速路与大海之间。这两个乎平行的地域性元素决定了（或者在某种程度上限制了）这地区的发展，并且随着时间的推移，它们引发了一些内部功性的问题，以至于在同地区内人类的过多的活动达到了周边境所能忍受的最大限度。

海大道是vinaroz都市交通系统的关键元素。通向滨海大道780米长的道路上矗立着近100栋建筑物及其所带来的具体通需求，这其中还包括至少35家餐馆所需要的日常补给运需求及其来访客人的交通需求。

就是为什么这一项目将优先权赋予了行人，我们需要使整个目的交通保持循环来走进大道旁边的房屋和餐馆。现有的车场被保留了下来，而通向它们的道路则经过了改善，我们保留了一定预留区域以应对未来可能产生的停车需求。

了适应大道位于不同都市地貌的南方（在这里我们发现平行街道）和北方（所有街道正交），我们在北部提出了地下停的设想（如果预算允许），它将作为一条交通隧道来保证项在任何时间下的交通循环。

间设置

市是一种经年累月不断循环地发展的有机体。地中海气候以种非常具体的方式保证了这一地区一年之中的正常运行，尤是在那些以旅游业为主要收入的地区。夏天到来的密集人在其余的三个季节里却戏剧性地回落了，然而好消息是，这巨大的反差正随着瓦伦西亚地区的市政设施建设不断地得缓解。然而，差异依然是存在的，这使得公共空间的模式不地从人流密集、充满活力的夏天转换到余下的季节中宁静祥的模式。这就是为什么建设一条适应一年中不同功能要和环境的滨海大道成为了一种必要的行为。这条步道并不仅是一个我们构筑的特殊环境，而是一条除了承担基本的功以外，更能够适应各种活动和随机需求的木质平台。

剖面设置

我们对这条步道的剖面提出了一种新的设置想法，使用棕榈树作为其重要的结构元素。这样我们所获得的就是一个将交通循环设置于中央的剖面，而南北两侧则可以选择性的被使用，以应对步行人流减少的情况。

都市网格

几何

这条大道长792米，总面积为30500平方米。我们创造了三米网格，并将整个地表面平分，以便于管理和使用。

动态

我们将一种新的动态赋予了这条大道。在这里，行人和机动车可以和谐共处。而城市中的人流集中地（学校、市场）、停车场、餐馆、广场以及通向海滩的道路，在成为人流密集区的同时也和谐地融入其周边的环境当中。

有轨电车

有轨电车与线性网格上的交通循环系统交叠在一起，为空间提供导向，并从功能上将这些空间的维度稳定在能够恰当配合地段内主要活动的程度上。我们为靠近海滩的行人提供小型的网格，而在逐渐接近城市的地方逐渐扩大这些网格。

元素

在步道上我们更根据功能的需要创造了一系列具有韵律性的元素（44个18米的单元）。这样他们就能够环绕在主要的建筑物周围，成为儿童、年轻人和家庭聚集的地方，从而成为城市的元素和通向海滩的一种方式。

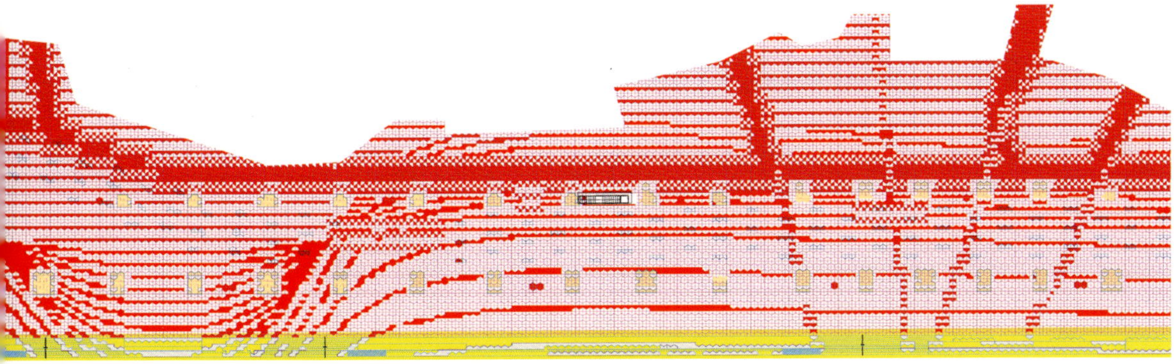

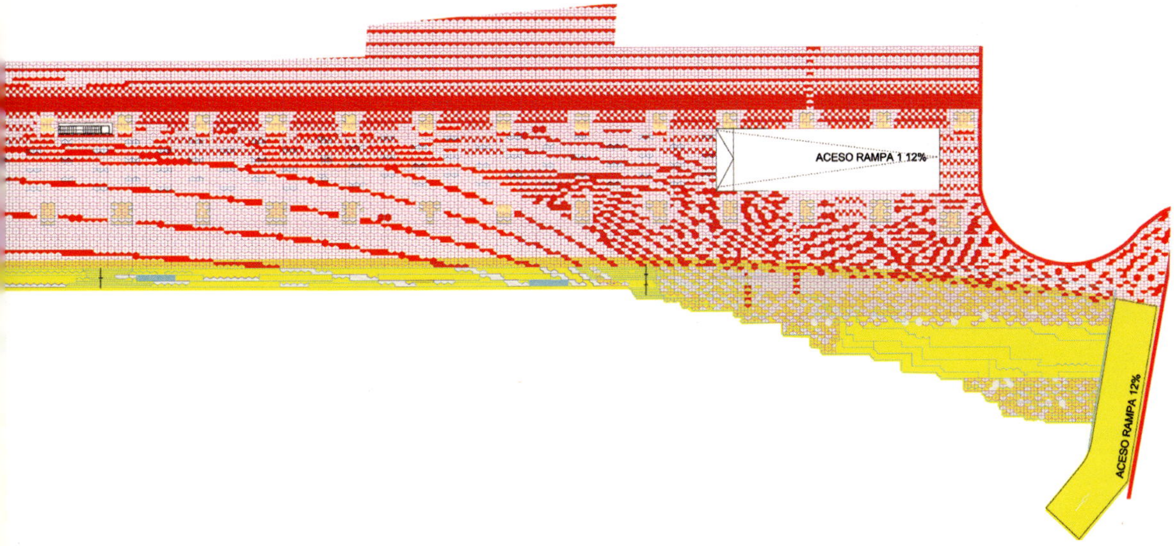
ACESO RAMPA 1 12%
ACESO RAMPA 12%

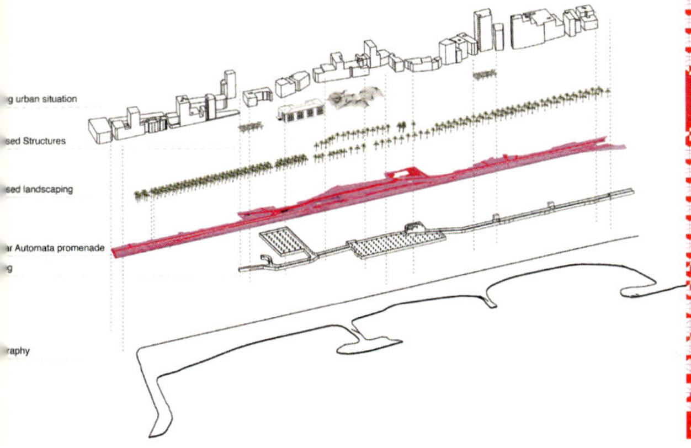
urban situation
Structures
landscaping
Automata promenade

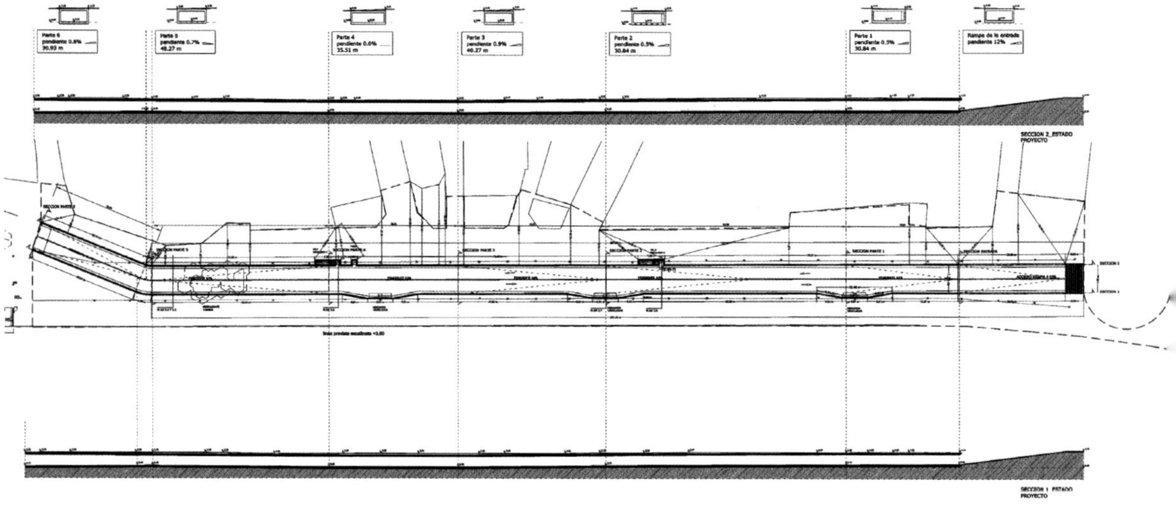
Parte 6
pendiente 0.8%
30.93 m
Parte 5
pendiente 0.7%
48.27 m
Parte 4
pendiente 0.0%
35.51 m
Parte 3
pendiente 0.9%
40.27 m
Parte 2
pendiente 0.5%
30.84 m
Parte 1
pendiente 0.5%
30.84 m
Rampa de la entrada
pendiente 12%
SECCION 2_ESTADO PROYECTO
SECCION 1_ESTADO PROYECTO

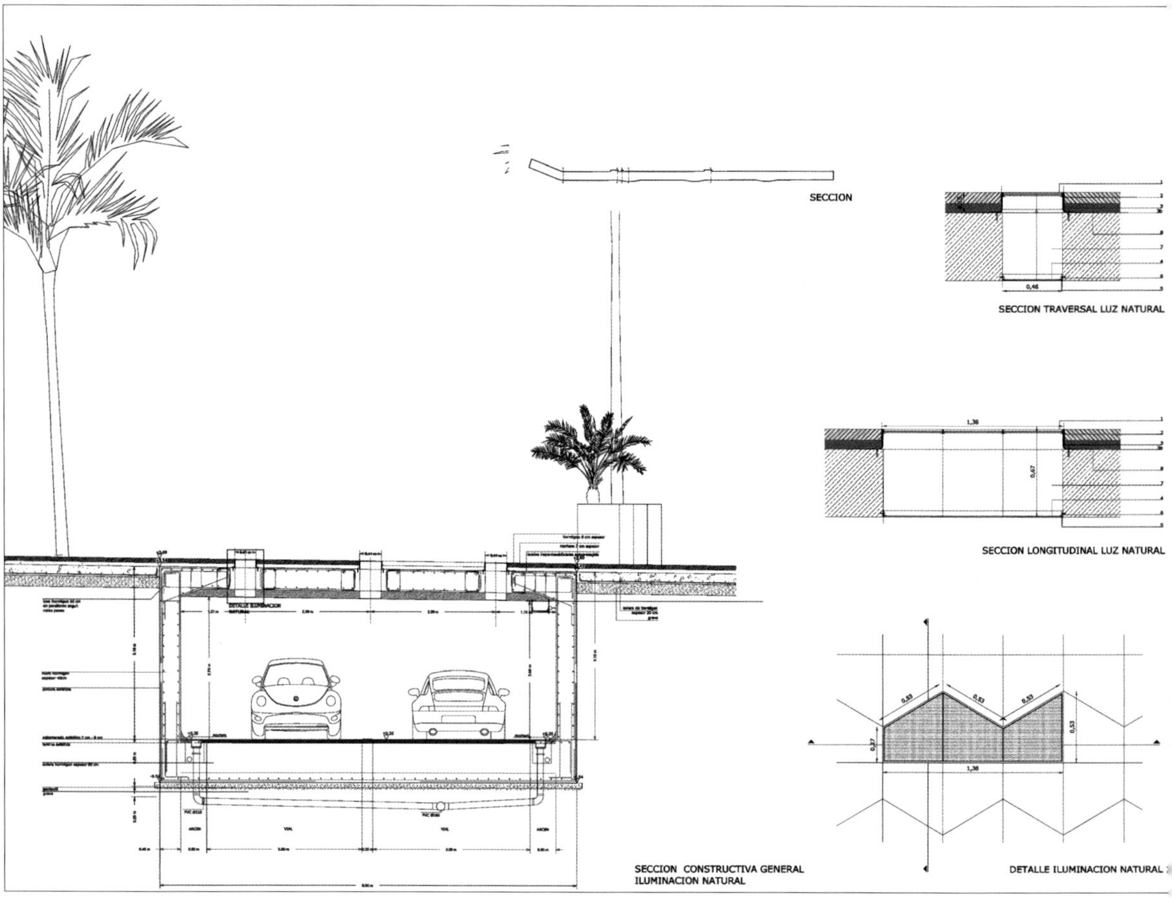
SECCION
SECCION TRAVERSAL LUZ NATURAL
0,46
1,38
0,67
SECCION LONGITUDINAL LUZ NATURAL
0,53
0,37
1,38
SECCION CONSTRUCTIVA GENERAL
ILUMINACION NATURAL
DETALLE ILUMINACION NATURAL

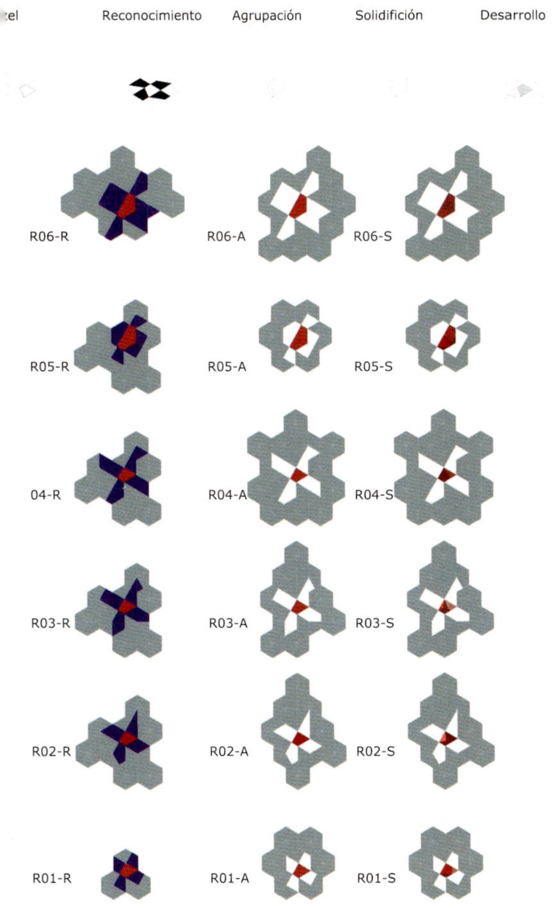
el
Reconocimiento
Agrupación
Solidifición
Desarrollo
R06-R
R06-A
R06-S
R05-R
R05-A
R05-S
04-R
R04-A
R04-S
R03-R
R03-A
R03-S
R02-R
R02-A
R02-S
R01-R
R01-A
R01-S

R-01
b=1
L=1
g=1a
nc=8
R-02
b=1
L=1
g=1a
nc=8
R-03
b=2
L=2
g=1a
nc=9
R-04
b=2
L=2
g=1a
nc=11
R-05
b=4
L=2
g=1a
nc=12
R-06
b=1
L=2
g=2a
nc=8
R-07
b=1
L=2
g=2a
nc=8
R-08
b=2
L=2
g=2a
nc=9
R-09
b=4
L=3
g=2a
nc=11
R-10
b=1
L=2
g=2a
nc=8
R-11
b=1
L=2
g=3a
nc=28

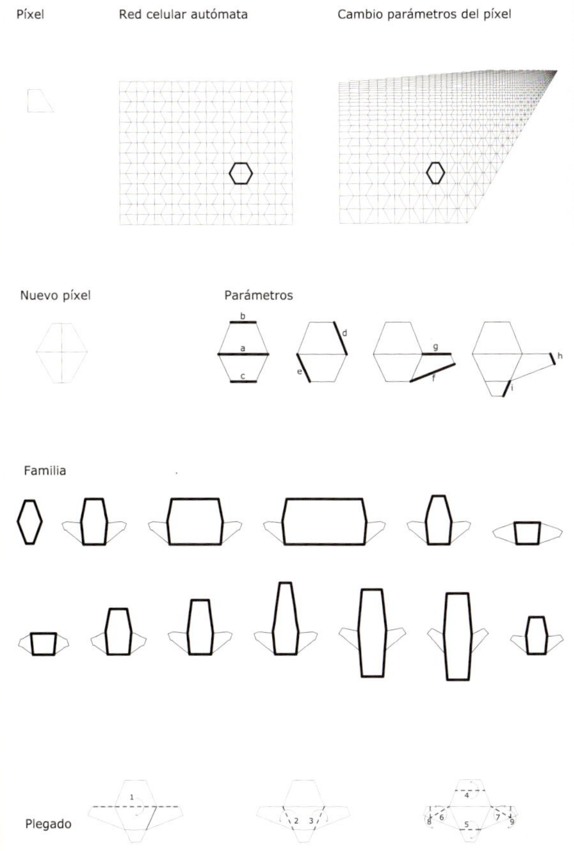
Píxel
Red celular autómata
Cambio parámetros del píxel
Nuevo píxel
Parámetros
Familia
Plegado

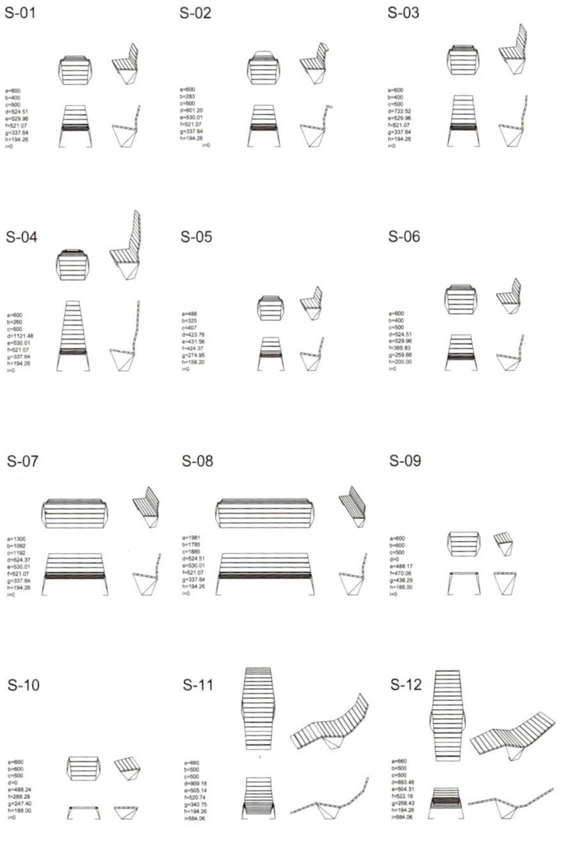
S-01
S-02
S-03
S-04
S-05
S-06
S-07
S-08
S-09
S-10
S-11
S-12

ELEMENTS IN THE PROMENADE

"Chiringuitos" and toilets

The "chiringuitos" are ephemeral constructions located along the promenade. They have little bars, stores or information stands. They are made of a metal structure and a wood covering and a textile top generates a huge shadow in its interior. These "chiringuitos" are located near the beach accesses to avoid constructions in the beach itself. Certain "chiringuitos" incorporate Toilets.

Terraces

In front of the existing restaurants the installation of terraces over a wood platform installed over the rock floor is allowed. On top of these platforms its possible to protect with umbrellas over the summer and the light plastic tents that allow the use of these terraces over the winter.

Relax islands

Over the hard floor surface, we placed a series of wood platforms, that act like islands in the dynamic promenade's surrounding. These islands (hexagonal wooden platforms) have different structures that depending to their adopted form they will allow the pedestrian to either sit down or lay down. These islands can be either found individual or in groups to reinforce the human interaction.

Kids playgrounds

We find these playgrounds in the same hexagonal forms but this time they have been subtracted from the surface, creating a soft surface (earth) over which we place modern wood and steal games that allow the dynamic interaction with the objects.

Accesses

The height from the promenade until the beach varies from 20 cm in the southern part and 2.10 m in the northern part. To overcome this distance we've placed 7 accesses made with wood platforms that allow different access ramps (specially conceived for handicaps and children) and stairs.

Showers / fountains

Next to the ramp accesses of the beach we place groups of showers, placed in a way that once you shower you will not have any contact with the sand again. They are places over hexagonal units and some other ones are equipped with fountains. We can find them next to the accesses or kids playgrounds.

Lightning

Following the palm trees alignment we placed a series of structures that incorporates the lightning of the promenade and the beach. It was on purpos that these structures were not placed in the borde of the promenade and the beach, we didn't want to block this limit we wanted it completely clear.

Structure

In the central part of the promenade, facing the old schools we located the bigger structure. A very light metallic structure that cover itself with a sem transparent textile cloth. This space permits the celebration of different events, concerts and show in an spontaneous and open way . As this structure is not closed and in absence of an spectacle, the pedestrian simply crosses this conceived space as another face of the promenade.

Materiality

The promenade will have an abstract and sensual materiality. The transition between the spontaneous and free beach space and the city, happens in the promenade, a space that allows being bare foo in the city, lay down, encounter a game of shadows and have the conditions to enjoy the Mediterranea climate.

Stone

We are using a basaltic volcanic stone of various sizes respecting the longitudinal joint. In the central part of the promenade they have a thickness o 7 cm to allow the passage of the vehicles and in th lateral parts a thickness of 3.5 mc.

Wood

We are using the Ipe wood with a special treatmen Measures 12 x 150 cm. Thickness 3 cm in the walkin parts and a thickness of 6 cm in the relax areas an benches.

海大道上的元素

Chiringuitos"和卫生间

Chiringuitos"是一种沿步道设置的临时构筑物。它们设小型的吧台、商店和信息问讯设施。这些设施采用金属结构木质外壳，其巨大的帆布顶棚在室内生成了巨大的阴影。这"Chiringuitos"布置在通向海滩的道路上来防止人们在滩上面胡乱搭建。某些"Chiringuitos"包含了卫生间。

地

岩石之上的木质平台上，现有餐厅的前面设置了一座台地装。在这些台地的平台表面上，夏天可以张开遮阳伞，而冬天可以在轻质塑料的帐篷下举办活动。

憩岛

们在硬质地面上设置了一系列的木质平台，使得它们宛如充活力的滨海大道周围散落的岛屿一般。这些小岛（六角形木平台）在不同的形态下有着不同的结构，而行人则可以在其或坐或卧，进行休息。这些小岛可能独立存在，也可能以组的形式来加强人们之间的互动。

童游乐场

些游乐场同样是六角形的，但是这次它们从地面上被提取出，创造了一种柔软的表面（土地），我们在上面铺设了现代的质和钢质的游乐设施，使得儿童与环境之间也充满了互动的能性。

通道

步道与海滩之间的高差从南边的20厘米变化到北边的2.1米。为了减缓这样的差异，我们设置了七条带有木质平台的通道通向不同斜度的坡道（尤其是为残障人士和儿童着想）和台阶。

淋浴/喷泉

紧邻通向海滩的坡道的地方，设置有成组的淋浴设施，这样游客在淋浴之后就不必再和沙子产生任何接触。这些提供淋浴的场地都是六角形的，同时其中一部分的淋浴设施被喷泉所代替。而这些设施的旁边就是通往儿童游乐场的通道。

照明

沿着棕榈树排布的走向，我们布置了一系列将滨海大道与海滩的照明结合在一起的装置。我们有意识地避免将这些装置设置在滨海大道与海滩的交界处，因为我们不想在两者之间创造任何阻碍。

装置

在滨海大道的中央部分面对学校的地方，我们设置了一个更大的装置——一个半透明材质包裹下的轻型金属装置。这个空间以一种独立而开放的形式来进行不同的庆典、集会、音乐会以及展示会等等。而当这个装置并未完全封闭或是不进行展示活动时，行人们则可以像穿过滨海大道的另一个部分一样从这里穿行。

物质性

滨海大道将具有一种抽象而感性的物质性。自由而自给自足的海滩与城市之间的过渡就发生在海滨大道，这是一个使在城市中赤足成为可能的空间，你将有机会随意地躺下与阳光、树影一起享受地中海气候。

石材

我们根据纵向交界点的情况使用不用尺寸的火山玄武岩。在海滨大道的中央，厚度为7厘米的石材令机动车的行驶成为可能，而两边的部分则只有其一半的厚度。

木材

我们用一种特殊的方式来使用Ipe木材。量取12厘米X150厘米，厚度为3厘米的木材来构成步行部分，而休息区域和休息椅则采用厚度为6厘米的木材。

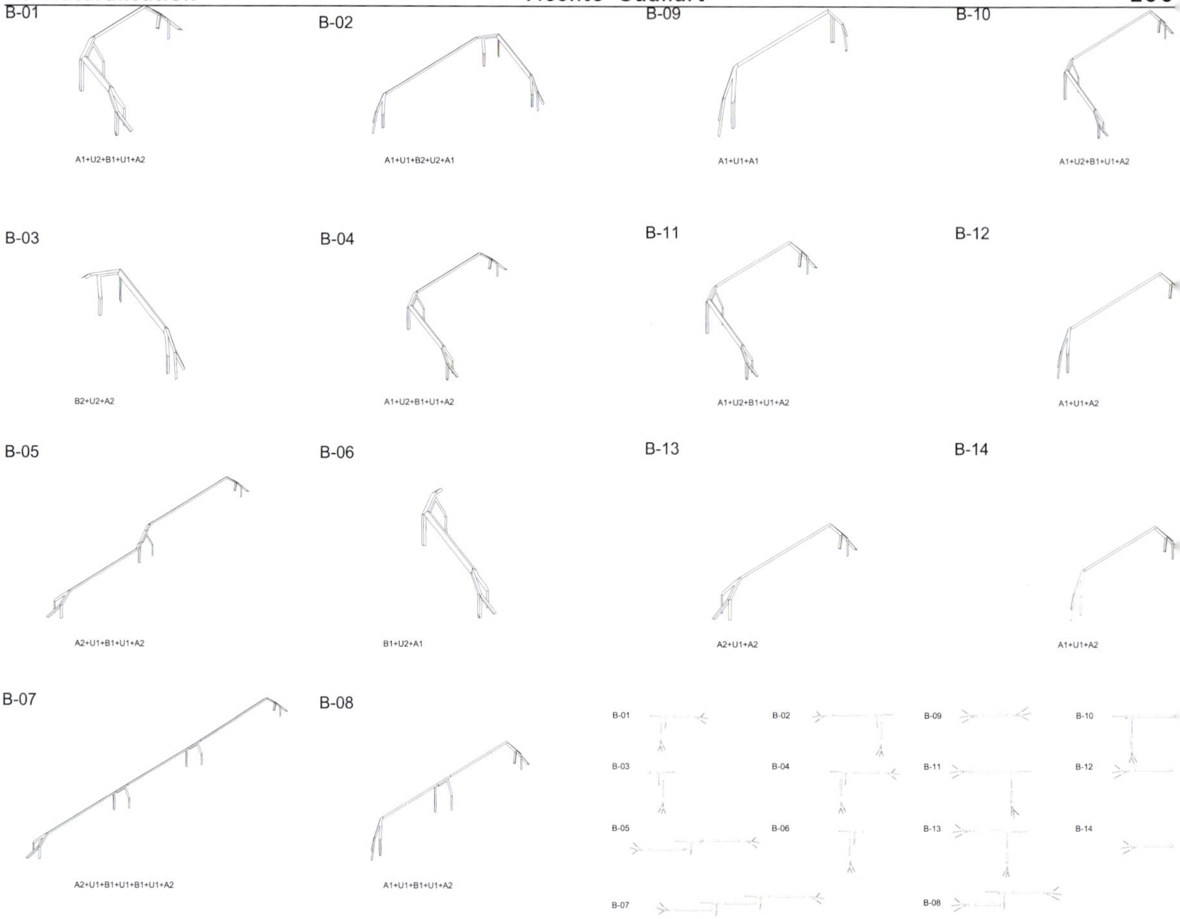
B-01
A1+U2+B1+U1+A2
B-02
A1+U1+B2+U2+A1
B-09
A1+U1+A1
B-10
A1+U2+B1+U1+A2
B-03
B2+U2+A2
B-04
A1+U2+B1+U1+A2
B-11
A1+U2+B1+U1+A2
B-12
A1+U1+A2
B-05
A2+U1+B1+U1+A2
B-06
B1+U2+A1
B-13
A2+U1+A2
B-14
A1+U1+A2
B-07
A2+U1+B1+U1+B1+U1+A2
B-08
A1+U1+B1+U1+A2
B-01
B-02
B-09
B-10
B-03
B-04
B-11
B-12
B-05
B-06
B-13
B-14
B-07
B-08

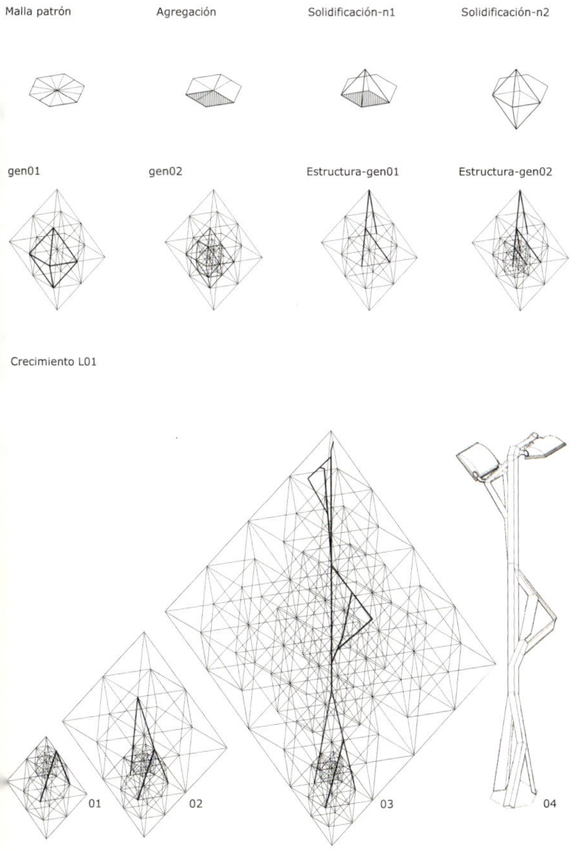
Malla patrón
Agregación
Solidificación-n1
Solidificación-n2
gen01
gen02
Estructura-gen01
Estructura-gen02
Crecimiento L01
01
02
03
04

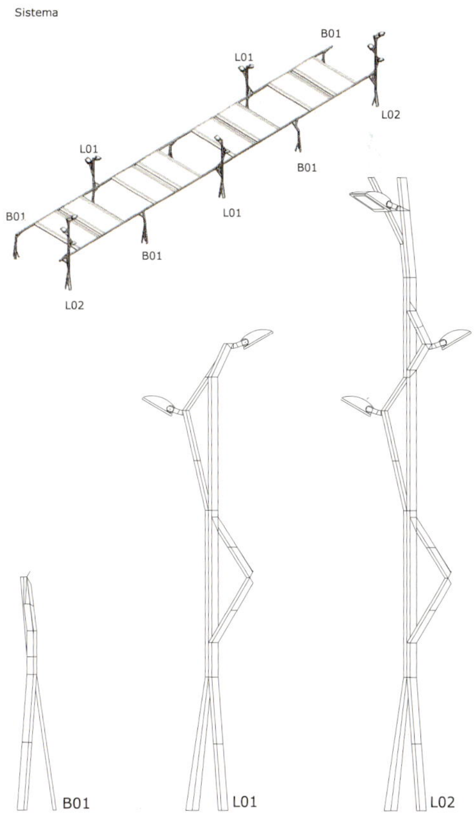
Sistema
B01
L01
L02
B01
L01
L01
B01
B01
L02
B01
L01
L02

But these two worlds form just one and must be constructed at one and the same time. This situation produces a new operative system in which the architect must be able to take part, by making the most of all the physical and digital means within his reach. The architect must consequently assume the fact of experiencing, today, a moment when new paradigms are being invented, a moment when we are defining new human activities, and changing existing situations, which require new constructed references.

但是这两种世界的形式只有一个，并且只能在同一个时间段内被建构成一个同样的世界。这样的情形制造了一种新的操作系统，在这个系统当中，建筑师必须能够通过在其能力可及的范围内尽量使用所有物质的及数码的建造方式来参与其中。在今天这样一个不断有新范例被创造出来的时代，这样一个我们不断地定义新的人类活动并改变现状的时代，建筑师必须能相应地假设感受，而这需要一种新的已建成的参照物。

HORTAL HOUSE
HORTAL住宅

In a hilly area some three kilometres from the Mediterranean coast we built this villa anchored in the terrain, projecting itself towards the sea.
The construction was conceived as a great concrete bracket, almost like a geographical feature that operates according to its own rules, in order to create a privileged position in its surroundings.
Given that buildings are far more expressive during construction than when they are completed, this building presents itself as deliberately unfinished, with the presence of the original texture of the concrete external walls, and the metal structure and the floor slabs forming the interior volume.
Nevertheless, the openings in the wall have been given a more delicate finish, with stained glass windows, like gems incrusted in a massive rock, creating personalized landscapes in the interior.
The house responds to its environment with four different scales: to the west, fronting the access road, it is a single-storey volume defining a domestic scale (characteristic of a small house) with a pitched roof covered with rocks. To the north the volume rises up from the access ramp to the garage in a flat, compact form to create a place for mobility. To the south, the pool (which can be partially closed in) slides under the stepped bracket with a sun terrace around it.
To the east, facing the sea, the villa is metallic and transparent, an almost urban building of three floors, with views of the sea.
We cannot say if this villa is a beautiful building or not. We would prefer to evaluate it using the same criteria with which we appraise a mountain or a rock: strong, hard, amorphous...

我们在距离地中海海岸大约三公里的山丘区域建造了这座别墅，它从地面上拱起并朝向大海。
整个建筑物由一个巨大的混凝土支架构成，它看上去几乎是一个以自己的原则运行的地表特征，以在其周边环境中创造出一个具有先机的地位。
在建筑的建造费用远远高于其建成后的使用费用的前提下，这座建筑物将自身以一种被迫为完成的形态呈现了出来，其外墙显现出的是混凝土原有的肌理，而金属的结构以及地板骨架则构成了内部空间。
然而，墙上的开窗或多或少地更加精细地被打造出来，使用磨砂玻璃窗，宛如嵌入岩石的宝石，在室内空间创造出了一种非常个性化的景观。
整栋住宅通过四种不同尺度对环境进行反馈：设置有连接到该住宅的路面的西面以一层的体量定义了一种家庭式的尺度（显著的小住宅特性），同时具有一个倾斜的岩石质屋顶。北面的体量则由通向住宅的斜坡上升至车库成为一个平台，压缩了本身的形体以创造一个具有活动性的空间。南面的游泳池（一部分能够被遮蔽到室内）沿着具有台阶的支架向下倾斜，并在其周围设置了阳光平台。在面对大海的东面，别墅带有金属质感且透明，采用了一种大部分都市住宅所采用的三层形式，并可以观赏海景。
我们并不能说这栋别墅是一座美丽的建筑物。我们更倾向于使用我们赞赏一座山或一块岩石的标准来评价它：坚固、硬质、无定形……

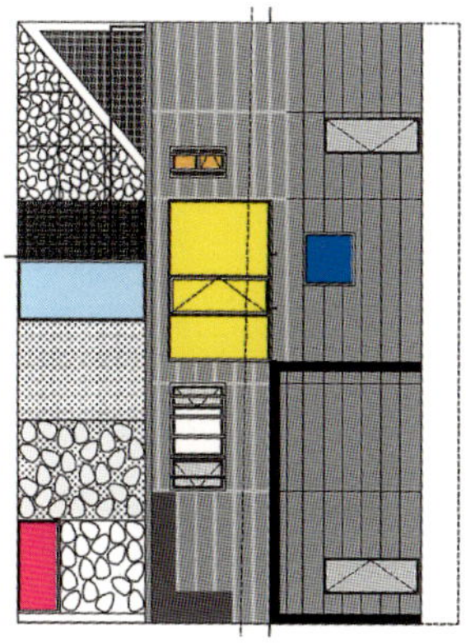

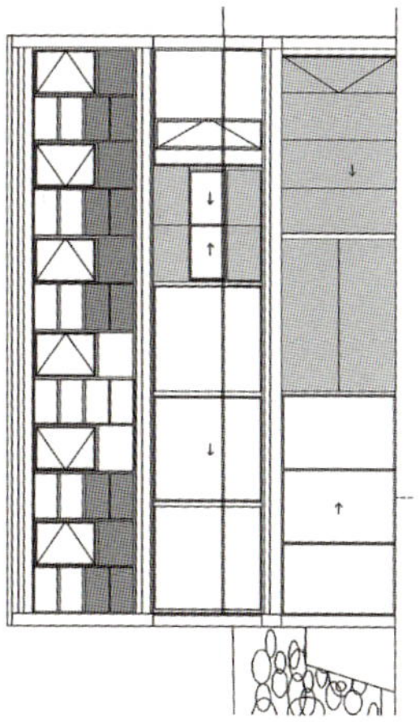

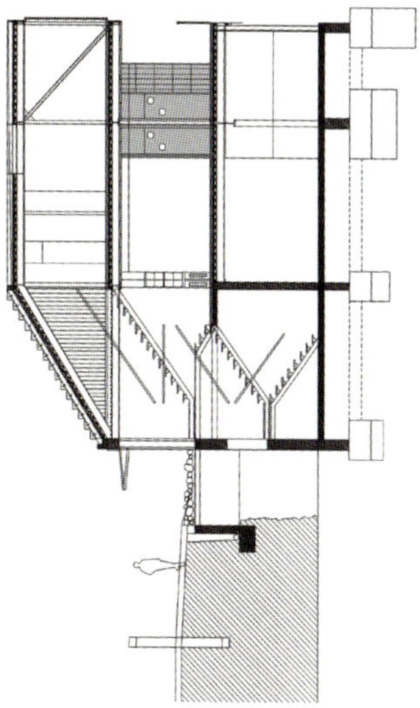

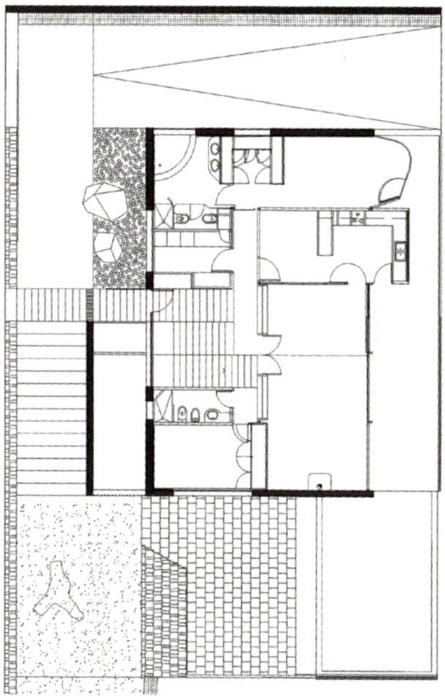

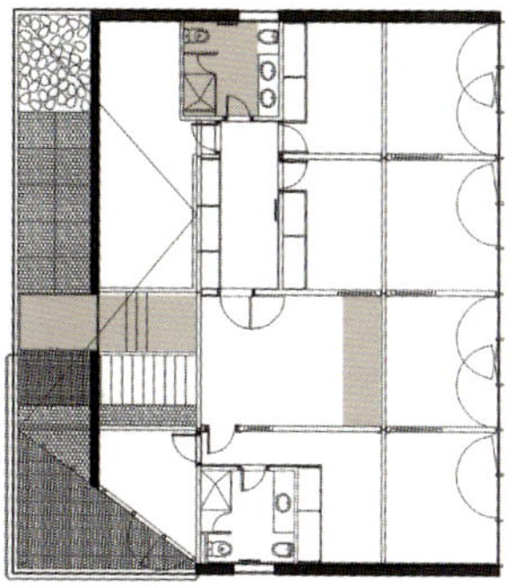

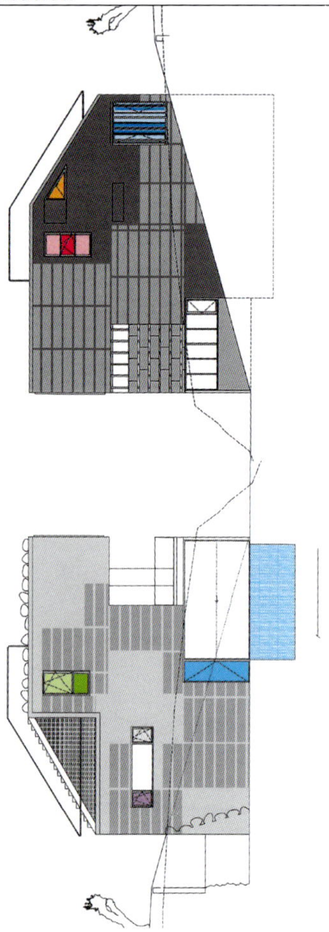

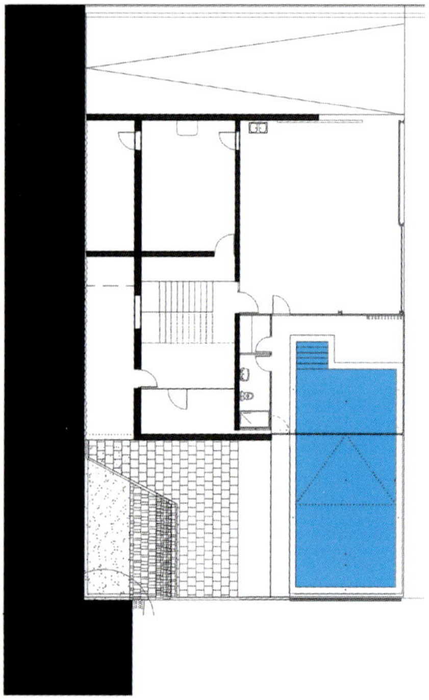

CAMBRILS SEA APARTMENTS
CAMBRILS海滨公寓

Haus am Meer

Haus am Meer

With all the respect that assumes building in the frontline of the sea, this buildings, of squared plan, Project themselves according to new space organizations and functions with the objective of profitin the maximum the sea location of the site. One of th differential factors has been the relation between the rooms and the big terrace that surrounds them conceived literally to enhance and enlarge each apartment. By folding doors and curtains we can have terraces follow essential elements for the circulation in the terraces. The kitchen and bathrooms look over exterior. The sea can be seen from the bathrooms. Another remarkable condition: the water elements are placed in the façade, reflect the surroundings and generate a relation between the interior skin and the exterior of colored glasses.

我们将这些建筑物设定在面朝大海的方向，这些方形平面的建筑通过新的空间和功能上的组织来最大程度地发挥地段中海的作用。而一个不同的因素则是房间与环绕它们的台地之间的关系，这种关系被用来加强和扩大每间公寓。通过将门和窗帘折叠令台地的走势随着其中的交通循环走势来流动。厨房和卫生间都能够俯瞰室外。浴室则能够看到大海。另一个显著的情况是：立面中加入了水元素，它将周边环境发射出来并在内表皮和有色玻璃的外表皮之间生成了一种关系。

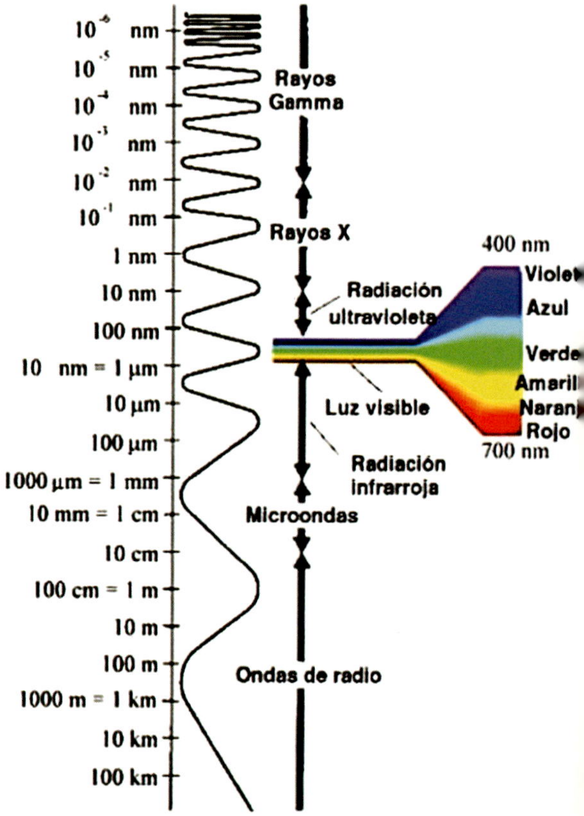

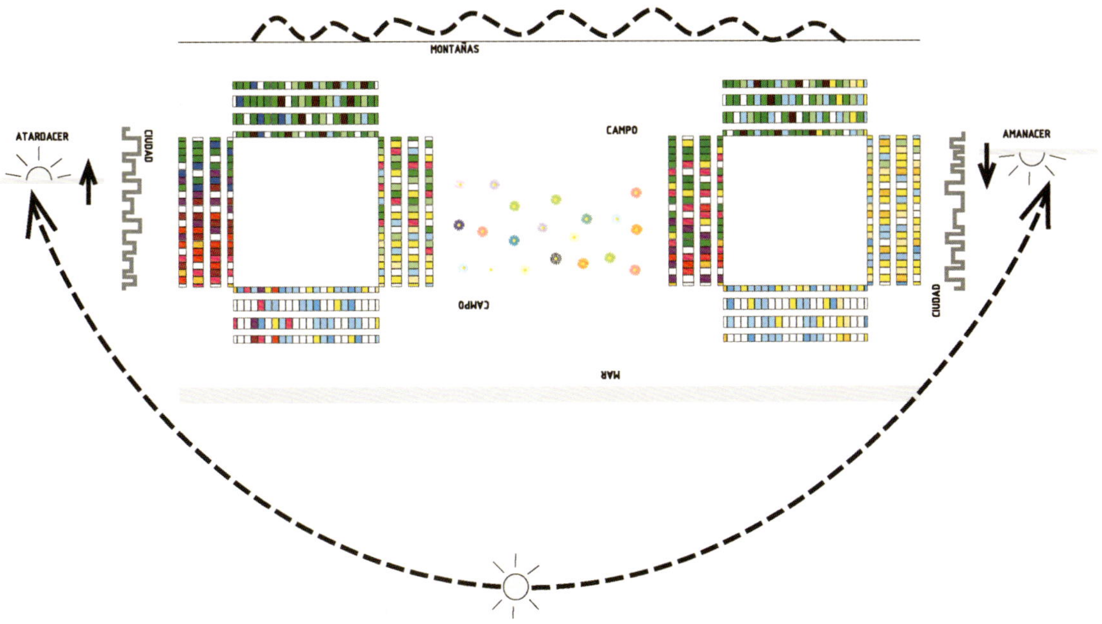
MONTAÑAS
ATARDACER
CIUDAD
CAMPO
AMANACER
CAMPO
CIUDAD
MAR

1
2
3
4
7
8

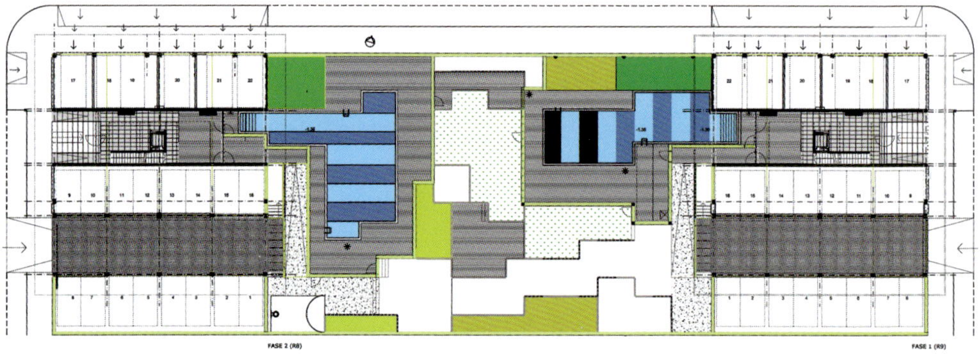

26
ANOIA

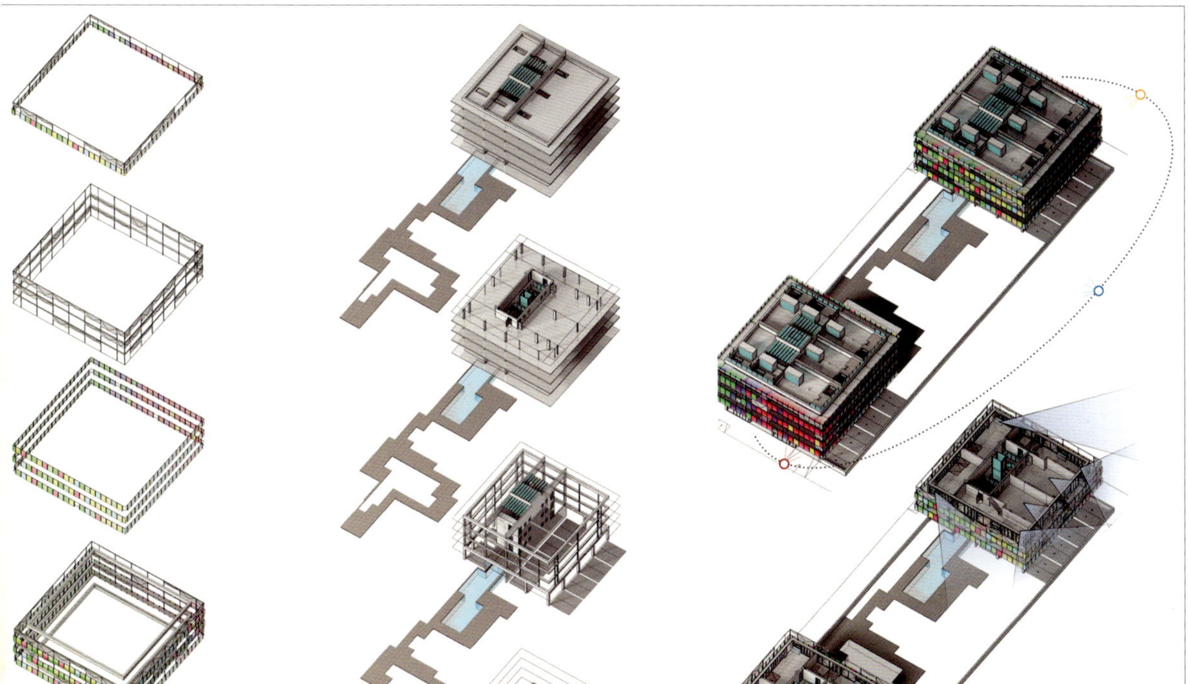

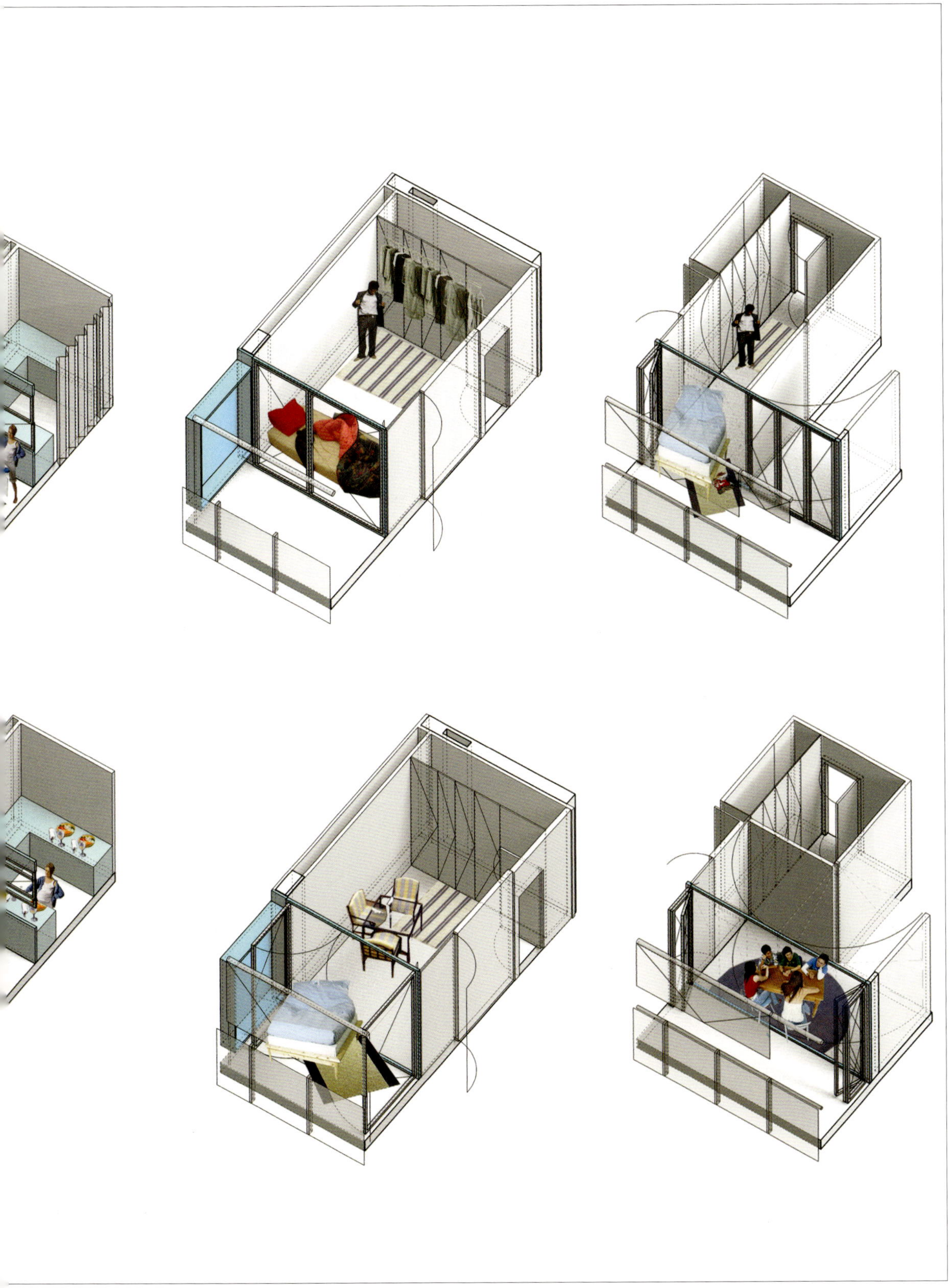

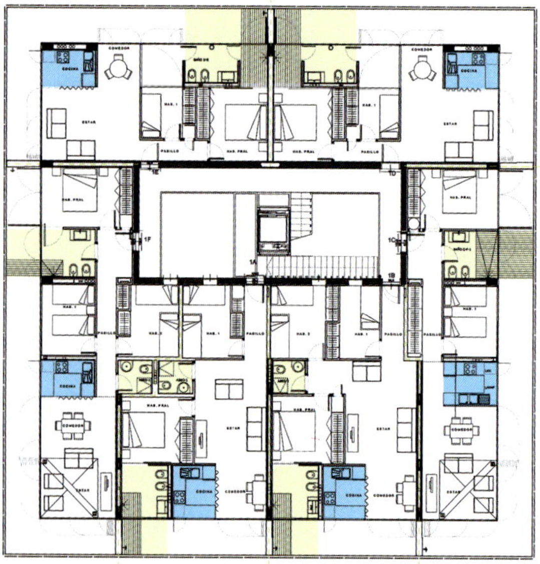

Level 01

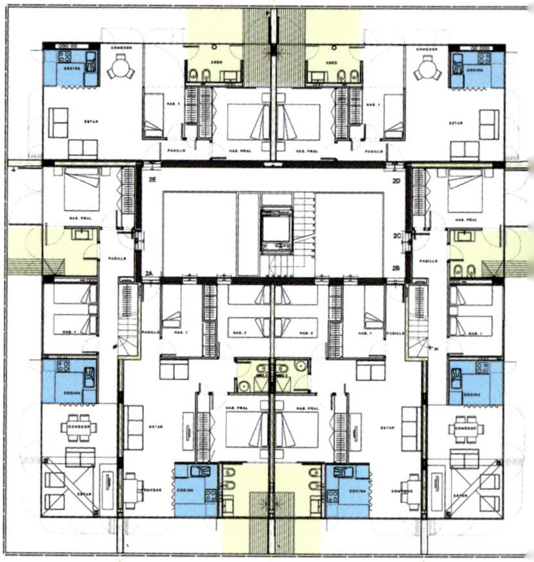

Level 0

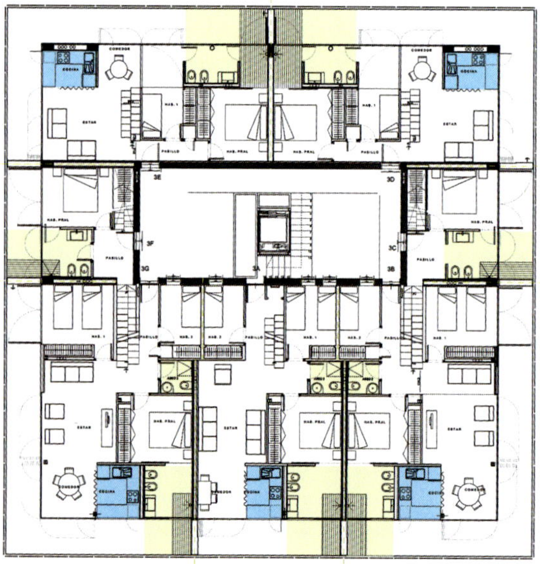

Level 03

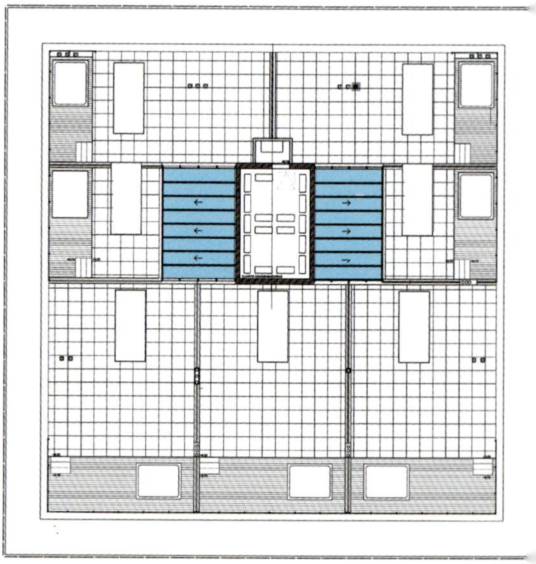

Roof

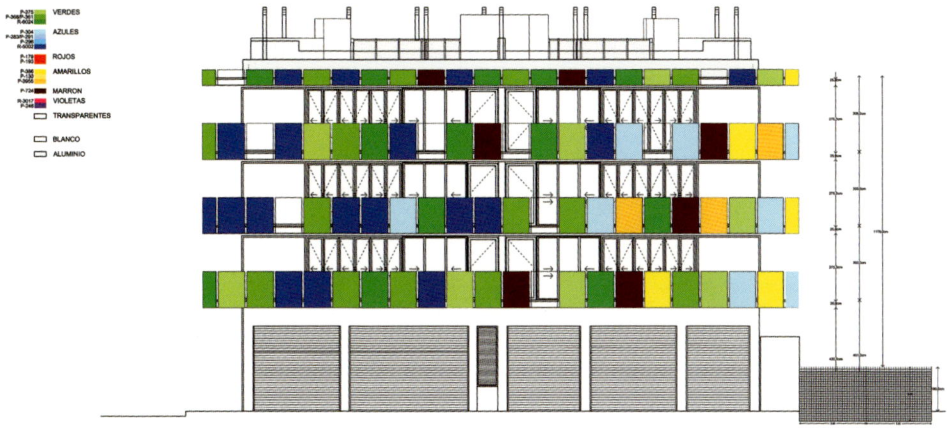
VERDES
AZULES
ROJOS
AMARILLOS
MARRON
VIOLETAS
TRANSPARENTES
BLANCO
ALUMINIO

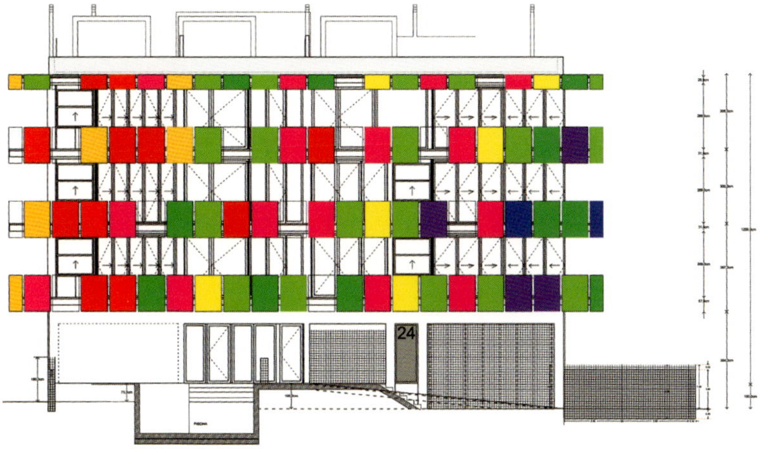
24

24

KIM HYNG YOON EDITING CO PAJU BOOK CITY-SEOUL

In association with Young Joon Kim/YO2 Architects

KIM HYNG YOON出版社 首尔PAJU图书城

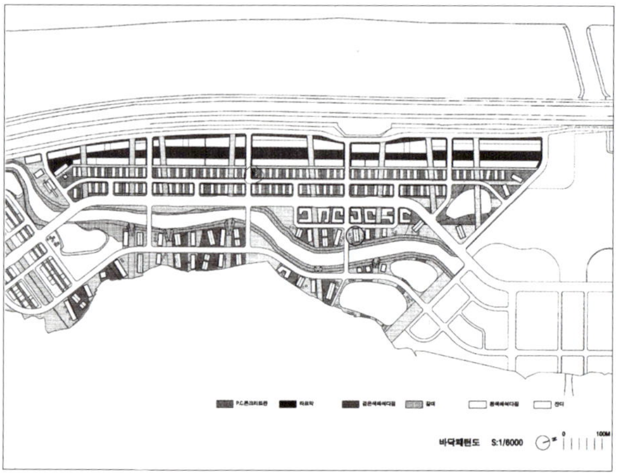

3d path

A building that could grow in the future

a structure that will be flexible

Given a fixed volume, we propose to build a structure that will allow the implementation of different configurations and the construction of the building in different phases. The building will thus have coherence, even if part of its surface area is not yet built. We have organized the possible programmes as a 3d path sequence, one for commercial purposes, the other for design functions. The building can be developed in phases: first building one promenade, and later building the other, using 6x3 m units. The building can even have two users, in different promenades.

In the basement there is a photo studio and the conference rooms. The bookshop is on the entrance level. Following the 3d path, each department will have its own level (and its own identity). _ the meeting rooms are in the centre of the building._ on the upper floor we have located the chief editor and the president's offices near two natural gardens.

Believing that the flexibility of the organization demands a flexible building methodology, we propose a metal structure (heb), with a metal frame for each floor slab (6x3m), finished with wooden beams and wood surfaces.

The facade is formed by an aluminium frame filled with transparent glazing for one promenade and coloured glass for the other: cyan, magenta, yellow and black.

三维通道

一个能够生长着进入未来的建筑

一个灵活可变的装置

以给定的体量为前提，我们提出了建造一种能够执行不同的装置命令，并以不同的语汇来建造建筑的结构。这种建筑将因而取得一致性，即使其表面区域的一部分尚未建成。我们将可能实现的程序以一种三维通道序列的形式组织起来，一部分用于商业用途，另一部分用于设计用途。建筑能够以通过如下步骤建立：首先建立一条大道，然后再以6米X3米作为基本单元建立其他部分。这栋建筑甚至可以允许两条大道上的两个不同使用者来使用。

地下室中是摄影工作室和会议室。书店在入口层。沿着三维通道，每一个部门都将拥有其独立的楼层（和独特的识别性）。会议室则位于整栋建筑的中心。在较高的楼层则设置有主编和会长的办公室，两者都临近两个室外花园。我们认为，组织的灵活性需要一个灵活的建筑来保证，因而我们提出使用一种金属结构（heb），并为每一层的楼板（6米X3米）配备金属框架，并用木制的梁架和表面最后完成。

立面由一种铝制框架中填充透明玻璃来构成一条大道，而另一条大道则以青色、红色、黄色和黑色的有色玻璃构成。

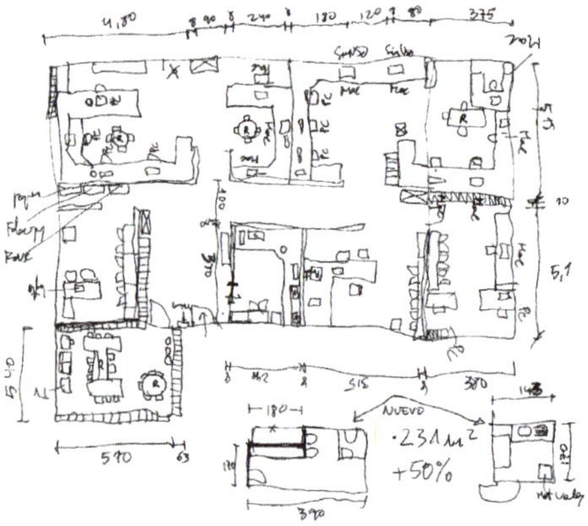

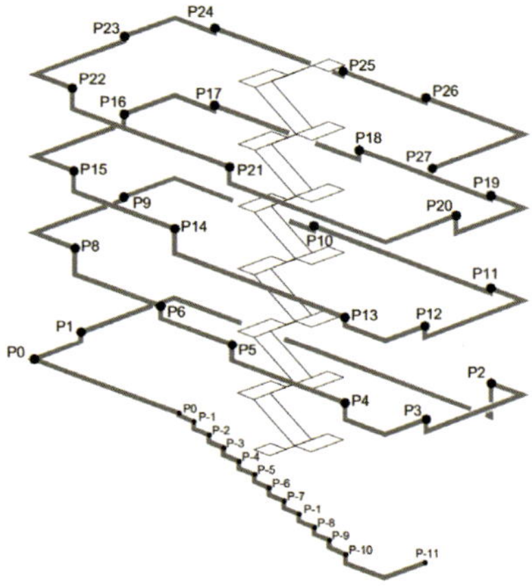
P23
P24
P25
P26
P22
P16
P17
P18
P27
P19
P15
P21
P9
P20
P14
P10
P8
P11
P6
P13
P12
P1
P0
P5
P2
P4
P3
P0
P-1
P-2
P-3
P-4
P-5
P-6
P-7
P-1
P-8
P-9
P-10
P-11

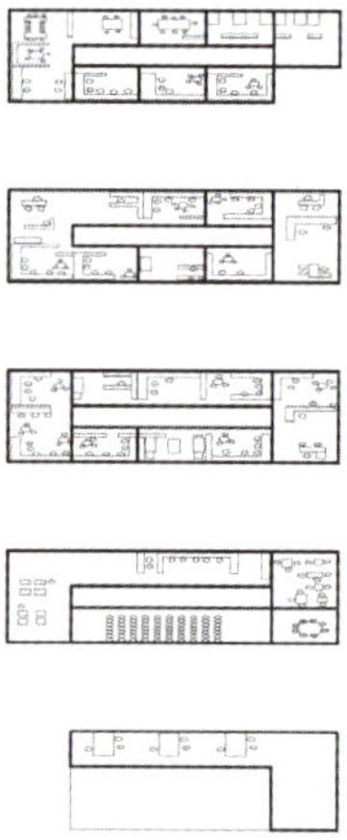

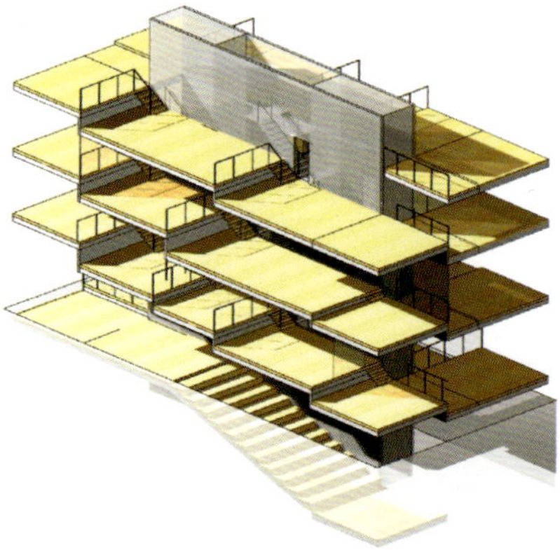

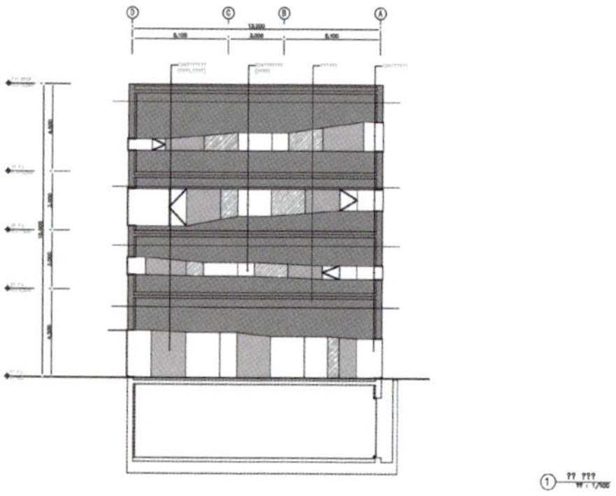

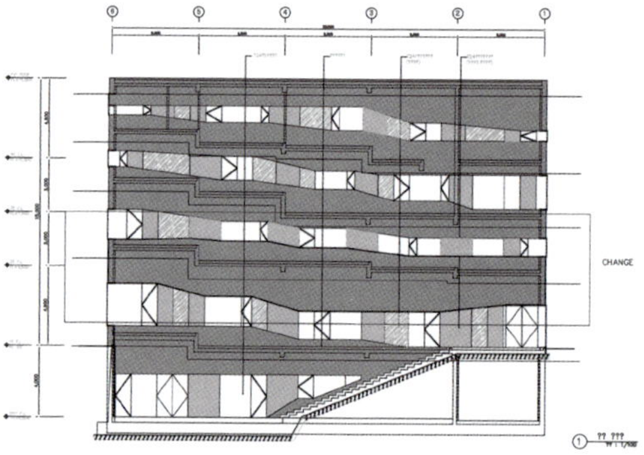
CHANGE

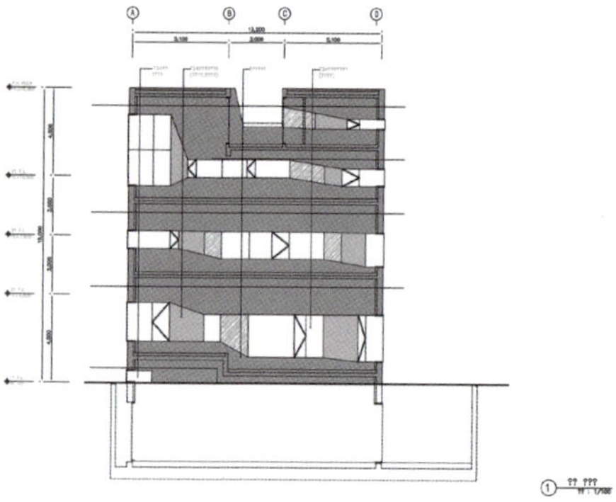

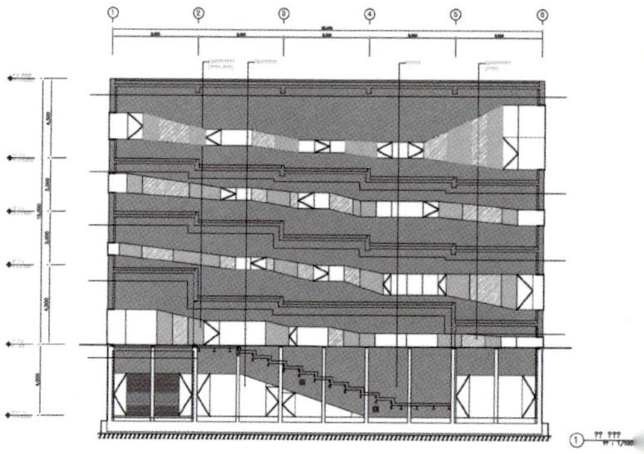

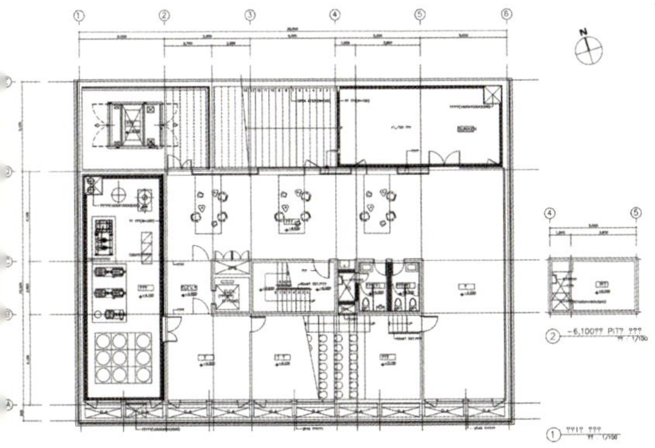

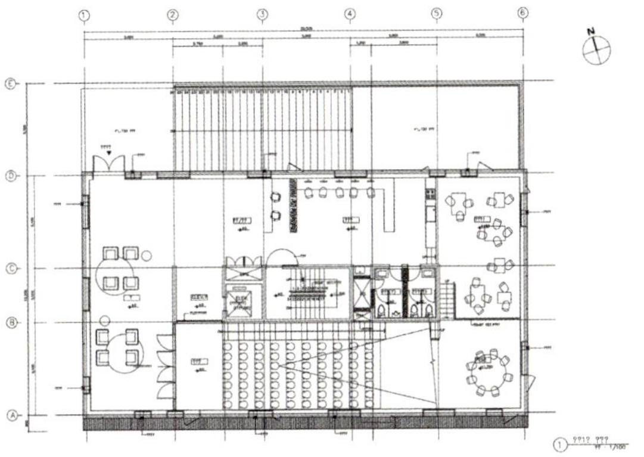

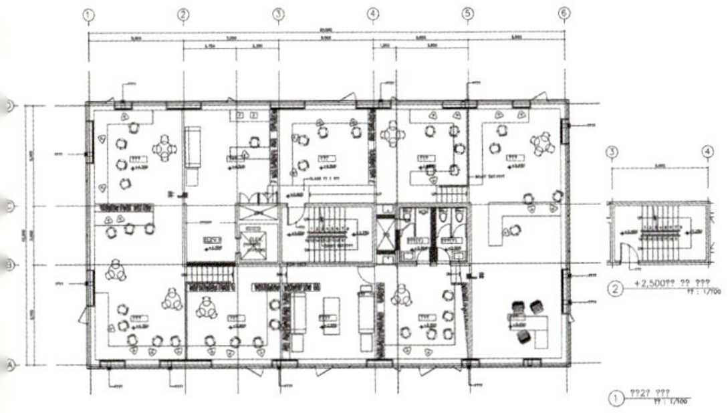

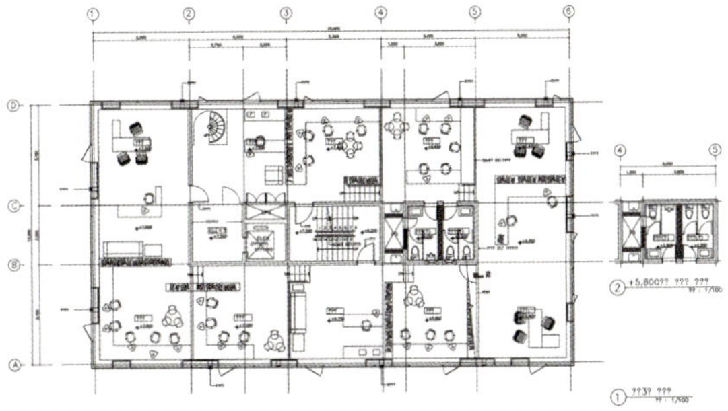

It is moment of linguistic fervour, of inventing and experimenting with new materials, of developing new constructive processes, incorporating natural and artificial forms of energy, defining viable growth, recycling and re-informing cities, a moment of getting acquainted with new relationships in the physical world through an experimental passage in the virtual world. It is thus important to think about this new hybrid condition of the world that is peculiar to our day and age. The natural, the artificial, and the digital. By way of media, mountains, and architecture.

这是一个语言热的时代；一个不断发明并试验新材料的时代；一个发展新的建造过程，将能量的自然与人工形式相结合，定义可持续增长，将城市循环利用并重组的时代；一个通过虚拟世界的试验性路径来不断适应物质世界当中的新关系的时代。因而，重要的是去思考这个对于我们的生活和时代来说独一无二的世界中那些交错混杂的环境——自然世界、人工世界以及数字世界——而这样的思考正需要通过媒介、山峦与建筑去完成。

SOCIOPOLIS
MASTERPLAN
社会学都市总体规划

SOCIOPOLIS

Sociópolis is a project fomented by the Generalitat Valenciana for the construction of a new model neighbourhood of accessible housing, in line with the modern tradition of investigating and proposing new habitable environments that reflect the social reality of the time.

The project promotes the construction of housing that responds to the needs of the new types of family unit (young people, the elderly, single-parent families, etc.), both owner-occupied and rented, in a quality urban setting in which the landscaped areas, social amenities and good architecture generate urban excellence.

Sociópolis was presented at the Valencia Biennial in 2003, as a project in which 13 international architecture firms took part. The project put forward a model of new urban development in which housing and multifunctional amenities were integrated in an agricultural environment, a continuation and updating of the model constituted by the Mediterranean hortulus.

Following the presentation of the project it was decided to construct a first neighbourhood of 2,500 homes in the La Torre district to the south of the city of Valencia, on a 350,000 m^2 site on the banks of the new course of the diverted River Turia; the same principles would be followed, but on a larger scale.

In this project the urban transformation is guided by a commitment to ensuring the maximum protection for the existing huerta (one of the traditional agricultural zones surrounding the city of Valencia) irrigated with waters from the River Turia by way of channels originally dug by the Arabs some 800 years ago. The new urban development reinforces the protection of the landscape and the environment while at the same time fulfilling a much-needed social function, making housing available at a controlled price to a great number of people.

Within the neighbourhood has four well-conserved historic farmhouses, and around these will be the focal points for 'urban farm' zones cared for by the local residents. The neighbourhood will also promote social interaction and a sense of community by means of sports facilities such as a soccer pitch, an athletics circuit, games areas and a skating rink.

All of the proposed buildings are oriented toward the central landscaped zone, which has a surface area of 120,000 m^2, with direct access from the peripheral traffic circuit running round the complex.

In addition to residential blocks and towers the neighbourhood will have amenity buildings accommodating a hybrid programme, around which the public life of the neighbourhood will be organized. The public buildings will have rental housing intended for young people under thirty and elderly people, and at the same time all of the buildings will fulfil their public vocation by means of programmes that encourage social relations, such as a arts centre, a kindergarten, a sports zone, a social centre, a youth centre and studios for artists.

The architects participating in the project include Vicente Guallart, Abalos & Herreros, Manuel Gausa Eduardo Arroyo, José María Torres Nadal, Sogo Arquitectos, Willy Muller, Antonio Lleyda / Eduard de la Peña, Toyo Ito, MVRDV, Greg Lynn FORM, Duncan Lewis, José Luis Mateo, Kim Young-Joon, JM Lin, Jose Maria Lozano and Maria Colomer.

SOCIOPOLIS

社会学都市是一个由瓦伦西亚加泰罗尼亚政府推动的住宅设项目，它以建立一种新的邻里关系为主要诉求，同时也希能够实现一种反应现实的社会实质的新的宜居环境。

该项目推行建设一种针对新的家庭（年轻人、年长者、单亲家等等）单位需要的住宅，这种住宅可以购买也可以出租，并位于具有景观区域、福利机构的高品质城市环境中，更将采出众的建筑设计。2003年，威尼斯双年展推出了由13个国际筑公司参与的社会学都市。这个项目推出了一种新的城市开模式，住宅和多功能公共活动设施被融入到一种农业的环境中，而这一模式的被地中海Hortulus不断延续和更新。

随着这一项目的展示，一个包含2500个住宅的邻里社区首次在瓦伦西亚南部城市的La Torre地区建成，占地面350000平方米，并紧邻Turia河转向的岸线地区。在这一住区中，同样的原则将以更大的尺度被推行。

在这个项目当中城市的变化由一系列的规则来主导，以保证有huerta（一个环绕瓦伦西亚市的传统农业地区）能够继通过800年前阿拉伯人开凿的水渠受到Turia河的灌溉。新城市开发项目加强了对于景观和环境的保护，同时也满足了会功能上的巨大需要，使住宅在能够控制的价格下成为市能够获得的产品。

在这个社区内，有四个完整保护的古农场房屋，而在这些房周围则是我们设置的并由当地居民管理的都市农场区域。这社区也将通过建立足球场、运动回廊、游戏区域以及一个市冰场的方式来推行社区之间的互动交流。所有预计建成的筑都朝向中心面积达120000平方米的景观区，同时拥有外交通环绕其周围。

除了居住性街区和大楼以外，这个住宅区还将建设能够容纳合功能的公共活动建筑，社区的公共生活将会围绕这个建展开。公共建筑将能够提供为30岁以下的年轻人和老年人备的出租性住宅，同时这些建筑又能够通过提供艺术中心、儿园、体育锻炼区域、社区中心、青年中心和艺术家工作室功能空间，来鼓励社会交往以满足他们的公共生活需求。

参加这个项目的建筑师包括：

Vicente Guallart, Abalos & Herreros, Manuel Gaus Eduardo Arroyo, José María Torres Nadal, Sogo Ar quitectos, Willy Muller, Antonio Lleyda / Eduardo la Pena, Toyo Ito, MVRDV, Greg Lynn FORM, Dunca Lewis, José Luis Mateo, Kim Young-Joon, JM Li Jose Maria Lozano 以及 Maria Colomer.

IPLANTATION STRATEGIES

ociópolis is situated in the district of La Torre, › the south of Valencia, on the banks of the new ourse of the diverted River Turia, in a sector of al-ıost 350,000 m² delimited by high speed express-'ays and a degraded urban periphery. The objec-ve addressed by the Plan is the implantation of lmost 3,000 protected housing units, both owner-ccupied and rented, respecting as far as possible ıe existing agricultural landscape and its historic rigation channels, integrating these into a central ark of 150,000 m² surrounded by amenities.

he sector has been laid out on the basis of an nalysis of the landscape of the huerta, the exist-ıg farmhouses and the age-old system of irriga-on channels, superposing on these a specialized etwork of circulation routes for cars, bicycles, ports and pedestrians, from which there is direct ccess to a sequence of towers and blocks distrib-ted around the perimeter. All of the streets are onnected with the existing urban fabric of La Torre ı order to ensure effective urban continuity.

he necessary urbanity will be achieved not on the asis of a predetermined urban form, but through he encouragement of social interaction in a se-uence of specific public spaces that offer sports, gricultural and cultural activities, all linked a major etwork of amenities.

RBAN ORCHARD-GARDENS

he conservation of areas of huerta in the interior f the cities, transformed into urban orchard-gar-ens, has entailed working in close cooperation vith the organizations that have historically per-ormed this function. To this end an agreement has een signed with the Acequia de Favara irrigation ssociation, whose water channels crisscross the ector, for the conception, development and sub-equent management of the largest urban orchard-ardens in any major Spanish city. Urban orchard-ardens are qualified as green zones, located within he city boundary, whose vegetation is agricultural, sed and maintained by the citizens themselves, nd accessible to all of the local population by neans of roads and paths. This means it is the lo-al people, cultivating small orchard-garden units from 25 to 100 m²) to produce fruit and vegetables or their own consumption, and creating for this urpose associations and groups that together to onserve the public space and the landscape, who lirectly foment urban cohesion and social interac-ion.

he participation of the Acequia de Favara irriga-ion association in the project ensures the future of hese zones, while the fact that they will be watered rom the Acequia makes them subject to the Water ribunal, thereby extending or developing its com-etency to a new form of cultivation, and injecting new life into it. The sector will have some seven kilometres of conserved, repaired or renewed water channels, while will be integrated into the design of the central park, and by guaranteeing the safety of the pedestrians will enable the culture of water to impregnate the design of the entire public space.

培植策略

社会学都市位于瓦伦西亚南部的La Torre区，同时该地区正是Turia河转道的新岸线区域，被高速公路和双向的都市主干路线划定的一块350000平方米的区域。这个总体方案欲植入近3000套可购买可出租的住宅单元，同时尽可能地尊重现有的农业景观和古灌溉水渠，将这些元素融入到社区中心面积为150000平方米的公共活动设施当中去。

该方案以给定的都市形态为基础，通过在一个具体的公共空间序列之内提供体育、农业和文化活动等与公共活动相关的设施来鼓励社会性交流互动。

都市果园

城市中间受到保护的huerta地区被变成都市中的果园，继续承担着从历史上就承担该功能的组织的作用。为了达到这样的目的，有水渠穿越该地区的Acequia de Favara灌溉协会与住宅区签订了协议书，从概念、开发和后续管理的方面参与到这个具有全西班牙大城市中最大的都市果园的开发工作当中来。在城市边界之内的都市果园被定义为绿化区域，其植被是农业性的，由市民负责养护和使用，并且对所有的本地居民开放。这就意味着本地居民可以通过种植小型的都市果园（25~100平方米）来生产自己所需的水果和蔬菜，并创立专著于这项事业的联合会及组织来一起保护公共空间和景观。这样的活动就直接地维护了都市的一体性并加强了社会性的交流互动。

Acequia de Favara灌溉协会对于该项目的参与保证了这些区域的未来——即这些区域将以获得Acequia de Favara灌溉的形式来成为决定水源利用的主体，并因此拓展和发展其对于一种新的种植形式的竞争力，为其注入新的活力。这个地区将会拥有达七公里的保护、修复或更新过的灌溉水渠，同时这些水渠也会被融入到新的中央公园的设计当中。并且，在保证行人安全的情况下，这些水渠也正是催生水文化所主导的新的公共空间的主体。

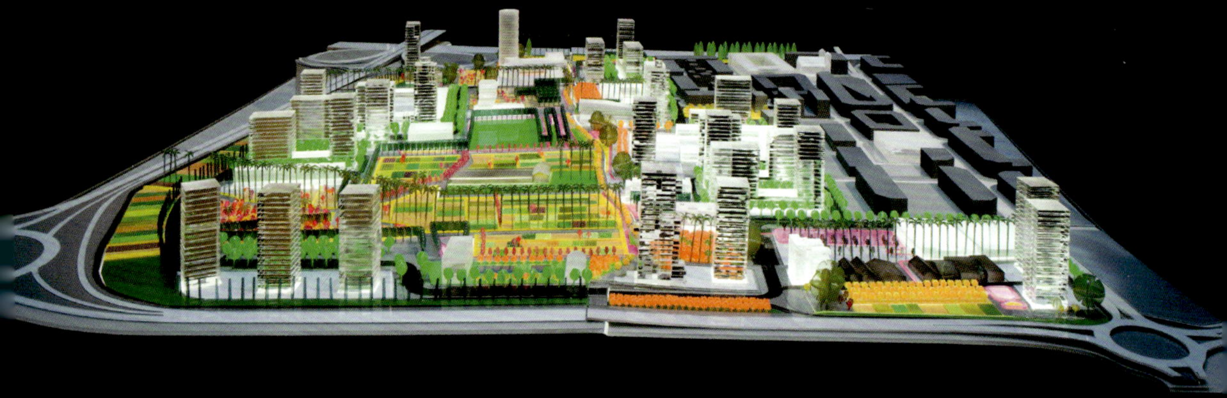

IGITAL NEIGHBOURHOOD

ne project promotes the use of the new infor-ation technologies within the home and in the eighbourhood, enabling people to share resourc-s, create support networks and develop a strong ense of community, thus promoting face-to-face ontact between people in the sector's various ublic spaces.

nis will be done by implanting a fibre-optic net-ork parallel to the traditional telecommunications etworks in order to create a great Intranet for the hole neighbourhood.

lso in development is the 'neighbourhood web-te', a collaborative project involving various bod-s in the construction of the digital urbanization f the neighbourhood so that the day comes for eople physically to occupy the homes and public paces there will be a network of digital spaces nd services to foment the urbanity of the neigh-ourhood on the basis of digital applications. This atform will be open to the development of new oplications by the residents themselves.

PORTS CIRCUITS

oort is an activity with a manifestly social char-cter that can be and is practiced by people of all nds and all ages. In view of this, a sequence of oorts facilities will extend the entire sector, includ-g a soccer pitch, a municipal swimming pool, ulti-sports courts, basketball courts and a mini-asketball area. These facilities will be connected y a sports circuit with sports stations for physical xercise set along a continuous jogging track.

ne soccer pitch of artificial grass will be in the entre of the sector and will be at the disposal f the existing local La Torre football team. The nanging rooms will be in the falls of an adjacent ervices building, and there will be a raised cafete-a overlooking the pitch and the tiered stand.

ne local council will also construct a municipal pen-air swimming pool on a plot located in the outh of the sector.

order to develop the plan a multidisciplinary eam was created: led by the architect Vicente uallart, the team included, among others, the ociologist José Michael Iribas, the ecologist Javier barti, the landscape architect Manuel Colominas, ne anthropologist Artur Serra, the biologist Javier uallart, the agronomist Ximo Sánchez and the wyer Maria Ángeles García Capdepon, together ith the technical staff of the Instituto Valenciano e Vivienda directed by Maria Jesús Rodríguez.

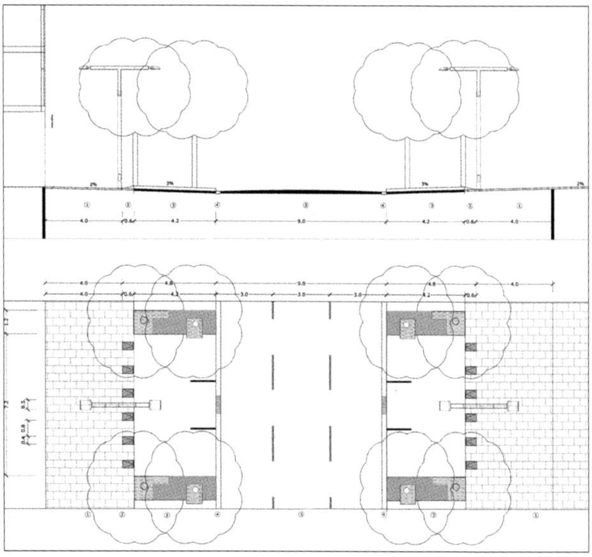

数字化社区

这一方案推广在家中和社区中使用新的信息技术，以保证人们能够进行资源共享，创造相互支持的网络来建立一种强大的社区意识，最终鼓励人们在社区各种公共设施中进行面对面的交流。

我们在社区中植入了一套与传统的电信通讯相平行的光纤网络来为整个社区创造一套不同凡响的交流网。与此同时我们也在开发的是"社区网页"，一个在社区的信息化建设过程中包含了多个主体的协作型项目。因而，当人们物质性地占有了住宅和公共空间后，一套相应的数字化空间和服务系统将会从信息方面加强这种社区的城市化特性。这一平台将会向社区居民开放以鼓励他们开发出新的应用方式。

体育回廊

体育是一种具有显著社会特性的活动，它适用于不同身份不同年龄的所有人。从这一点出发，一系列的体育运动设施将会被扩展到整个社区当中，其中包括足球场、公共游泳池、多功能运动场、棒球场和迷你篮球场。这些设施将会布置在用以进行跑步活动的步道旁，被一条用以容纳各种体育锻炼的运动回廊连接起来。

植有人工草地的足球场将会是这个区域的中心，并被配给本地区现有的La Torre足球队。更衣室则位于一座临近的服务型建筑下面，同时还配有一个可以俯瞰整个足球场和站立阶梯的咖啡厅。地方政府也会在该区域的南面建造一个露天的游泳池。

为了实现这一计划，一个多学科的工作队伍被组建起来。在建筑师Vicente Guallart的领导下，这个工作组包括社会学家José Michael Iribas，生态学家Javier Obartí，景观建筑史学家Manuel Colominas，人类学家Artur Serra，生物学家Javier Guallart，农学家Ximo Sánchez和律师Maria Angeles García Capdepon，并与来自Instituto Valenciano de Vivienda 的在Maria Jesús Rodríguez指导下的技术队伍进行合作。

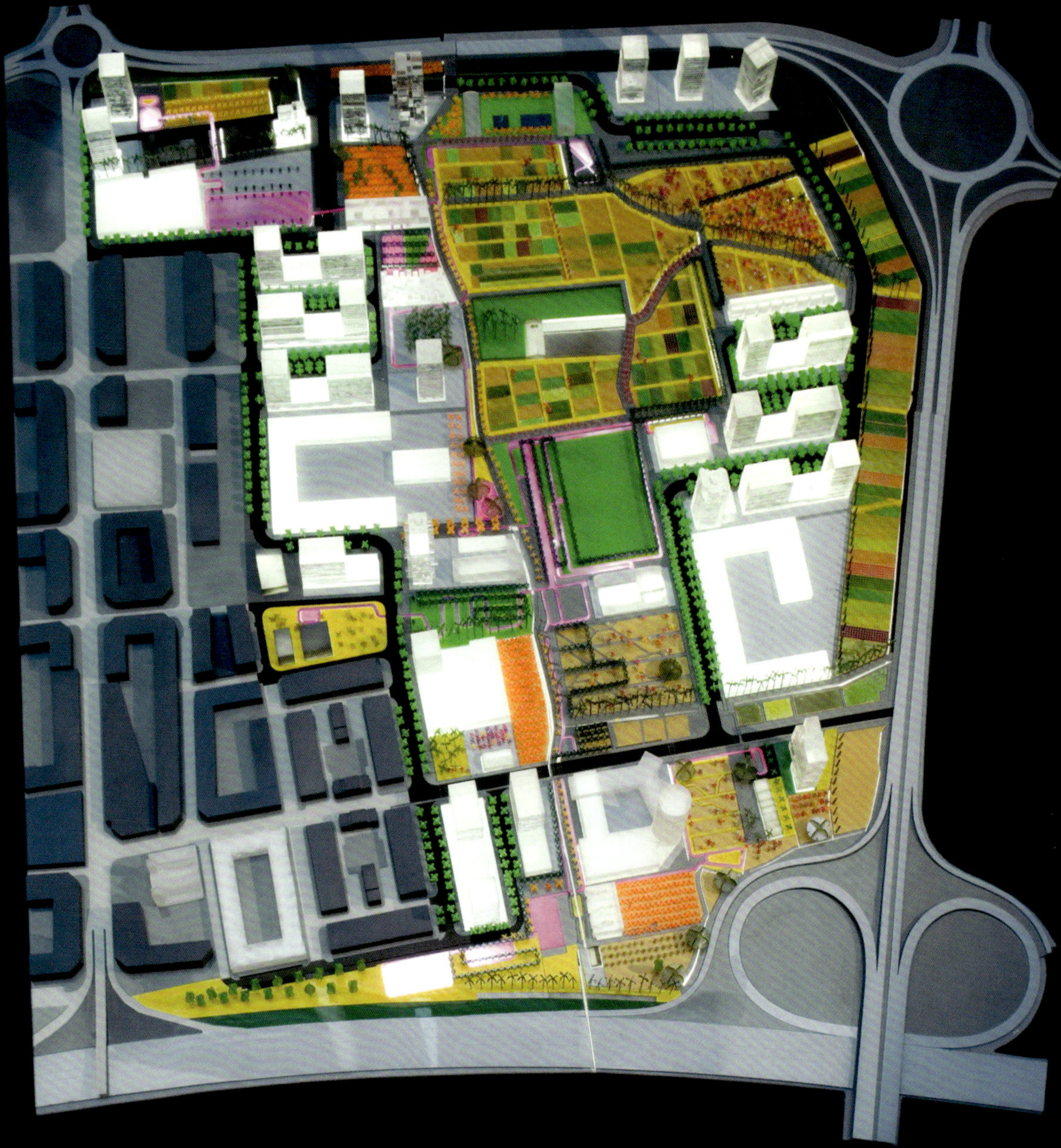

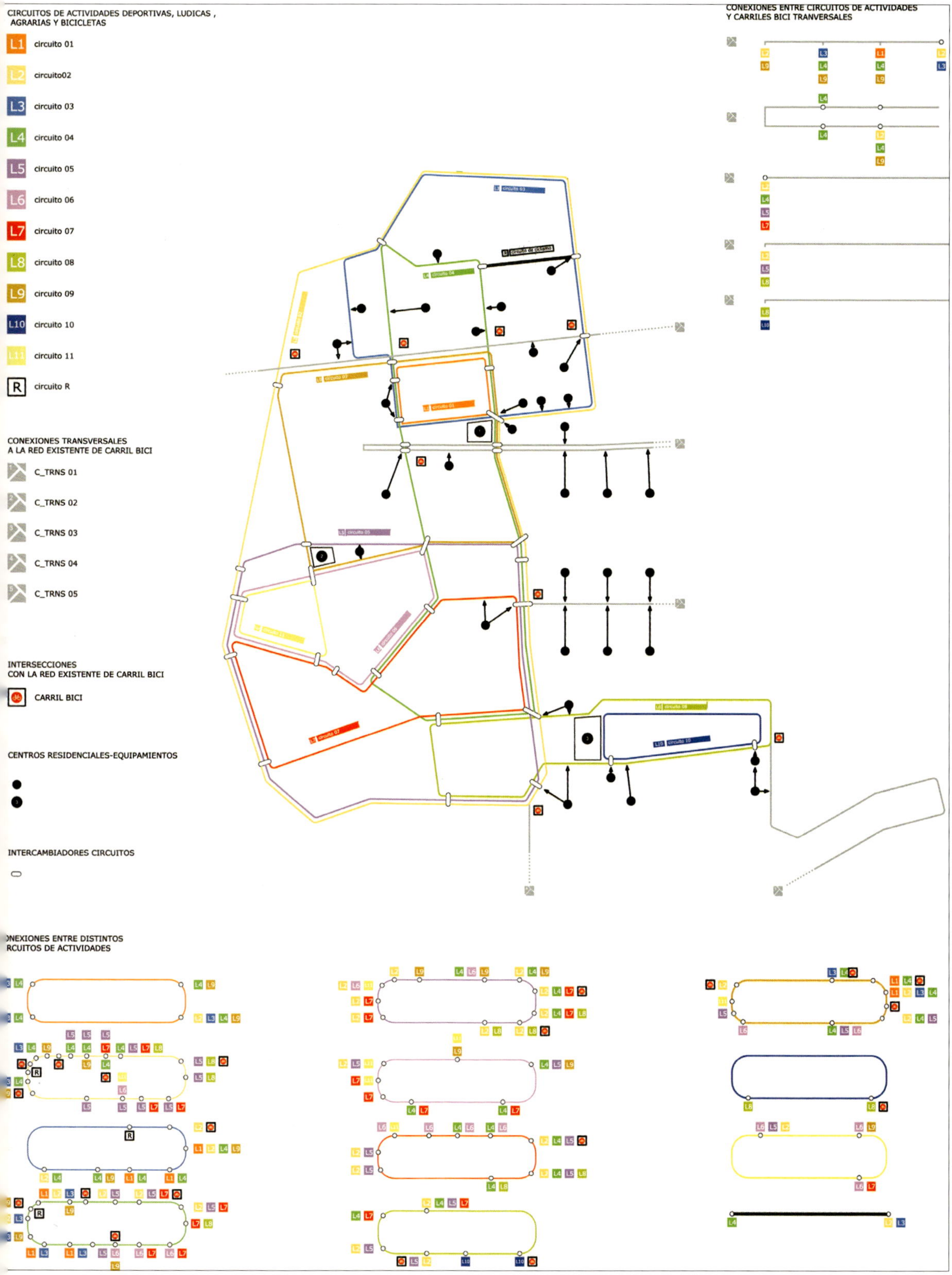
CIRCUITOS DE ACTIVIDADES DEPORTIVAS, LUDICAS , AGRARIAS Y BICICLETAS
L1 circuito 01
L2 circuito02
L3 circuito 03
L4 circuito 04
L5 circuito 05
L6 circuito 06
L7 circuito 07
L8 circuito 08
L9 circuito 09
L10 circuito 10
L11 circuito 11
R circuito R
CONEXIONES TRANSVERSALES A LA RED EXISTENTE DE CARRIL BICI
C_TRNS 01
C_TRNS 02
C_TRNS 03
C_TRNS 04
C_TRNS 05
INTERSECCIONES CON LA RED EXISTENTE DE CARRIL BICI
CARRIL BICI
CENTROS RESIDENCIALES-EQUIPAMIENTOS
INTERCAMBIADORES CIRCUITOS
CONEXIONES ENTRE CIRCUITOS DE ACTIVIDADES Y CARRILES BICI TRANVERSALES
ONEXIONES ENTRE DISTINTOS RCUITOS DE ACTIVIDADES

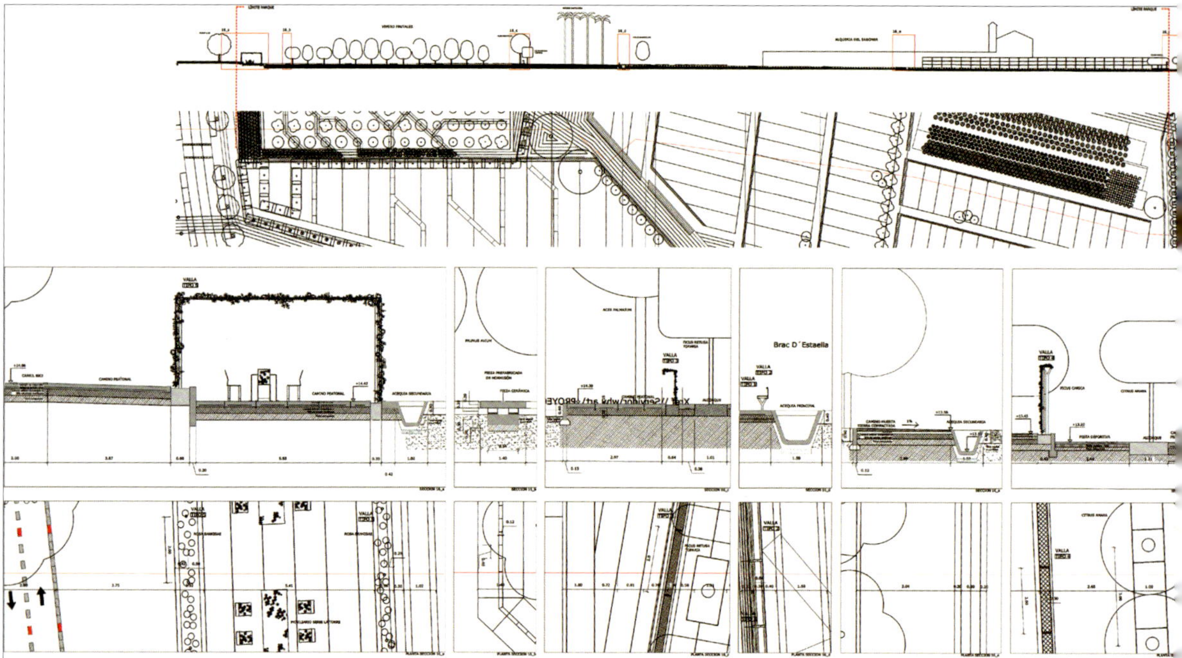
Brac D´Estaella

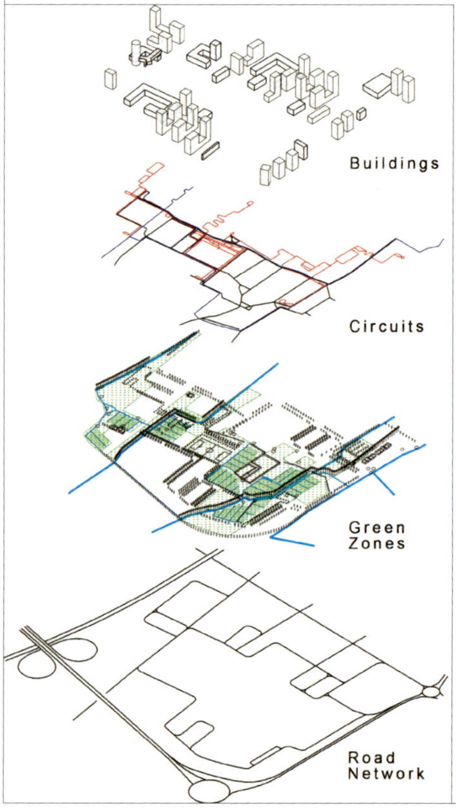
Buildings
Circuits
Green Zones
Road Network

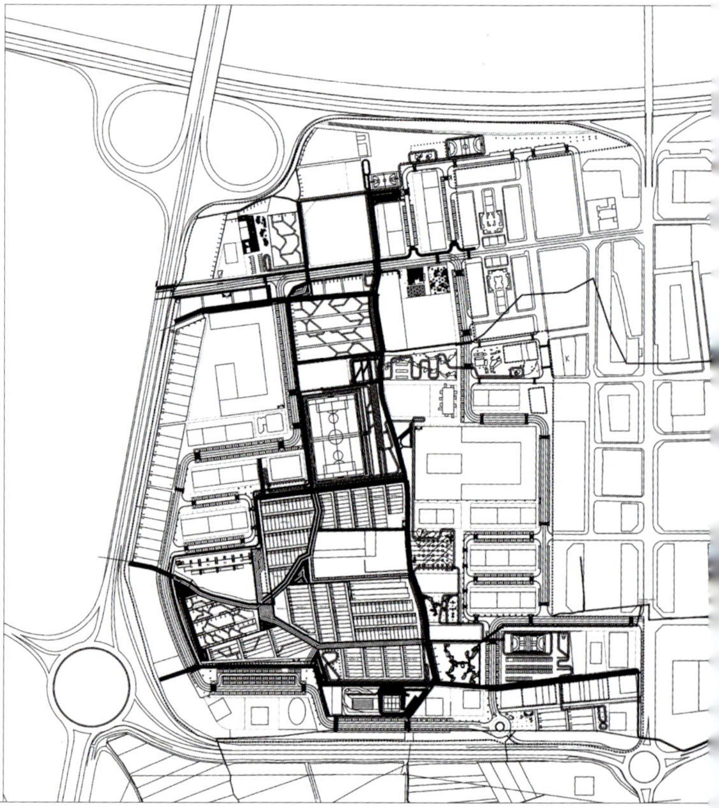

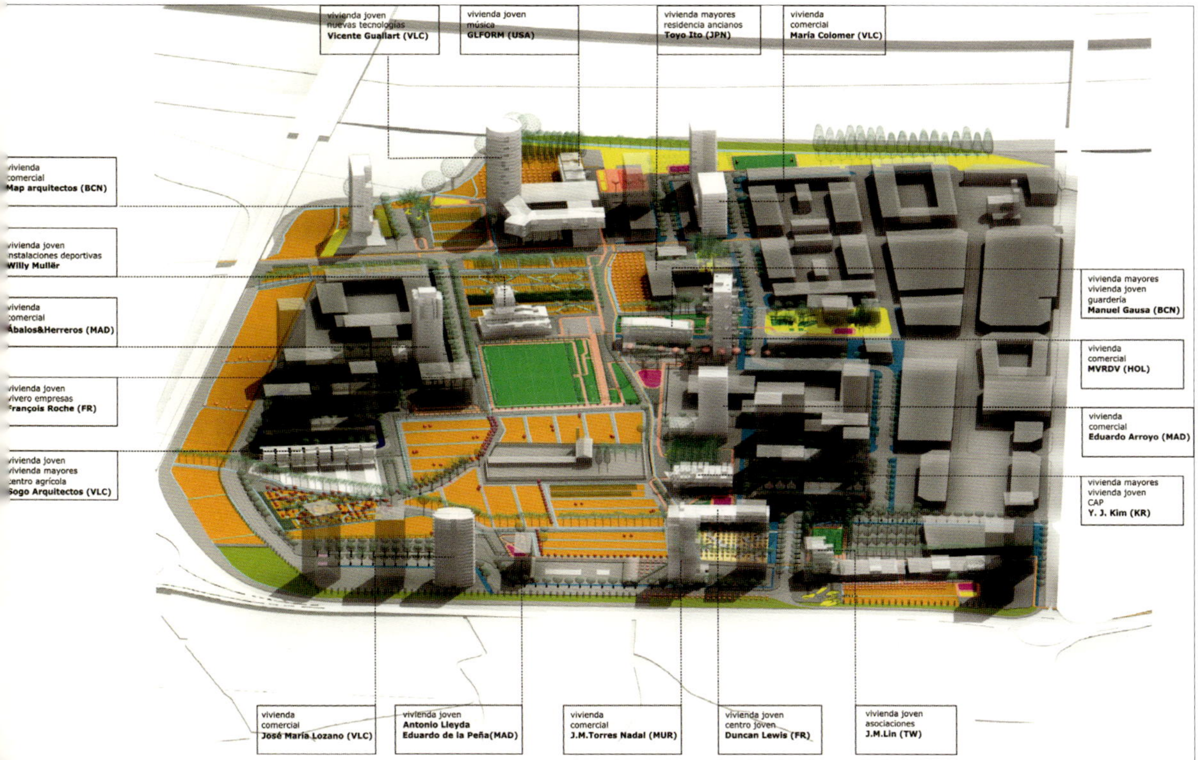

yo Ito
partamentos para mayores

Willy Muller Arquitectos
Pile Tower

Scape Archiecture. Duncan Lewis
Huerta m3

YO2 Architects. Young Joon kim
Urban Mat

R&Sie(n)
Cajas en obra

NO.MAD Arquitectos. Eduardo Arroyo
101373

EA Arquitectos
Dom Extica

Manuel Gausa
Topografía Residencial

Vicente Guallart
Sharing Tower

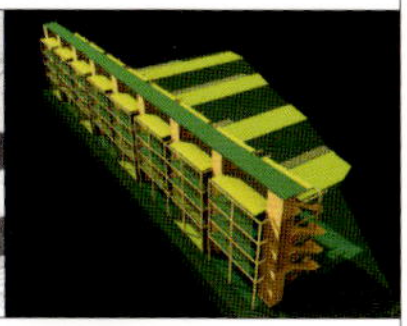
Sogo Arquitectos
Bloque con Centro Agrícola

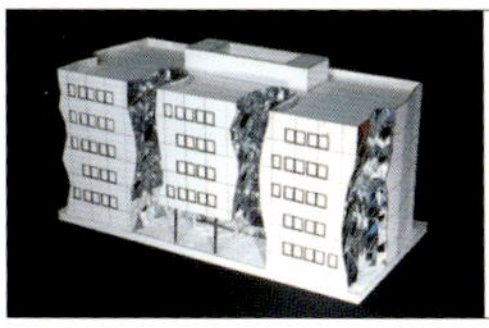
Greg Lynn
Geode Block

Abalos & Herreros
Torre Solar

Jose Luis Mateo. Map Arquitectos
Torre de viviendas

Colomer & Dumont
Tob

Arquitectura Mediterránea
Torre Radial

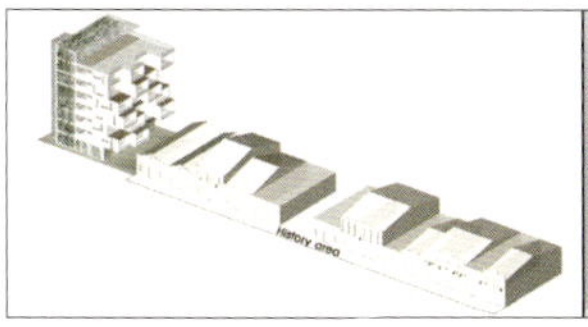
The Observer Design Group. Jm Lin
Valencia Block

Arquitecturas Torres Nadal
Tower

MVRDV
Huerta Tower

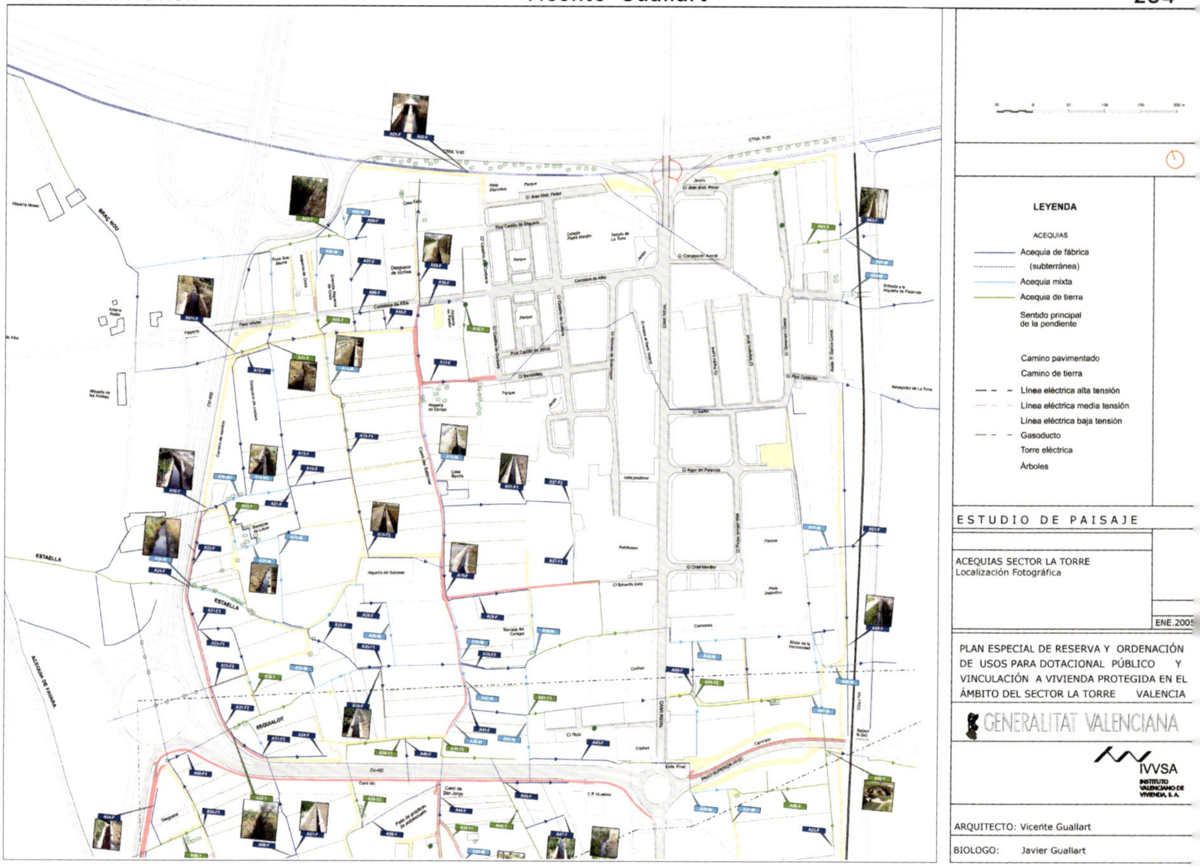

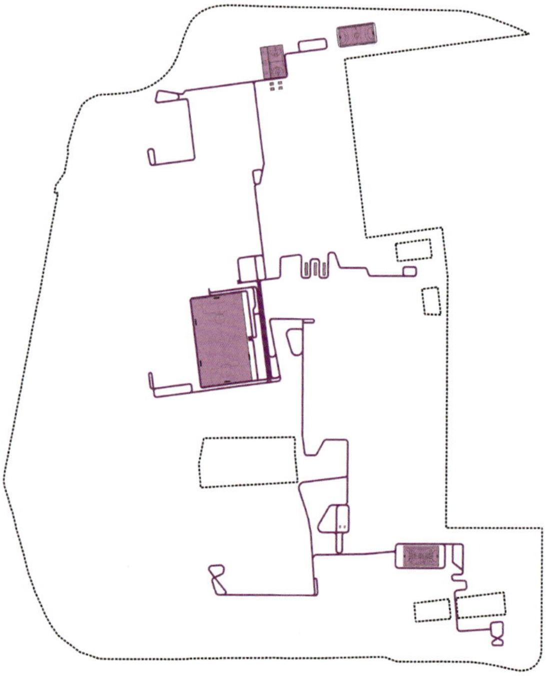

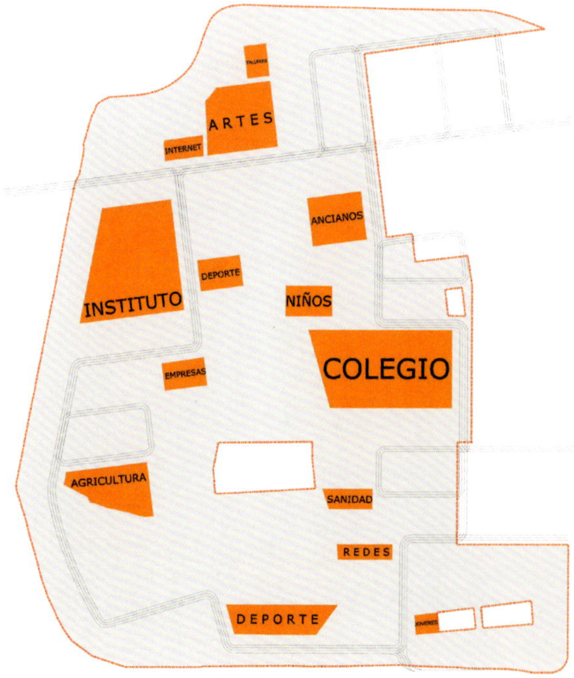

EQUIPAMIENTOS

LIMITE DEL SECTOR

01

Acequia A03 (izquierda) y A03a (derecha). En A03 se observa la parada, para dar agua a A03a. En A03a está puesta la compuerta que impide que el agua de riego llegue a la zona del regador correspondiente a la parcela de "e1" en barbecho.

f002

Acequia A03, en nudo con A03c (izquierda) y A06 (derecha, tras el pontón). En A03 se observa tras el pontón el lugar de la parada para dar agua a A03c y A06.

f009

Partidor con torno para desviar el agua desde A11 (losa abajo izquierda) hacia A14 (bajo derecha). En la foto un agricultor a última hora de la tarde abre el partidor para dejar de desviar agua por la A14.

f010

Acequia A03. A la derecha se observa los tubos de desagüe de la estación de bombeo del paso inferior de la carretera de Alba bajo la CV-400.

03

omo f001 pero desde diferente ángulo. Acequia A03 (derecha) y 03a (izquierda). Al fondo, en la parte superior, la CV-400. A03c ontinúa hacia la izquierda siguiendo el muro de la autovía.

f004

Nudo entre las acequias A03 (principal, de tierra) y A04 (derecham, de fábrica). Se observa la compuerta para toma de agua de A04; la parada para que entre agua por ésta se observa en f005.

f011

Acequia A12. Se observa cómo se ha limpiado la acequia para traer agua desde A18 (discurriría al fondo de la imagen) para regar una parcela en "e16".

f012

Acequia A12, desde el mismo punto que f011. Al fondo se observa la parte en desuso de A12, junto a un desguace de coches abandonado.

05

onfluencia entre A03 (bajo derecha) y A07 (entubada, ontinuación de A03) y A05 (izquierda). está puesta la compuerta n la parada al origen de la A05 para impedir que pase el agua a sta. Se observa el lugar de la parada al incio de la zona entubada, que sirve para desviar el agua para A04 y A05.

f006

Acequia A03b, en el sector que riega la parcela "e9". Al fondo se observa el partidor de A11, zona donde vierte las sobrantes A03b.

f013

Confluencia entre las acequias A13 (arriba), A19 (derecha), A21 (abajo) y A18 (izquierda). Las distintas paradas permiten varias opciones para el curso del agua.

f014

Partidor bajo trapa en el cruce del camino deel Saboner y la calle Benidoleig. El partidor permite desviar el agua del Braç Nou (A15a y A17) hacia A15 para el riego de "e19" y "e26".

07

Partidor bajo trapa, que se utiliza para desviar agua hacia A09 (se observa en la foto, paralela a la carretera de Alba) y regar así las parcelas que se observan a la derecha (en barbecho y viveros).

f008

Partidores bajo trapa en la esquina del camno del Saboner y la carretera de Alba. Entre ambos se controla el desvio del agua a A08, A16 y A16a.

f015

Acequia A18 junto al camino de servicio de la CV-400. A la izquierda se observa un pequeño regador de tierra para el riego de una pequeña parcela en "e20"; en este caso se realiza la parada en el punto que se observa junto al tocón del árbol.

f016

Confluencia de las acequias A18, A20 y A20a. Se observa un partidor al fondo que regula el desvio de agua a A20.

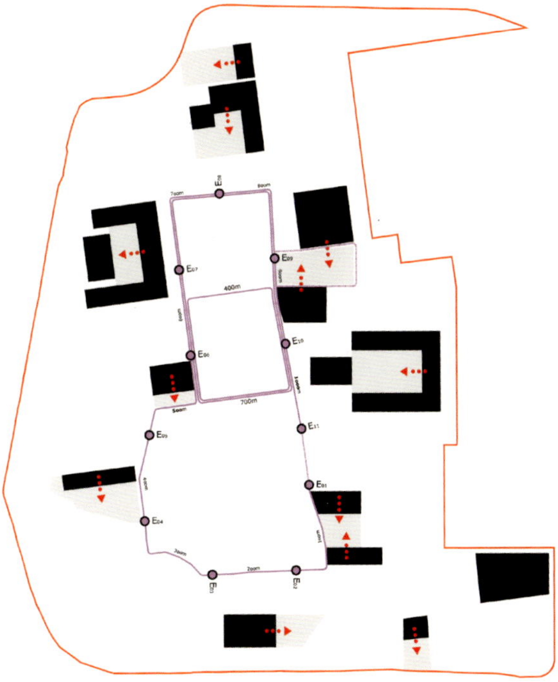

CHINCHILLA BIOCLIMATIC VILLAGES

CHINCHILLA生态气候村

This is a project of a new neighbourhood in the Albacete metropolitan area, with 2,800 subsidized homes, cultural facilities and technology incubators. Designed by Spanish and international architects, the district will use renewable energy sources —an experimental hydrogen system— in an urban setting of great landscape and environmental quality.

The project grew out of research into the creation of self-sufficient habitats for a new generation of citizens who want homes at an affordable price in a complete urban setting in which they can live, work and relax, in keeping with the lifestyles of the society information.

The district is intended for young people from the Albacete metropolitan area, the whole of Castilla-La Mancha and even from Madrid or Valencia, whose way of life is centred on networking or teleworking, taking advantage of Albacete's excellent communications thanks to the AVE advanced passenger train. The new district, laid out on 100 hectares near Chinchilla station, will be organized in sections defined by the circulation systems, which will also include pedestrian, cycle, sports and agriculture circuits. There will be more than 20 hectares of green space, sports facilities and agricultural plantations.
The houses are organized in 'villages' with self-sufficient management systems and energy production, urban allotments, homes and workplaces. There will also be business incubators and offices for innovative companies, especially in the fields of new technologies, sustainability and logistics.
The district will use on an experimental basis, for the first time in Europe, a domestic hydrogen micro-network, connecting buildings with different functions (homes, offices and cultural amenities) in such a way that each generates hydrogen cleanly and can share it with other units when needed. This will mark the beginning of the introduction of hydrogen networks, which the European Union has identified as the future of energy. The district will also have its own hydrogen-fuelled transport, thereby implementing the introduction of a public hydrogeneration system on the A-31, one of Spain's busiest motorways.
In addition, a high-speed Intranet will connect all of the homes, enabling the residents to organize themselves by way of the neighbourhood web, an innovative browser that will promote social interaction.
From an urban perspective, an advanced neighbourhood is a compact nucleus developed on the basis of an integral project, with sufficient density and critical mass to generate the kind of social and economic relations associated with a city. As a result, the advanced district enjoys functional autonomy, generously equipped as it is with amenities that enhance the quality of life and accommodating a productive tertiary sector that provides jobs. A place to live in, not just to sleep in.

PLANTA ORDENACION

ARQUITECTOS INVITADOS

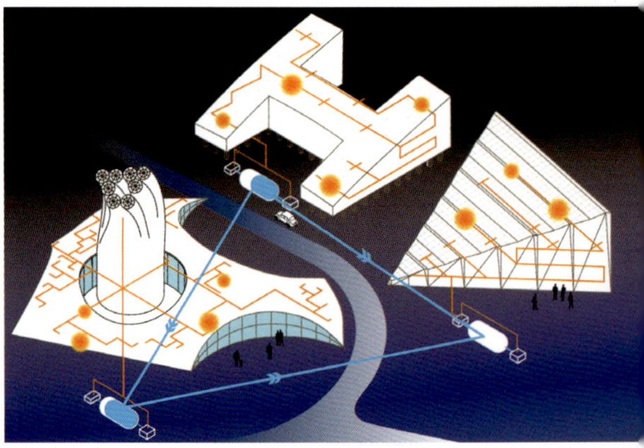

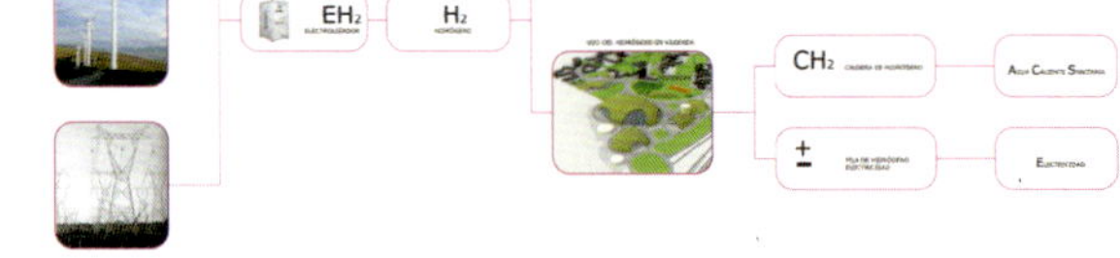

The technological conception of such a neighbourhood is defined by a number of different strategies:
The architectural, which incorporates an avant-garde design that embraces new housing typologies in line with new kinds of home and new ways of living, generating accessible spaces by means of a design without exclusions.
The telecommunicational, given that these are neighbourhoods with a digital infrastructure directly related to the information society and to the new ways of working currently being developed in the most advanced parts of the world.
And finally, the environmental (with the application of the most rigorous criteria of sustainability), integrating the proposed interventions with the natural landscape in which the neighbourhood is located and introducing the technology required to ensure maximum savings in water, energy and transportation. In short, the achieving of maximum internalization of environmental costs.

The project incorporates various bioclimatic strategies, which guarantee all of the traditional standards related to the production of hot energy by means of passive systems and the intelligent use and recycling of water.

Provision has also been made for the setting up of cooperatives to manage solar farms that will ensure the clean generation of energy.
There will also be a pilot project for the use of hydrogen in the home. In addition to producing hydrogen in situ by the hydrolysis of water and storing it, this internationally pioneering experience will create a micro-network with three different units —housing, workplace and leisure centre—generating and sharing hydrogen according to the needs of each one.

这是一个位于Albacete都市区域内的一个新的邻里社区项目，其中包含2800套补助性住宅、文化设施以及技术孵化器。这个由西班牙建筑师同国际上的其他建筑师一起设计的区域将会使用新的可再生能源——一种新的实验性水系统——并存在于一个都市背景下良好的景观和环境质量当中。

这一项目来源于对于自给自足的住宅的研究和创造，旨在为想要在可负担的价位获得一套包含了所有生活必需设施的住宅的新一代市民提供一个他们能够在其中生活、工作和休闲的地方，同时也使得他们能够时刻接收到与生活方式相关的社会信息。

这个居住区的目标人群是来自Albacete都市区域的年轻人，Castilla-la Mancha的所有人，甚至于来自马德里或瓦伦西亚的那些生活方式以网络工作和电信工作为中心的人，使得他们能以ave高级乘客列车所提供的服务为契机来利用Albacete便捷的交流方式。

这个新的住宅区位于临近Chinchilla车站约100,000平方米的土地上，它将会被分割为不同区域，而不是由道路的环线系统来定义。这里将会拥有不少于20，000平方米的绿化空间，体育设施以及农业植被。
住宅以村落的形式被组织在一起，并带有自给自足的管理系统和能源生产、都市土地配给、家庭、工作场所等等。这里还会拥有商业孵化器和创新型公司的办公场所，尤其是那些与新技术、可持续发展和后勤相关的企业。
这个住宅区将会使用一种在欧洲首次尝试使用的试验性基础设施——一种内部微型氢网络，将不同功能（住宅、办公以及文化休闲）的建筑联系在一起，并使得每一种建筑都能够生成干净的水并在需要的时候同其他单元共享。这标志着被欧盟定位为未来能源的氢网络引入的开始。该住宅区同时还将拥有自己的氢能源交通工具，从而在a-31这条全西班牙最为繁忙的道路上完成公共氢系统的引入。
此外，一种高速的互联网将所有的住宅联系在一起，使得居民能够通过使用邻里网络这种用于推广社交互动的浏览器自己组织活动。
从都市的角度来看，一种高级的邻里社区应当是在综合项目的基础上开发出来的密集核心，它应当拥有足够的密度和严格的质量来生成与城市之间的社会和文化关系。因此，高级的居住区享受功能上的自治，总体上具备能够保证生活质量的娱乐休闲设施，并且拥有具有生产力来提供工作机会的第三产业。总而言之，它应当是人们生活的地方，而非人们睡觉的地方。
这样一个邻里社区的技术概念是由一系列不同的策略决定的：
建筑方面与倾向于新的住宅原形前卫设计合作，同时也带来了新的住宅种类以及生活方式，通过一种无差别的设计方式来达到生成友好空间的目的。
电信通讯方面的前提则是住户均配备有能够直接介入信息社会的数码基础设施，并且一种新的工作方式正在世界发达的地区之一被不断开发。
最后，环境方面（伴随着最为严格的可持续发展标准的应用），将设计与住宅区所在的自然景观的干预以及为保证最大程度节约水资源、能源和交通而引入的技术相结合。简而言之，即成就最大程度的环境消耗内部化。

这一项目将多种生态气候策略结合在一起，通过消极系统以及水资源的智能使用和回收，保证了其符合所有与放热能源相关的传统标准。

为了建立合作管理太阳能农场的准备工作也正在进行当中，该农场将会保证能源的清洁生成。
同时，一个指导家庭使用氢的方式的计划也会被推出。除了通过水的氢离析和储存来现场制造氢能源，这一国际性的先锋体验将会创造一种包含三个不同单元的微型网络系统——包括住宅、工作场所以及休闲中心——来生成并根据每部分的需要共享氢能源。

MOTRIL TROPICAL ECO-NEIGHBOURHOOD
MOTRIL热带生态社区

Motril, capital of the Costa Tropical area, in the province of Granada, is experiencing a major process of urban development as a result of its strategic geographical location on the Mediterranean coast between Málaga and Almería. Its role as the port of Granada, its position between the sea and the mountains, and the arrival of the Autovía del Mediterráneo motorway make the town a key site for the development of the region.

In the case of the MOT-3 project addressed by the present Plan, this is a key site for the development of the town. In its configuration, Motril gives a very clear message in terms of the models of urban growth that it wishes to promote, and especially with regard to the relationship between the population nuclei and the surrounding natural environment. This relationship between growth and interaction with the environment is the key issue to be resolved by urban planning at the start of the 21st century.

Concentric Growth or Radial Permeability
MOT-3 poses the choice between concentric growth and radial permeability on the northern edge of Motril. The approved general plan posited that a large environmentally protected area should be created between the port and the urban nucleus in the form of a sugar cane plantation and a park route. This large-scale green facility connects up with the town's Parque de los Pueblos de América and the Rambla de los Álamos avenue, which runs east to the sports centre and the municipal cemetery.
The MOT-3 sector is capable of opening up the town to the north, to the hills and the Sierra, reproducing in a symmetrical (but initially on a smaller scale) the environmental sugar cane plantation and giving concrete physical form to the sea-Motril-mountain environmental relationship.

Local and Global
The scheme thus identifies the opportunity to transform the green zones of a residential sector into a park, which would become the town's primary network, thereby benefiting at the local scale the residents of the new neighbourhoods adjoining this major amenity, and at the global scale benefiting the municipality as a whole by ensuring its openness to the natural networks around it.

The Layout
The project addresses a location with a complex topography, with hills and differences in level. It is envisaged that the sector will be bisected by a new ring road bordering the historic town centre, with a new roundabout to facilitate connectivity between all the new sectors of the town.

Transfer of Traces

In the light of the foregoing reflections the design team decided to transfer the traces of the existing paths and plots and conserve the open spaces in their present state.

This is the basis not only for an environmental continuity in the territory, but also for a cultural continuity, in preserving the lines of tracks that in many cases date from ancient times and —small though they are— have served to structure the use of the territory.

Planning Criteria

The criteria adopted in the laying our are:

•To guarantee the creation of a major green space, as large as possible, running north-south through the entire sector.

•To create the maximum number of open spaces and amenities, well above the established minimum standards, compatible with the creation of the necessary residential plots for the real-estate development of the sector.

•To lay out the circulation routes according to the actual topography and the existing roads and paths.

•To provide an optimum mix of subsidized and private-sector housing.

•To ensure the proximity of all the housing plots to the large landscaped space.

•To distribute the amenities throughout the sector so that there are urban attractors in all directions.

•To occupy all of the land eligible for development in plots of similar size.

•To promote the inclusion of buildings that contribute to the definition of the urban space.

•To introduce the smallest possible number of traffic routes.

•To conserve the Rambla de los Álamos and create an environmental corridor around it.

位于格兰纳达省滨海热带区域的首府Motrial由于其在地中海沿岸Malaga与Almeria之间的战略性位置，正在经历一次巨大的城市开发。它位于海与山之间的位置，以及Autovia del Mediterraneo机动车道都令这一市镇成为了该地区开发的重点地段。

在现有规划所布置的Mot-3项目是整个市镇开发的重点区域。在Motrial本身的构成当中，它清晰地传递了自身所期望推崇的都市生长模式，尤其是有关人口核心与其周围自然环境之间的关系。这种生长和与环境互动之间的关系正是城市规划在21世纪初期需要解决的最为重要的问题。

集中生长或放射式渗透

Mot-3将集中生长或放射式渗透的选择放在了Motril北部边缘上。已经得到承认的总体规划中明确要求在港口和城市核心之间设置一块以甘蔗种植园和一条停车路线为主要形式的大型环境保护区域。这种大尺度的绿色设施将市镇的Parque de los Pueblos de América与Rambla de los Alamos大街连接起来，并向东连接体育中心和市民公墓。

Mot-3区域能够面对市镇向北面的山和Sierra开敞，重新以对称的方式模仿环境原有的甘蔗种植区域（事实上尺度稍小），同时也给当地的海山环境关系赋予更为坚固的物质形式。

本地的与全球性的

这一设计因而认知并利用了这一将居民区内的绿地转化成为公园的机会，而这种转化将会成为该市镇最为基本的网络，因而从本土的尺度上惠及了与该设施相关的新社区的居民，同时也从全球的角度为人类整体带来了良性的影响，即保证了人类文明对于其周边自然网络的开放态度。

布局

这一项目通过山以及层次的不同在场地上制造了一种复杂的原型。计划实施后，该区域将会被一条与古市镇中心相毗邻的新环路平分开来，同时也将拥有一个新的环状交叉路口来为市镇所有的新区之间的连接服务。

追踪转换

在前面叙述的反馈下，设计组决定将现有道路和节点转化，并以维持其现有状态的方式保护现有的开放空间。对于保护那些多数自古就已存在的道路来说，同时对于控制该地区设施的使用来说——尽管它们中有些规模微不足道——这不仅是整个区域环境可持续性的基础，更是一种文化持续性的基础。

规划标准

在规划布局当中我们采用的标准如下：

保证创造一个主要的绿化空间，规模尽可能地大，同时要南北向地贯穿整个区域。

创造最多的开放空间和休闲公共设施，其设置标准应当远高于现今建立的最低标准，同时与该区域内房地产开发设置的必要居住节点相互兼容。

根据实际的原型和现有道路路径安排环路的布局。

提供一个补助性住宅与私有住宅恰当混合的住宅区。

保证所有居住节点都易于到达临近的大型景观空间。

将休闲娱乐设施分散在整个区域内，以使得在所有的方向都有吸引人的都市景观。

将适合开发的整片土地的更小规模的开发来进行推进。

推行将为都市空间这一概念作出贡献的建筑纳入范围的行为。

推行数量最少的交通道路。

保护Rambla de los Alamos，并在其周围创造一种环境性的廊道。

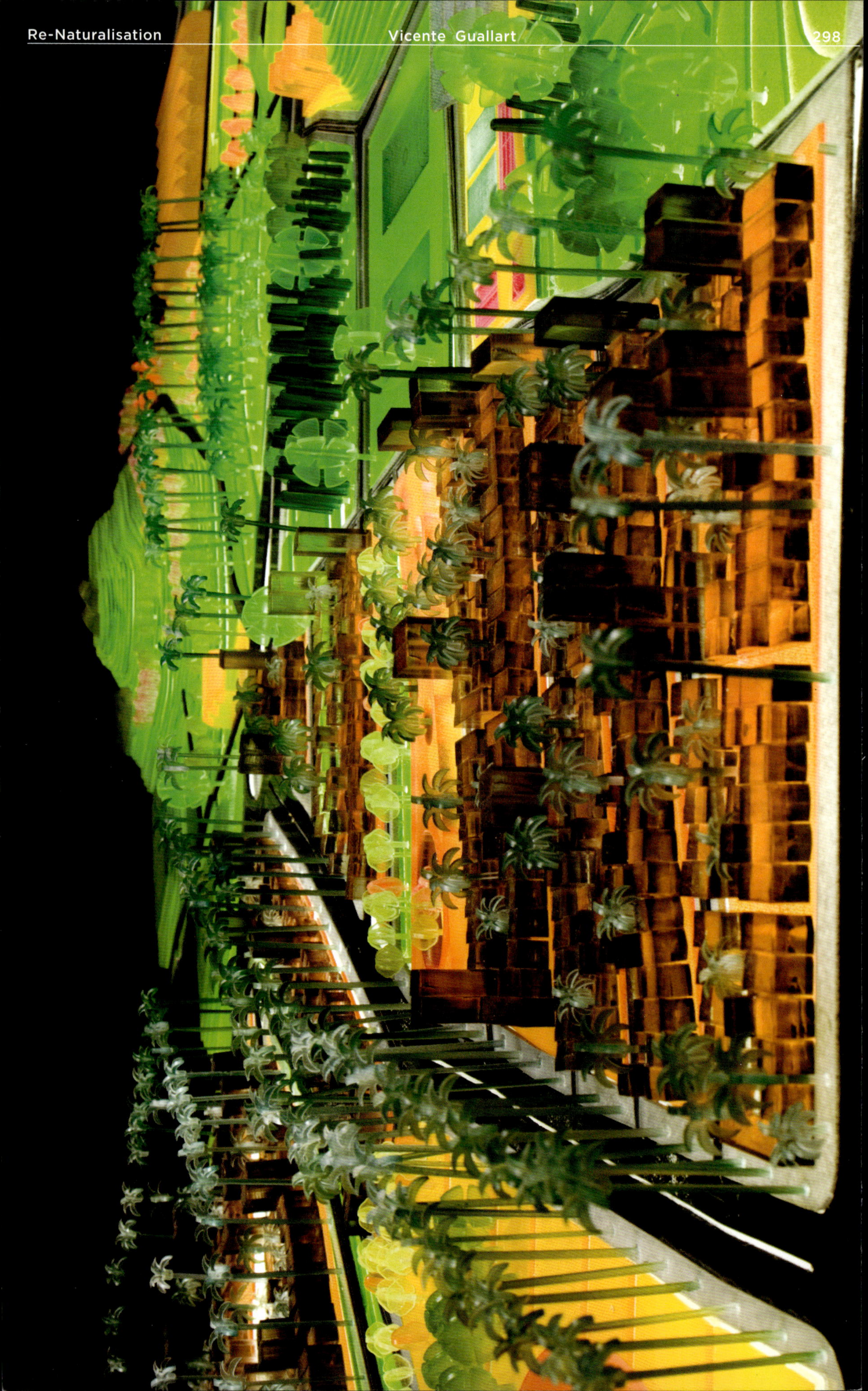

SHARING TOWER
共享大厦

Sharing Tower: a social information centre, car park and 99 student apartments. According to the Factor Four principle, we should be seeking to obtain twice the benefit with half the space, or with half the resources, in order to enhance the efficiency of our environment. This means that we need to construct more efficient homes and buildings that enable their occupants to share resources. The functions of the home —sleeping, washing, storing food, cooking, eating, washing clothes and relaxing— are organized around specific objects — bed, shower, fridge, cooker, table, washing machine, television. It is not logical for every small apartment to have all of the appliances required for these functions, which in many cases are in use only 10% of the time. The alternative is to establish dynamic relations between apartments on the same floor, sharing spaces in groups of two, four or eight. This can allow a 25 m2 apartment to have an effective area of 75 m2. At the same time, the communal use of the space encourages social interaction.
The project area, on the boundary between agricultural land and the city, has an underground car park that serves as a speed interchanger between the urban city and the garden city proposed by Sociopolis: between an environment for driving around and the pedestrian city, between the fast city and the slow city.

The network society makes it possible to share resources so that people can do more with less. Between the public and the private space there is a shared space. The physical resources contained in the home should provide for the individual functions of the act of dwelling; each individual should have at least these resources in a minimum area of 25 m2. If resources are shared by two, four or eight people, each of them will have a larger usable area at their disposal. Each person has the use of the sum of his or her individual space plus the collective space.

When it comes to sharing resources, there may be incompatibilities between certain activities because of different individuals' preferences in terms of how things look, sound and smell. In a sequence of different floors the sharing of resources can be organized according to the residents' compatibilities, with each floor functioning like a large house whose communal space is accessible from each personal space. The skin of the building is constructed like an X-ray of its internal functioning.

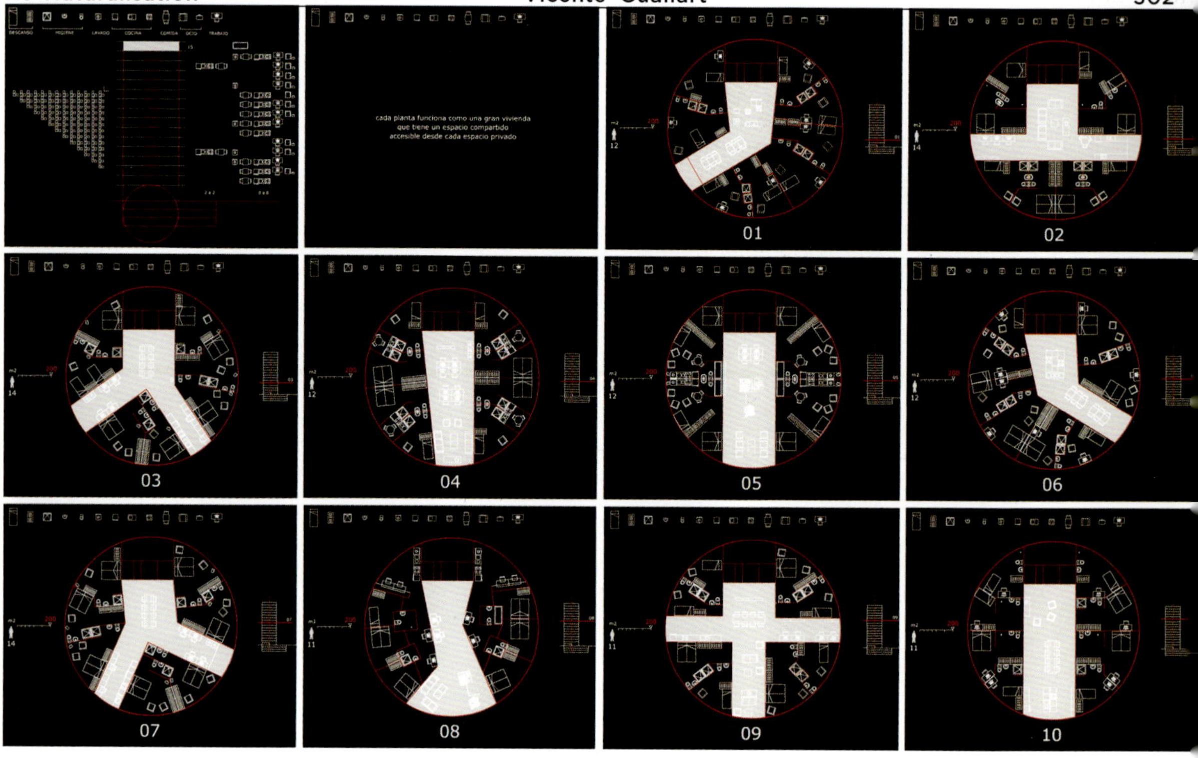
cada planta funciona como una gran vivienda
que tiene un espacio compartido
accesible desde cada espacio privado
01
02
03
04
05
06
07
08
09
10

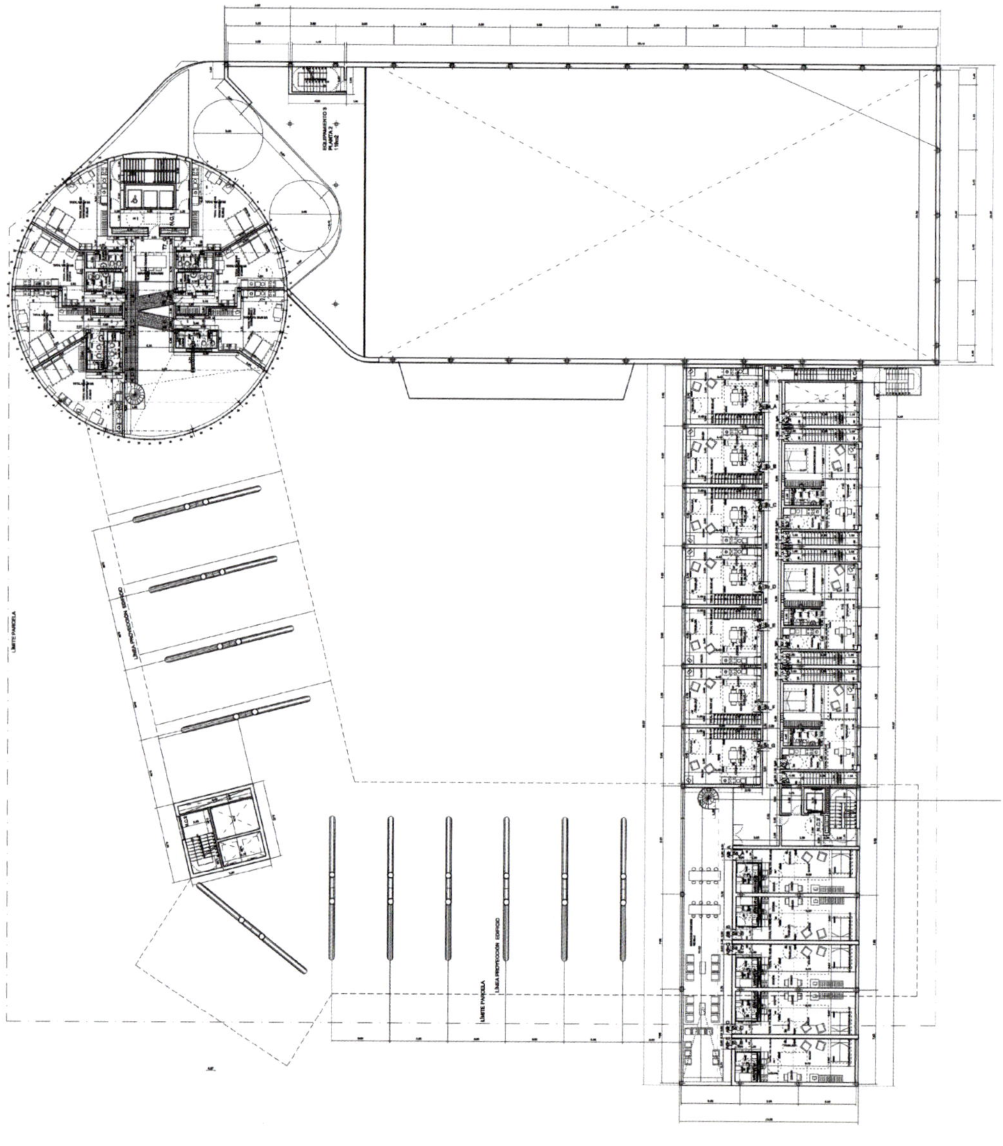

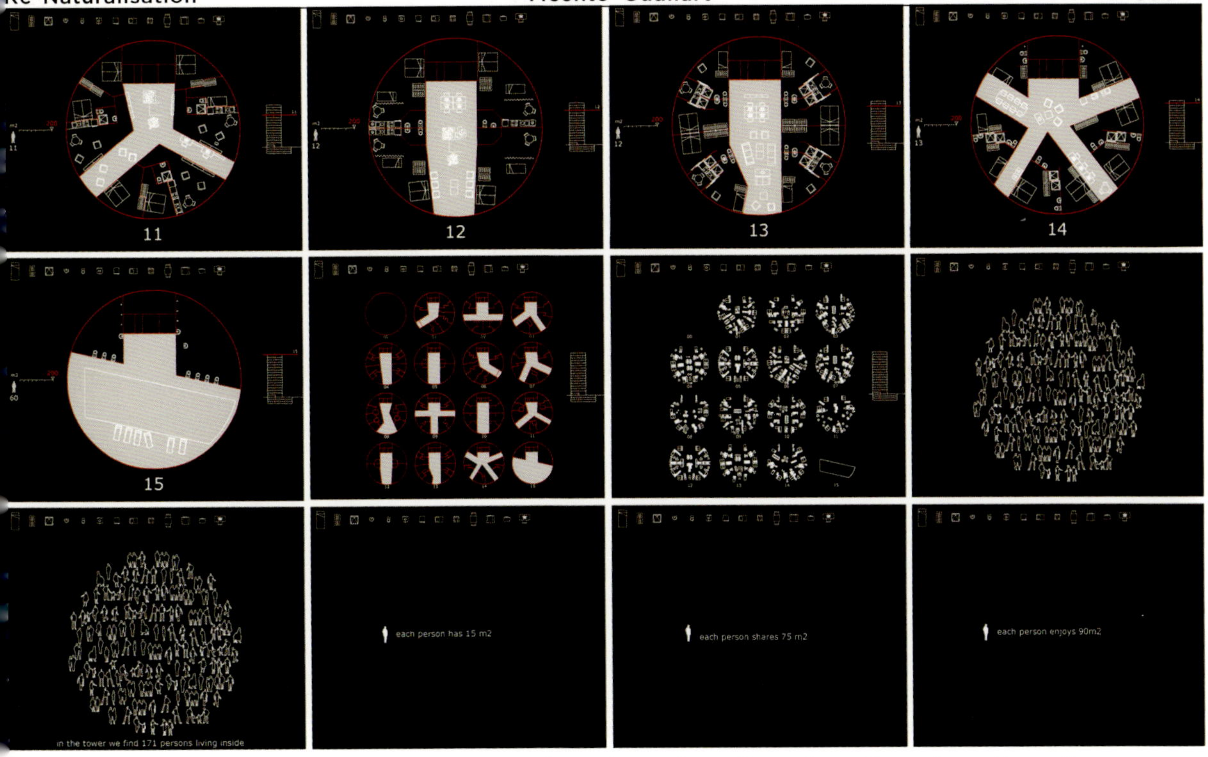
11
12
13
14
15
in the tower we find 171 persons living inside
each person has 15 m2
each person shares 75 m2
each person enjoys 90m2

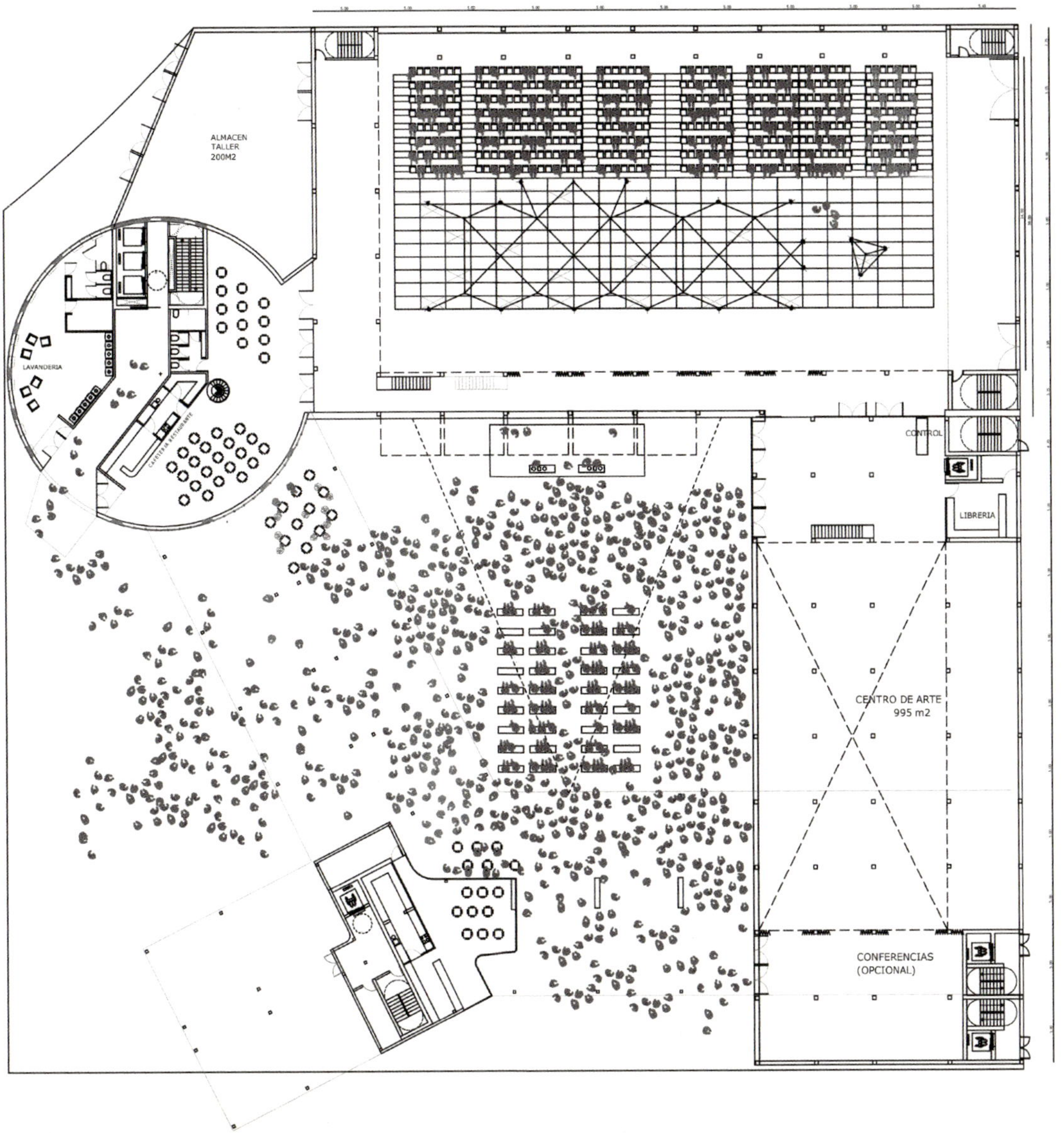
ALMACEN
TALLER
200M2
LAVANDERIA
CONTROL
LIBRERIA
CENTRO DE ARTE
995 m2
CONFERENCIAS
(OPCIONAL)

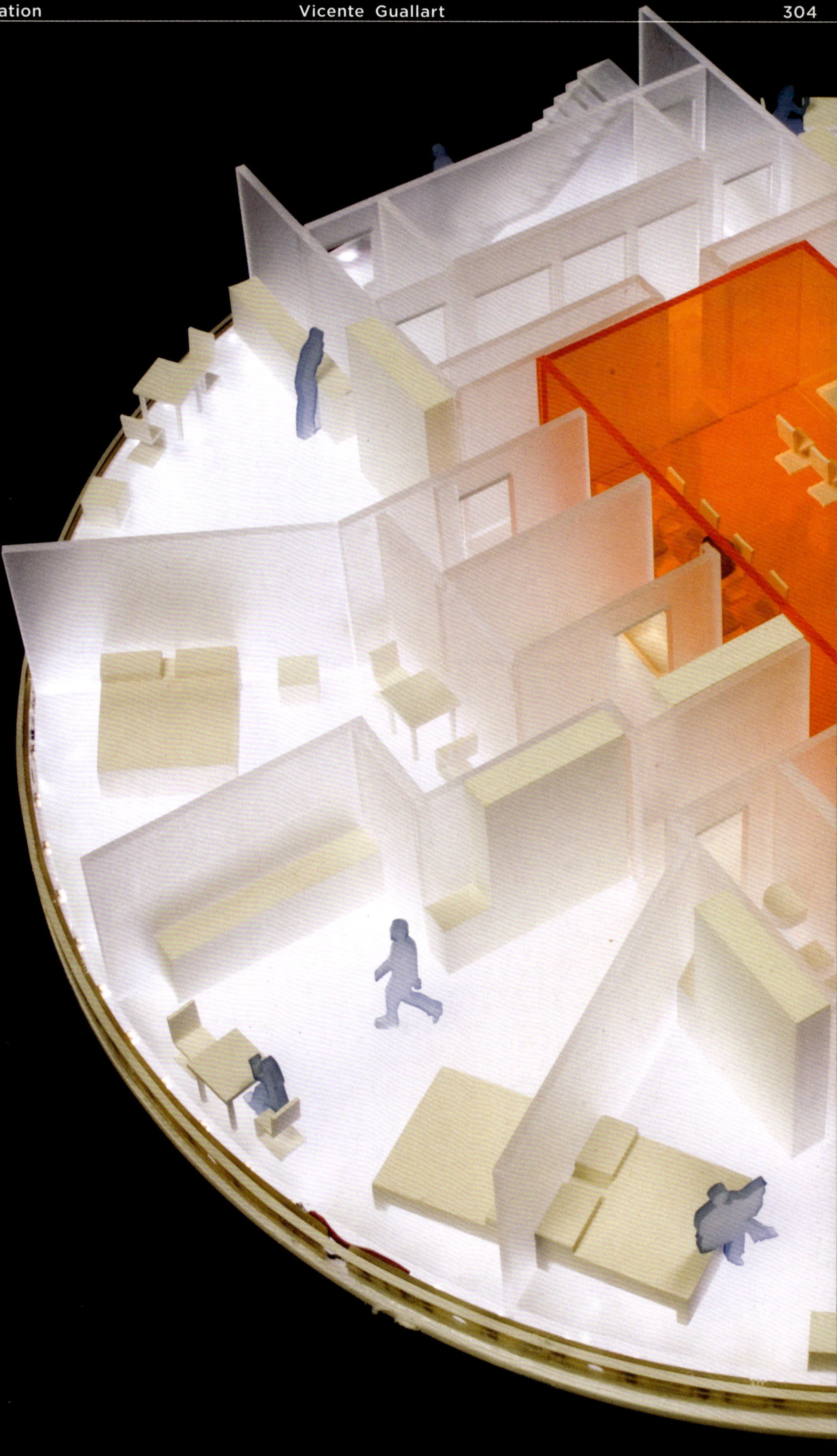

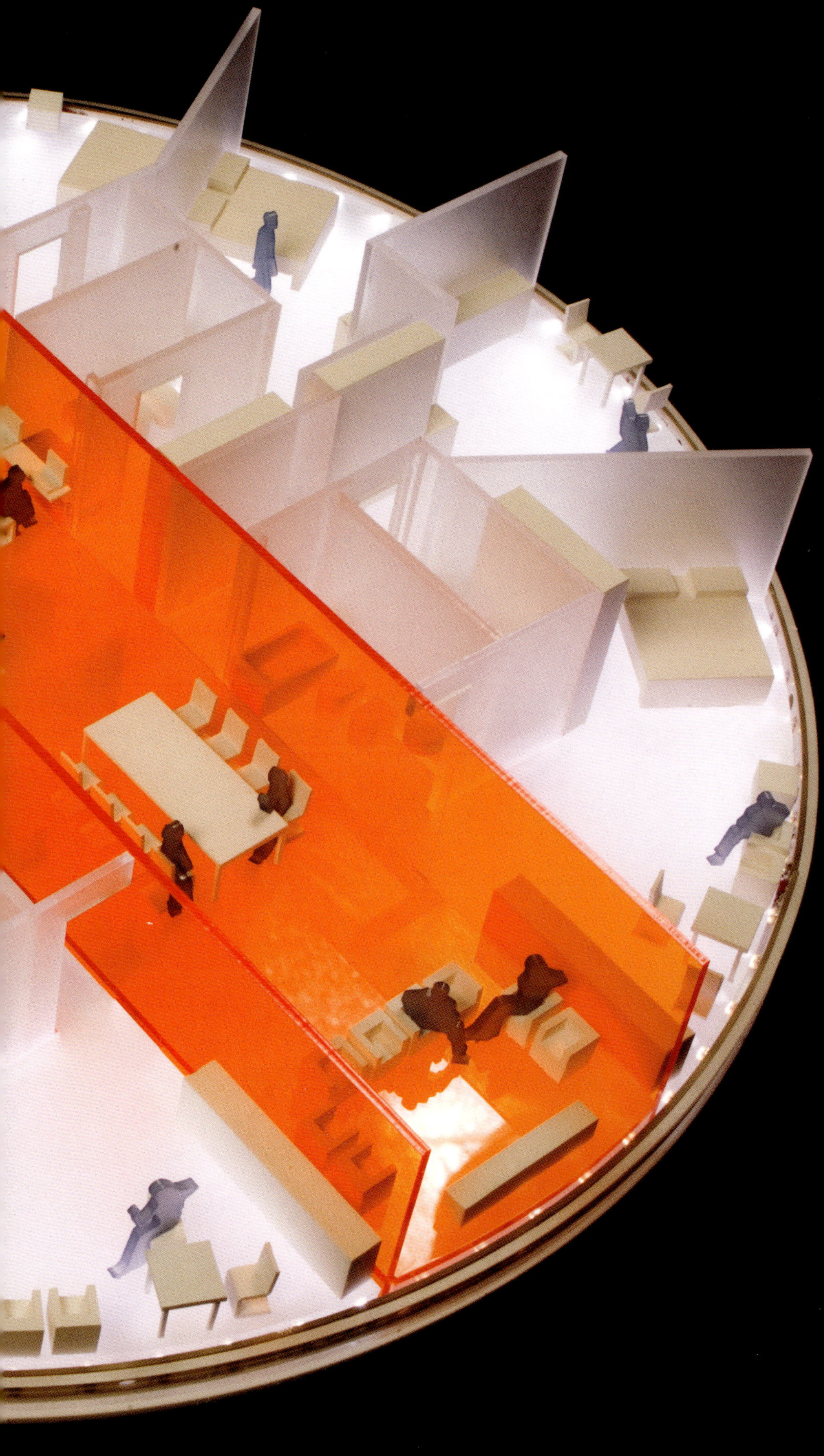

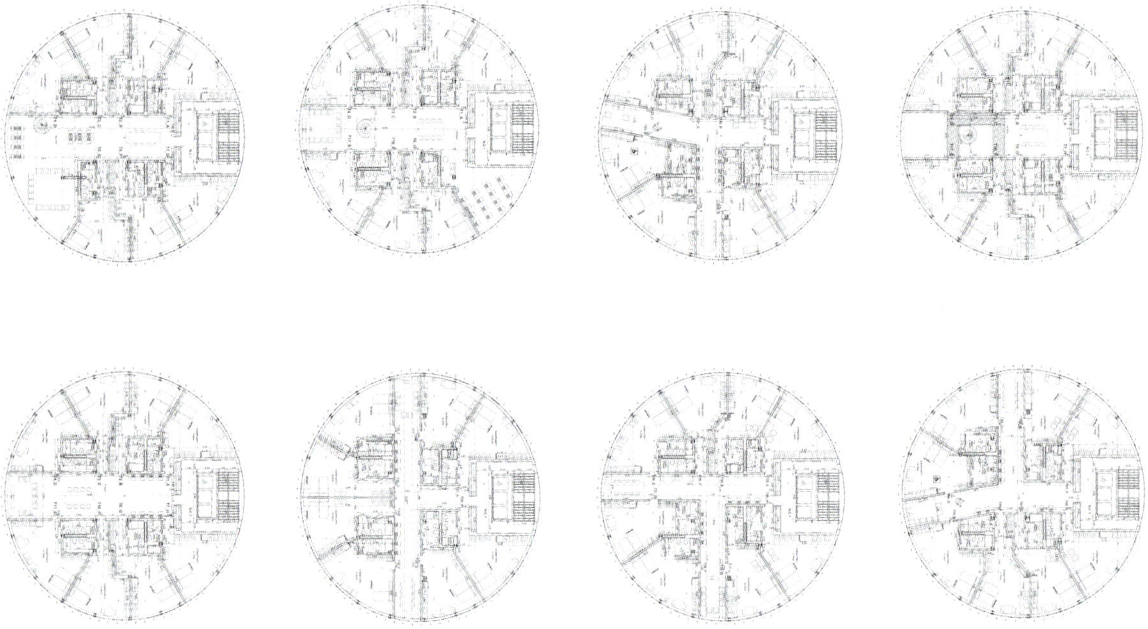

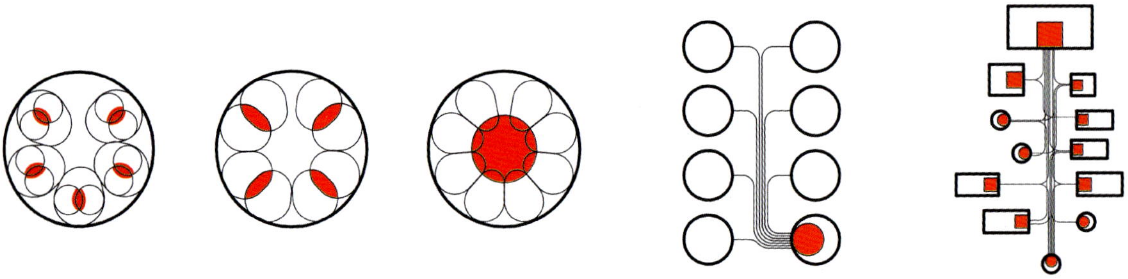

1 dwelling 2 dwellings 1 floor 1 building 1 neighbourhood

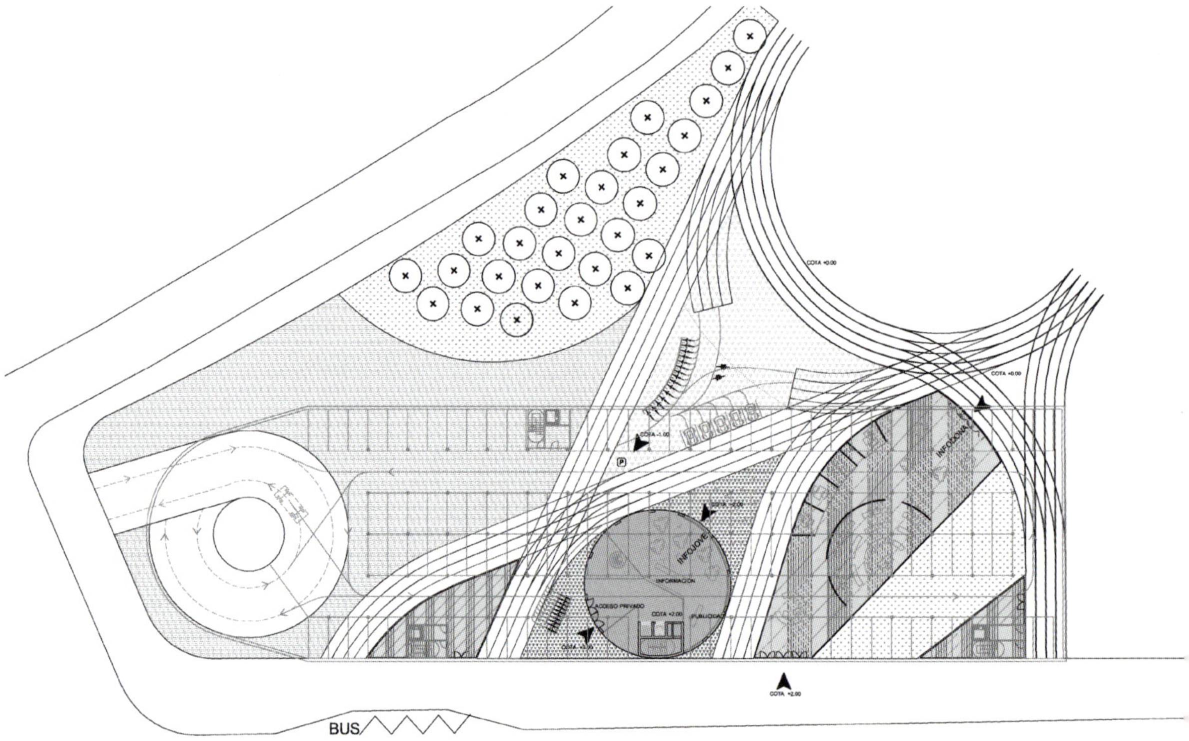

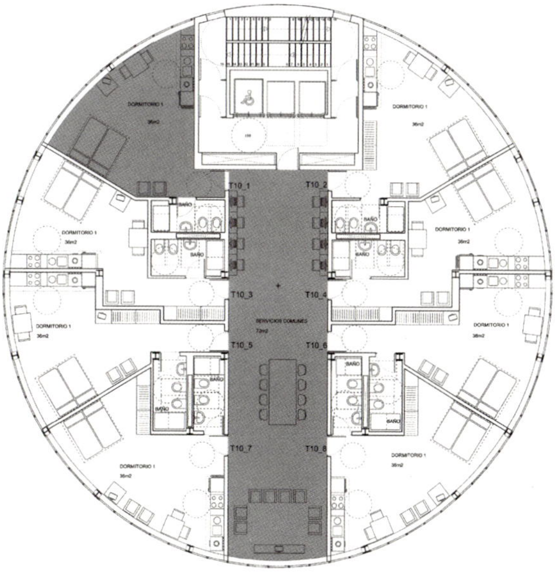
DORMITORIO 1
36m2
BAÑO
T10_1
T10_2
T10_3
T10_4
SERVICIOS COMUNES
72m2
T10_5
T10_6
T10_7
T10_8

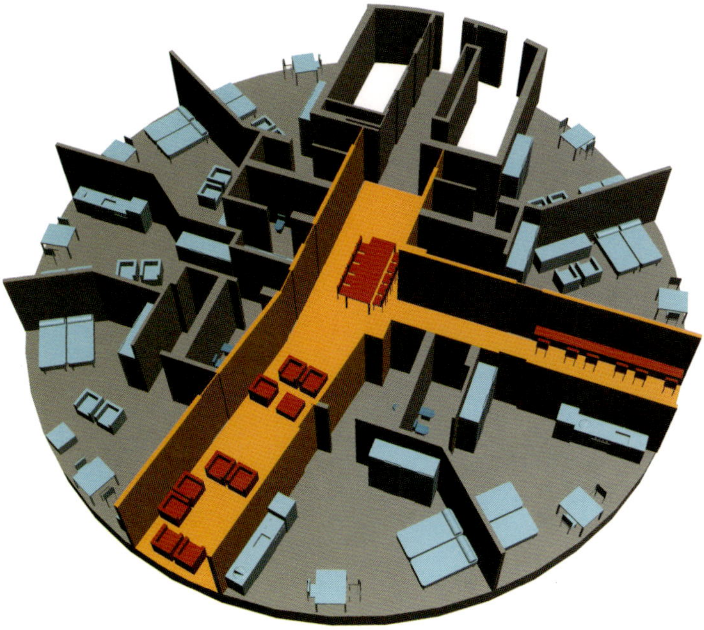

共享大厦：一个社会性的信息中心、停车场以及99个学生公寓。根据四倍原则，为了加强我们环境的效率，我们应当寻找事半功倍的空间或能源使用方法。我们需要建造更多的高效率住宅和建筑，使拥有者能够进行资源共享。住宅的功能——休息、洗浴、储存食物、烹调、吃东西、洗衣服以及休闲娱乐——都被组织在具体的物件——床、淋浴器、冰箱、厨具、餐桌、洗衣机、电视——等等的周围。对于小型的公寓来说，每一家都拥有他们享受这些功能所需的全部设施是不合理的，并且在众多的案例当中，这些设施只有10%的时间处于被使用的过程中。替代性的方案就是在同层的公寓及之间建立一种具有活力的关系，使得他们能够以两户、四户或者八户的规模来进行空间共享。这就允许一套25平方米的公寓能够具有一套75平方米的公寓所拥有的有效面积。同时，空间的日常使用也鼓励了社会性的互动。这一项目区域位于城市和农业土地之间，拥有地下停车场来作为社会学都市所提倡的都会城市和花园城市的转换器：一种在机动车驾驶环境和步行环境之间的转换，一种在快速城市和慢速城市之间的转换。

网络社区是人们能够进行资源共享，以更少的资源获得更高的效率。在公共空间与私有空间之间也同样存在着一个共享空间。包含在住宅之内的物质资源以提供居住条件的方式为每一个人提供需要的功能；而每一个人则至少需要在一个25平方米的空间内容纳这些资源。如果资源被两个、四个或者个人共享，他们中的每个人就会拥有更多个人可支配的空间。每个人都能够使用其私有空间与共享的共有空间的总和。

而在共享资源的时候，可能会出现由于个人所进行的活动所产生的观感、声音和味道的不同而产生活动之间不能够协调进行的情况。不同的楼层所组成的贡献空间序列可以根据居民活动的兼容性组织资源的共享，每一层都发挥着一座大住宅的功能，其交互空间与每个人的个人空间是相互连通的。整栋大厦的表皮就像X射线一样能够反映出内部功能。

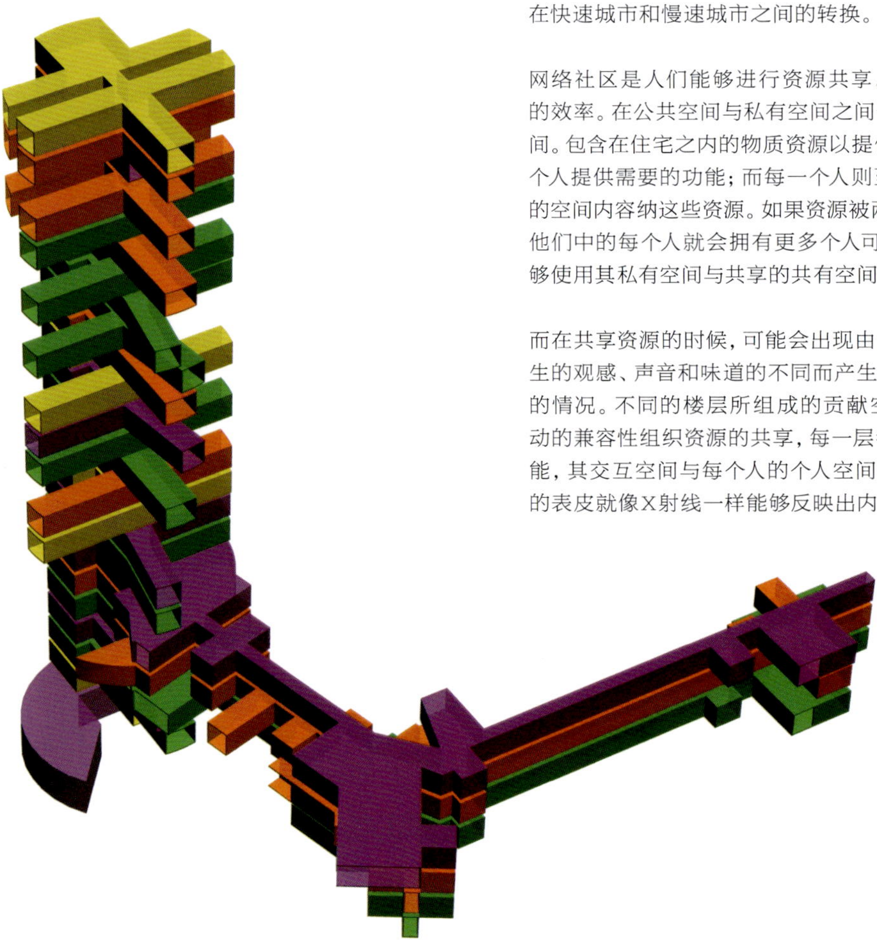

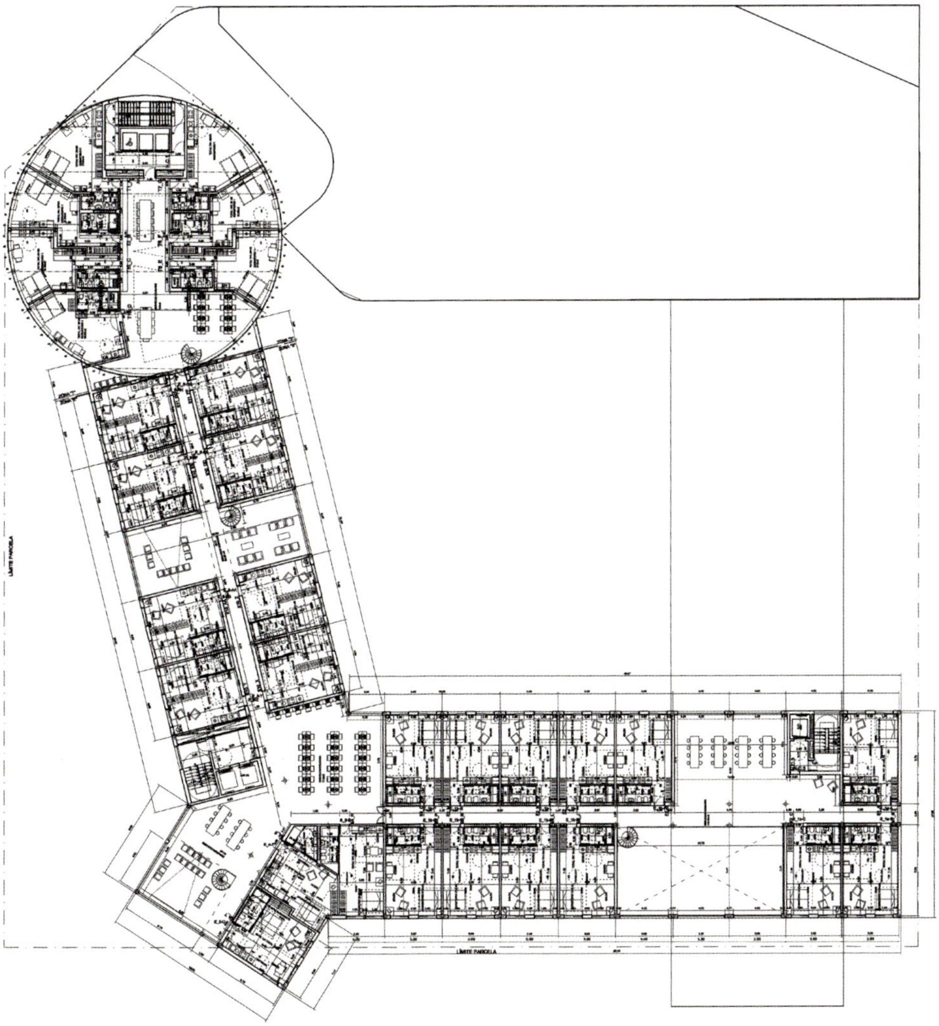

BOWLING
COLUMBINE
MICHAEL MOORE

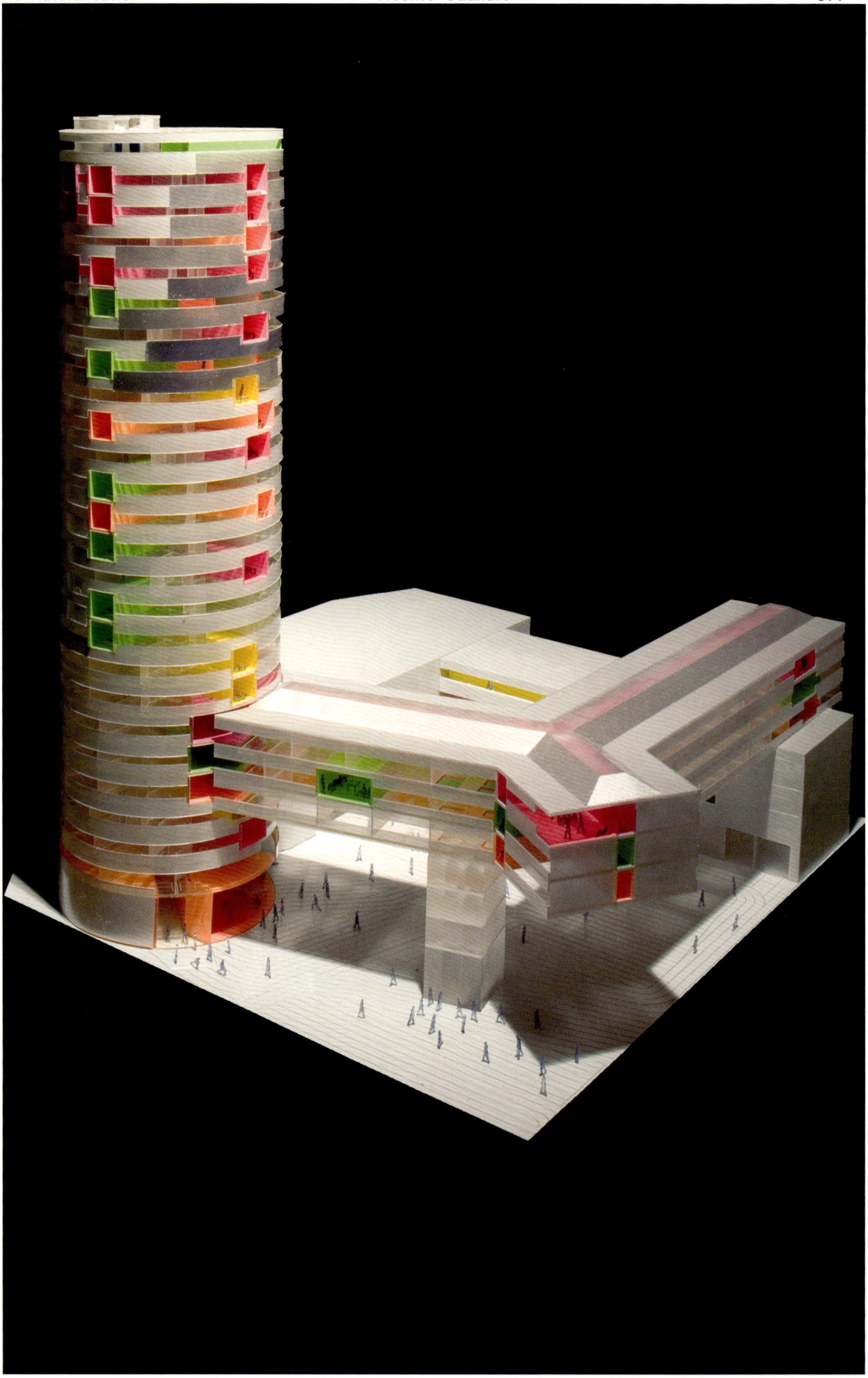

BIBLIOGRAPHY
建筑师年表

GUALLART ARCHITECTS

Vicente Guallart (Valencia 1963) is one of the most relevant emerging architects in Spain. He opened his studio in Barcelona in 1992.
He works in the confluence of architecture, nature and new technologies. Some of his more relevant projects are the Denia Mountain (reconstruction of an old quarry, in an environment that hold and old Arab castle, selected for the Venice Biennale 2004), Three ports in the North of Taiwan that will improve the tourism in the island (winners of two international competitions), or the Sharing Tower in the neighborhood of la Torre (Valencia).
He spends part of his time with research and educational projects as director of the new Institute for Advanced Architecture of Catalonia (Iaac), in Barcelona, that hold a international postgraduate program on architecture and urbanism working in many scales "from Bits to geography" (www.iaacat.com). In 2000 he co-directed with MIT's Media Lab and other research Center, the Media House Project, the prototype of an informational house, based, as first time, in distributed computation.
He is author of the master Plan of the project Sociopolis, an innovative housing project in Valencia that allows the urban development and the protection of the agriculture environment. He coordinates a group of international architects that will build different equipments and housing projects in this neighborhood. This project was shown in the Valencia Biennale, or the Arkitertur Zentrum in Vienna. It will be built for the year 2009.
He is co-author of the Metapolis Dictionary of Advanced Architecture (Actar),
HiperCatalonia, or Media House Project. The Korean publishing house DD dedicates the monograph "Intelligent realities" to his work. Actar publishing house is preparing his new book "GeoLogics".
He have given lectures in different Universities and Cities around the world as Princeton University, Upen, Architectural Association, UCLA, MIT, Space (Seoul), Taipei and others.
His work has been exhibited in the Biennale di Venezia 2002 and 2004, 2008, São Paulo International Architecture Bienale 2007, MACBA, IVAM, MoMA Archilab, Gratz, and others.

MAIN BUILDIGNS AND LANDSCAPES

-Sharing Tower. Sociopolis-Valencia. 2005-2007
-Kim Hyng Yoon editing Co Headquarters. Paju Book City. Seoul. Korea (2001-2005)
-34 apartments in Cambrils. Tarragona (Spain) (2002-2006)
-Fugee Turistic Harbour. Taiwan 2004-2007
-Batoutz Touristic harbour. Taiwan 2004-2007
-Keelung ocean plaza. Taiwan 2004-2007
-Vinaroz Fish Market 2004-
-Hortal house. Commarruga. Barcelona. Spain (2001-2004)
-Denia Mountain Project. Denia. Alicante Spain (2003-
-Web Hotel. Fabrications at MACBA, (Barcelona) 1998
-Metapolitan Loft (2001-2004)
-House in the historic center of Barcelona. 1992.
-Project for the urban environment of the Plaza del Arbol in Valencia. EUROPAN II. 1991

URBAN PLANS

- Bioclimatic Villages in Chinchilla. Spain. 2007
- Ecobarrio Tropical. Motril, Spain. 2007
-Sociopolis Master Plan. 2003-2005-Sociopolis
- Project of Urbanization for Sociopolis. 2004-2005
-Xativa Biogolf. 2005-
-Burjasot 2005-
-Plan of Uses for the Port of Vinaroz, 2001-2002

-Museum of Fashion in Valencia, 2002-2005

OTHER PROJECTS

-Project for the construction of a multifunctional complex (Hotel, Car Park, Auditorium, Commercial Center) in the castle quarry in Denia, 2002-
-Construction project for the south beach and services building adjacent to the Port of Vinaroz, 2001-2002
-Special Plan for Protection and Interior Reform of the UA6 and UA7 in the environs of the Arabic city wall in Valencia
-Project for a Telecenter and laying-out of the main square, including the partial covering-over of the river.
Ribes de Freser, 2000-2003
-36 apartments in Cambrils (Tarragona), 2001-2004
-C/ Cristòbal de Moura. Scheme design, 1998-
-Grade 1 camp site in Pola de Gordón (León), 1999
-Competition for the covering-over of the railway line and housing in L'Hospitalet
-House of the Seven Summits, La Puebla de Vallbona, Valencia 1997
-Scape House. Housing for agricultural environments. Prototype, 1998
-Study for a plastic house. European Association of Plastics Manufacturers / European Space Agency, 1999

EVENTS, PRIZES, EXHIBITIONS

PRIZES

-1st Prize in the Competition "New Taiwan by Design": the re-design of Fugge Port and Batouz Port .
-1st Prize in the Competition Taiwan Gateways, for the re-design of Keelung Water front.
-1st Prize in the Competition for the design for the waterfront of Vinaroz (Castellon. Spain)
-1st Prize in the Competition for the design of 2 km of the south Coast (Castellon. Spain)
-1st Prize in the National Projects Competition for the multifunctional Center in the castle quarry in Denia.
-1st Prize for the construction of the Hotel HUSA Reus.
-Finalist of the FAD Prizes 2008 with the Microcoasts
-Finalist of the European Prize of Urban Space with Microcoasts
-Award Ciutat de Barcelona Prize 2002, by the Media House.
-FAD Award 1992 for an apartment in Barcelona
-Finalist in the IberFAD Prize with a house in Liria in 1996.
-Finalist in EUROPAN IV with a housing project in L'Hospitalet.
-Mention in EUROPAN II with a project for the Plaza del Arbol and environs in Valencia 1991.
-Möbius Prize for the best Spanish CD-ROM 2000.
-Möbius Prize for the best Spanish CD-ROM 1995 for Mateo at ETH.

ONE-MAN EXHIBITIONS

Lógica Natual. IVAM. Valencia, Spain. 2007
The House of the Digital Man, Fundación COAM 1998.
La Beauté à la Nature, Avignon 2000.

COLLECTIVE EXHIBITIONS

-Veniced Biennale 2008. « Out There : Architecture Beyond Building ». Hyperhabitat : Reprogramming the World
-International Bienale of Contemporary Art of Seville. Spain. 2008
-ArchiLab 2008. Orleans, France.
-7th International Architecture Bienale of São Paolo, Brazil. 2007
-Sociopolis, Bienal de Valencia 2003.
-Hipercatalunya, MACBA 2003
-Invited to the official selection of the Biennale de Venezia 2000 with Barcelona Metapolis.
-Invited to the official selection of the Biennale de Venezia 2000 for Ecity.
-Archilab 2001.
-Archilab 2000.
-Exhibition HLM, Ministerio de Fomento. 2001
-Against Architecture, EACC 2000.
-Urban Natures, EACC 2001.
-Barcelona Metapolis
-36 modèles pour une maison, Bordeaux 1998.
-Fabrications, MACBA, MOMA, SFMOMA, WEXNER 1998.
-Europan II

TEACHING

Director of the Institute for Advanced Architecture of Catalonia (www.iaacat.com) since 2003
Co-Director of the Advanced Architecture Course, Fundació Politècnica de Catalunya since 2000
Visiting lecturer at: UCLA, ETH, AA, Princenton, Upenn
Arts Director Club, New York 2001
Media Lab, MIT, Boston 2001.
ICA, London 2000.
Biennale de Venezia 2000.
MACBA 1998.
Berlage Insitute 1994.
Lectures at the Official Architecture Associations in Catalonia, Madrid, Valencia and Seville.

PUBLICATIONS

Author or editor of:
Lógica Natual. Exhibition Catalogue. Spain, 2008.
Geologics. Actar. Spain, 2008.
Self Sufficient Housing. Actar. 2007
Sociopolis. Arkitertur zentrum, Actar Ed 2004
Media House Project. Iaac.2003
Geocat. Iaac. 2003
El teletreball i els telecentres com impulsor del reequilibri territorial, Generalitat de Catalunya 1999
Centros Históricos del siglo XXI, EUROPAN II 1992.
El efecto Albacete, Actar. 2007
Metapolis Dictionary of Advanced Architecture, Actar Ed. 2000-2003.
HiperCatlonia. Actar Ed 2003.
BARCELONA Metapolis, Actar Ed. 1998.
Festival Met 2.0.

BOOKS

Barcelona Plus, Actar Ed. 2001
10x10, Phaidon 2000.
ARCHILAB, 2000.
Housing. New Alternatives, Actar Ed. 1999.
Against Architecture, EACC. 1999
36 modèles pour une maison, Périphériques 1998.
Fabrications, MACBA 1998.

MAGAZINES AND NEWSPAPERS SUCH AS:

Pasajes, Arquitectura Viva, On Diseño, Valencia City, Lapiz, Descubrir el Arte. Domus, Quaderns, El Croquis, Werb, Wauen&Whon, Monocle, Wallpaper, Architectural Digest, Artpress, Mark, Loft Publications, etc.
El País, El Mundo, La Vanguardia, El Periódico, Levante, ABC, etc.

TV PROGRAMS

Special program in the series Creadores on TVE2 : 'The adventure of knowledge'.
Special program in the series SCAPE on Canal Satélite Digital.
Participation in the series Millenium on Canal 33: 'The house of the future'.

图书在版编目(CIP)数据

西班牙当代建筑实践：重构自然–山峦 媒介 建筑：Vicente Guallart作品 / 蓝青主编.
–武汉：华中科技大学出版社，2008.9
ISBN 978-7-5609-4861-4
I.西… II.蓝… III.建筑设计—作品集—西班牙—现代 IV.TU206
中国版本图书馆CIP数据核字(2008)第135642号

西班牙当代建筑实践：重构自然–山峦 媒介 建筑：Vicente Guallart作品
主　　编：蓝青
美国亚洲艺术与设计协作联盟(AADCU)

出版发行：华中科技大学出版社
地　　址：武汉市珞喻路1037号(邮编：430074)
出 版 人：阮海洪
责任编辑：张亦男
责任监印：张正林
制版印刷：北京画中画印刷有限公司
开　　本：889mmx1194mm　1/16
印　　张：20
字　　数：221千字
版　　次：2008年9月第1版
印　　次：2008年9月第1次印刷
ISBN 978-7-5609-4861-4/TU·414
定　　价：268.00元

销售电话：022-60266190, 60266199(兼传真)
网　　址：www.hustpas.com